I0115066

The Formidable Book of Flavonoids
Understanding Flavonoids and Synergy in Cannabis
Researched and written by Russ Hudson
Edited by Jacqueline Graddon, MBA
Images by Gloria Fuentes, Ph.D.

Copyright © 2025 Russ Hudson
Copyright © 2023 Front and rear cover photographs by Jeff Scheetz, AKA Doobie Duck

All rights reserved. No portion of this book may be reproduced in any form without permission from the publisher, except as permitted by U.S. copyright law. For permission, please use the Contact page on CannabisChemistry.com

ISBN: 978-0-9863571-5-2

This book was meticulously produced. Nevertheless, neither the author nor editor warrant that the information contained herein is free of errors. Readers are advised that statements, data, illustrations, procedural details, or other content may unintentionally be inaccurate. We encourage readers to independently verify studies presented or cited in this text.

Never Stop Growing

A Tribute to Jeff Scheetz

The front and rear cover photographs for this book, as well as the image of the hummingbird above, were taken by Jeff Scheetz, known in the cannabis industry as "Doobie Duck." Jeff's contributions to the cannabis community include decades of support for medical patients, thousands of exceptional cannabis macro photographs, his longstanding work with cannabis genetics and breeders, and his fascinating and unique cannabis dioramas.

Jeff passed away on December 26, 2023. The author wishes for his work to live on in this book.

You can see Jeff's work on Instagram at https://www.instagram.com/doobieduck/, or on Facebook at https://www.facebook.com/doobie.duck.

TABLE OF CONTENTS

FOREWORD

Silver bullets are fictional, existing only in lore for the control and eradication of equally fictional werewolves, vampires, and the occasional witch.

However, for decades, silver bullet theories were precisely how we in the cannabis industry disseminated information about this plant. First, we spent the better part of 40 years propping THC up as the most valuable component of cannabis. Then came CBD, ushering in more than 15 years of claims that this particular cannabinoid is the medical holy grail of cannabis. Incredibly, wars were waged between THC and CBD camps, while others in the industry seemed to collaborate, shouting from their rooftop pulpits that the two chemical constituents needed to be combined at a 1:1 ratio to have any real therapeutic value.

While that gawdy world spiraled out of control in the global cannabis markets, another contender abruptly entered the ring: terpenes. Terpenes were touted as being responsible for everything good about cannabis – from the way the plant smells and tastes, to the potency and effects of flower and other products. Another spammy world was born with this compound class silver bullet theory; a world of misinformation, pushy sales tactics, and purified terpene products that few in the industry understand.

And now, discussion and marketing efforts are moving on to flavonoids. Are these carbon-based compounds the next candidate for the cannabis industry's collection of rapidly outdated silver bullet theories? Is the market about to be flooded with flavonoid-dense strains and products developed with these compounds reintroduced after chemical stripping?

Fortunately, the conversation about terpenes shepherded in the concepts of synergy and the entourage effect. People started talking about "whole plant" or "full spectrum" cannabis products. Even laypeople understand at a basic level that several compounds in cannabis probably "work together," but in general only for positive or beneficial outcomes. Cannabis compounds – in the mind of the public – rarely if ever "work together" for negative or detrimental outcomes. So, while we're inching slowly away from silver bullet theories, our industry is still flooded with shifty brands, bro science, and Chads trying to hawk the next trendy cannabis products in what we've come to recognize as the classic cannabis industry pump-and-dump. Silver bullet products sell unfortunately well in this type of environment.

But cannabis is not a silver bullet plant, nor will any of its constituents ever be. Cannabis is both a social and chemical *community* plant – one that brings together seemingly limitless combinations of people and molecules. In fact, in 2016 cannabis scientists dubbed it "the plant of one thousand and one molecules[1]" in a paper that detailed the many different compounds produced by the plant. These compounds likely contribute to the effects, colors, flavors, and aromas of cannabis in a way that is far more synergistic than single-molecule based concepts. In addition to cannabinoids, terpenes, and flavonoids, other potentially important cannabis constituents include esters, alcohols, aldehydes, sulfurs, saponins, ketones, thiols, alkaloids, lipids, and other compounds. The medical, therapeutic, or recreational value of these compounds has been and will always be determined by the individual consumer and their desired outcomes, and cannabis offers a cornucopia of chemical constituent profiles that eventually will suit virtually every taste and need.

Consequently, The Formidable Book of Flavonoids is a small but significant part of the chemical story of cannabis. Consider this a companion text to The Big Book of Terps, with other books to follow in a textbook series called The Cannabis Chemistry Collection, hosted at CannabisChemistry.org. Together, these textbooks will offer the best refutation to the silver bullet theories of cannabis and will focus instead on potential synergies between these and other compounds.

---

[1] Andre CM, Hausman JF, Guerriero G. Cannabis sativa: The Plant of the Thousand and One Molecules. Front Plant Sci. 2016 Feb 4;7:19.

PROLOGUE

This book was written for researchers, scientists, educators, cannabis industry professionals, medicinal cannabis patients, and for everyone interested in flavonoids, and other chemical constituents of cannabis. Significant efforts were expended to make the work presentable and sustainably interesting to all demographics, including laypeople. For this reason, all citations in the text appear in-line for immediate reference. This format allows readers to instantly reference and verify information without breaking continuity by referring to an independent index or bibliography. Readers are encouraged to review the studies and papers referenced in this book in their entirety where necessity or special interest compels.

Please note that most studies cited here are not specific to cannabis. This is primarily because we simply do not have sufficient research available after decades of prohibitionist and research-restriction policies around the world, particularly in the United States. However, it should be understood that the molecules discussed herein are the same from plant to plant; for instance, the orientin that occurs in cannabis is the same orientin that occurs in buckwheat and barley. The rutin that occurs in cannabis strains is the same rutin that occurs in asparagus and celery. The flavonoid quercetin that occurs in hundreds of plants is the same molecule that occurs in cannabis.

We can presume that most plants use these compounds in a comparable way, and for the same or similar purposes, with exceptions for specialist plants and anomalies across species. Nevertheless, this presumption should be short-lived; we must continue to fight for legalization and scientific/research freedoms related to cannabis. Future editions of this book will be updated with emerging research, and readers that have unique or new research to share should contact the author directly at info@cannabischemistry.org.

Finally, it is important to consider that laboratory tests conducted as part of research for this book may be only partially accurate. This is because as of 2025, consistency in varieties or strains is sorely lacking. The Blue Dream acquired at a dispensary in California may be chemically entirely different than the Blue Dream acquired in Barcelona, Spain, or even in Oregon, Cali's northerly neighbor. Unfortunately, the power of strain name recognition often skews labeling and marketing efforts, making it extremely difficult to ascertain whether a particular batch of cannabis branded as any strain is in fact that stabilized strain. In most parts of the world, assurance of cannabis variety is challenging to impossible.

Because of these inconsistencies, breeders, geneticists, cultivators, dispensaries, legislators, investors, healthcare professionals, and all others involved in cannabis policy should advocate for more and better laboratory testing. It is not enough to only test for contaminants and cannabinoid content. Terpene, flavonoid, and other phytochemical content is important and can directly impact the therapeutic effects of cannabinoid and other constituents such as alcohols, ketones, thiols, aldehydes, sulfurs, and esters. Cannabis users need to know the phytochemical profile of a product to make sound decisions for their health and well-being. Testing and subsequent dissemination and education is the only method of obtaining this confidence.

INTRODUCTION

I sit and gaze, and wait with poise;
To dream in colors by plants employed;
Of yellows, purples, yet greens devoid;
We heed the call of flavonoids.

To see the daisy, petals deployed;
A splash of gold in fields conjoined;
And the purple bud, that calls so coy;
It speaks to us in flavonoids.

If we think of terpenes as contributing aroma and flavor to plants like cannabis, we can think of flavonoids as their visual counterparts; they are often responsible for the yellow, red, purple, and blue colors that we see in many plants, including cannabis. Flavonoids may also contribute to, alter, or modulate sensory perception of other molecules, resulting in their ability to influence flavor including bitterness, sweetness, and astringency. Like terpenes and cannabinoids, flavonoids may also contribute to the beneficial therapeutic effects of some cannabis products.

As pigment molecules, we can literally *see* flavonoids with the unaided eye; consider the produce section of a supermarket, and know that its vibrant colors are mostly attributed to an accumulation of specific flavonoids in each plant. Consider further the biological activity and substantial medical value of flavonoids as detailed herein, and it becomes clear how important these molecules are to the survival of plants, and to human health and nutrition.

How Do Humans Interact with Flavonoids?

Humans interact with flavonoids in two general ways: 1. Visual perception in the color of plants, and, 2. Direct consumption by eating plants or plant components. While the visual perception of flavonoids can cause pleasant feelings or feelings of nostalgia, indifference, or even distaste, the effect that these molecules have when consumed by humans is far more direct. This is important to know considering that flavonoids are ubiquitous in the human diet – it would be difficult to NOT consume flavonoids. For most people, this is excellent news, as the nutritional and therapeutic value of flavonoids in the human diet is critical to maintaining good health; perhaps more so than we know.

Unlike some cannabinoids and terpenes, flavonoids are generally not psychoactive, so consumption of these molecules typically won't modify behavior or cause sedation or euphoria, however, several flavonoids detailed in this text appear to exhibit antidepressant activities. Unfortunately, despite their significant nutritional value, some flavonoids are readily bioavailable, while others are not.

Flavonoids in Commerce

Flavonoids are found in hundreds of natural food products like fruit, vegetables, nuts, roots, and seeds, but they are also found in many value-added food products. Flavonoid content in packaged or prepared foods is often simply related to other ingredients in the item; for instance, a food item that contains tomato may also contain the flavonoids present in the tomato ingredient. Other times, flavonoids are added to products to enhance their nutritional value or provide health benefits, such as fortified juices, energy drinks, herbal teas, yogurt, protein bars, chocolate products, baked goods, gummies, and breakfast cereals.

Flavonoids are often used in supplements, particularly in traditional Chinese herbal-based supplements and products such as Xiang-lian Pill, Deduhonghua-7, Huangqi Guizhi, and Chaihu Shugan San, but also in other products including supplements like CocoaVia, Ginkgo, and Soyfit, among others.

Flavonoids are also commonly included in some types of skincare products because of their reported beneficial properties, such as antioxidant, anti-inflammatory, and skin-soothing effects. These products include serums and moisturizers, sunscreens, anti-aging products, and brightening treatments.

Flavonoids in Medicine

In the cannabis industry, there is much discussion about the medical value of terpenes, and how products can retain or even boost these compounds. While many of the primary terpenes found in cannabis are well-studied and offer significant medical or therapeutic value, flavonoids tend to dwarf terpenes in the scope of their actual and potential medical value. Some flavonoids are so heavily studied from a medical standpoint that thousands of papers exist on their use in a wide range of applications. In fact, the literature regarding some flavonoids presented here is so vast that in some cases entire textbooks could be written about a single molecule.

Flavonoids offer the potential to prevent and/or treat a massive variety of diseases and conditions, including many types of carcinomas, neurological disorders, injuries, heart attacks and strokes, viral, bacterial, and fungal infections including those that are resistant to traditional antimicrobial agents, joint and bone disorders, and many more. Notably, flavonoids can also synergize the effects of many other drugs, especially chemotherapeutics.

Flavonoids in Cannabis

Flavonoids appear to be the next chemical constituent of cannabis to generate excitement and interest in the industry, particularly for clinical and therapeutic use[2]. However, as of early 2025, most cannabis testing laboratories do not test for flavonoids, and those that do only quantify around 6-10 individual compounds. Regulatory bodies do not require testing for flavonoids, so most commercial cannabis operations do not request this type of testing. Included in this text are the top flavonoids known or hypothesized to occur in cannabis, although it is likely that the plant produces flavonoids or closely related compounds that are not covered here. The author recommends testing for any or all of the flavonoids detailed herein.

Ultimately, whether cannabis products are a logical choice to obtain therapeutic levels of flavonoids or not is a query that we seek to answer in this text and with ongoing research and collaboration. Until there is a proven answer, **to acquire the highest levels of flavonoids via a cannabis product, the author strongly recommends eating or juicing raw cannabis buds and leaves.**

---

[2] Hughston, L., Conarton, M. (2021). Terpenes and Flavonoids: Cannabis Essential Oil. In: Cital, S., Kramer, K., Hughston, L., Gaynor, J.S., Springer, Cham. (eds) Cannabis Therapy in Veterinary Medicine.

FRIENDS OF FLAVONOIDS

Introduction

Although humans have been consuming these molecules for millennia, flavonoids are relatively new to us as a concept, having been discovered less than a century ago. This chapter recognizes and summarizes the work of some of the most important flavonoid and polyphenol researchers and scientists in the world.

TIMELINE OF FLAVONOID FRIENDS

► Dr. Albert Szent-Gyorgyi - 1920s–1970s
► Dr. Richard Kuhn - 1930s–1960s
► Tom J. Mabry - 1960s–2000s
► Norman G. Lewis - 1970s–current
► Lester Packer - 1960s–2018
► Harold H. Draper - 1970s–1990s
► Dr. Helmut Sies - 1970s–current
► Bruce A. Bohm - 1970s–2000s
► Jeffrey Bland - 1980s–current
► Michael H. Gordon - 1980s–current
► Brenda S.J. Winkel - 1980s–current
► Erich Grotewold - 1990s–current
► Catherine A. Rice-Evans - 1990s–2010s
► Paul Kroon - 1990s–current
► Alan Crozier - 1990s–current
► Vincenzo Fogliano - 1990s–current
► Dr. Edward Giovannucci - 1990s–current
► Birgit Waltenberger - 2000s–current
► Mark J.S. Miller - 2000s–current
► Li Tian - 2000s–current

Albert Szent-Gyorgyi, M.D., Ph.D. – "Father Flavonoid"
University of Szeged, Hungary; University of Cambridge, England

Considered "the father of flavonoids," Dr. Albert Szent-Györgyi's work with flavonoids began when he was investigating the properties of vitamin C and searching for other biologically active compounds. While studying the health benefits of citrus fruits and paprika in the 1930s, he discovered a group of compounds that played a crucial role in maintaining capillary strength and vascular health. He initially isolated these substances from Hungarian paprika and citrus peels, finding that they reduced capillary fragility and permeability, which led him to name them "vitamin P." Gyorgyi won the Nobel Peace Prize in Medicine in 1937 for his discovery of Vitamin C; he has additionally been credited with the discovery of rutin.

Szent-Györgyi observed that vitamin C alone was not always sufficient to prevent capillary bleeding disorders, but when combined with these newly discovered compounds, the effects were significantly enhanced. This led him to hypothesize that *flavonoids worked synergistically with vitamin C*, likely by acting as antioxidants and strengthening blood vessels. His research demonstrated that flavonoids helped prevent small blood vessels from becoming too fragile, reducing issues like bruising and bleeding.

Richard Kuhn, M.D., Ph.D. - Biochemistry
University of Heidelberg, Germany

Dr. Richard Kuhn's work with flavonoids primarily focused on chemical structure, classification, and biological activity in plants. As a prominent chemist, Kuhn was one of the early researchers to systematically study the flavonoid family of compounds. Kuhn was instrumental in elucidating the chemical structure of flavonoids and their various subclasses, such as flavones, flavonols, and anthocyanins. His work provided a clearer understanding of their molecular makeup, helping to distinguish flavonoids from other plant compounds.

Kuhn's research also explored how flavonoids contribute to pigmentation in plants, particularly anthocyanins, which give flowers, fruits, and leaves their vibrant colors. His studies showed that flavonoids play an important role in attracting pollinators and protecting plants from UV damage. In addition to their role in pigmentation, Kuhn's research also highlighted the broader functions of flavonoids in plants, including their involvement in plant defense mechanisms against pathogens and herbivores.

Tom J. Mabry, Ph.D. – Botany
University of Zurich, Switzerland; University of Texas at Austin, USA

Dr. Tom J. Mabry's work with flavonoids focused on chemical structure, classification, and biosynthesis, as well as their role in plant biology. He made significant contributions to the understanding of how flavonoids are produced in plants and their ecological functions. Mabry was a key figure in uncovering the biosynthetic pathways of flavonoids in plants. He studied the enzymes involved in the production of various flavonoid compounds and explored how these pathways are regulated at the genetic level, and his work helped clarify the complex network of reactions that lead to the formation of different flavonoids.

Mabry also contributed to the chemical classification of flavonoids, identifying new compounds and detailing their structures. His research helped establish a more comprehensive framework for understanding the diversity of flavonoids in the plant kingdom and their variations across species.

Mabry's research extended beyond the molecular level to the ecological role of flavonoids. Mabry explored how flavonoids contribute to plant defense mechanisms against herbivores and pathogens and how they affect plant-pollinator interactions through pigmentation and scent.

Norman G. Lewis, Ph.D. - Biochemistry and Plant Physiology
Washington State University, USA

Dr. Norman G. Lewis's research focused on the biosynthesis, structure, and function of flavonoids, particularly in terrestrial plants. His work helped illuminate how flavonoids contribute to plant health and ecological interactions. Lewis also studied the biochemical pathways involved in the biosynthesis of flavonoids in plants. He investigated the enzymes and genetic factors responsible for producing flavonoids, including proanthocyanidins, flavonols, and flavones. Lewis's research also explored how flavonoids play a role in plant defense mechanisms, examining how these compounds help plants defend against herbivores, pathogens, and environmental stress, including UV radiation and oxidative damage.

Lewis's work extended to the ecological roles of flavonoids, particularly their involvement in plant-pollinator interactions. He investigated how flavonoid pigments and scents attract pollinators and facilitate reproduction in flowering plants. His studies also contributed to understanding how flavonoids regulate various aspects of plant growth, development, and reproduction, including their effects on seed formation and fertility.

Lester Packer, Ph.D. - Biology
University of California Berkeley, USA

Dr. Lester Packer's research on flavonoids focused on their role as antioxidants and their interactions within the antioxidant network of the human body. He demonstrated that flavonoids scavenge free radicals, protecting cells from oxidative damage and reducing inflammation. Packer's work highlighted how flavonoids enhance the effects of other antioxidants, such as vitamin C and vitamin E, by regenerating them and prolonging their protective functions.

Packer also investigated how flavonoids influence cellular signaling, gene expression, and mitochondrial function, linking them to potential health benefits in cardiovascular protection, neurodegenerative disease prevention, and immune support. His studies suggested that flavonoids improve circulation, protect neurons, and support brain health by modulating neurotransmitters and reducing oxidative stress.

Harold H. Draper, Ph.D. - Nutrition and Biochemistry
University of Illinois Urbana-Champaign, USA; University of Guelph, Ontario

Dr. Harold H. Draper's research focused on the biochemical properties and health benefits of flavonoids, particularly their role as antioxidants and their impact on oxidative stress. Draper studied the antioxidant properties of flavonoids, examining how they protect cells from oxidative damage caused by free radicals. He investigated how flavonoids help maintain cellular integrity by neutralizing reactive oxygen species (ROS) and preventing damage to lipids, proteins, and DNA.

Draper's research extended to the potential health benefits of flavonoids, including their role in reducing the risk of chronic diseases such as cardiovascular disease and cancer. Draper explored how flavonoids contribute to anti-inflammatory processes, support vascular health, and potentially inhibit the growth of cancer cells through their influence on cellular signaling pathways.

Draper's work also delved into the molecular mechanisms by which flavonoids exert their protective effects. He studied how flavonoids interact with enzymes, modulate gene expression, and influence the activity of proteins involved in oxidative stress and inflammation.

Helmut Sies, M.D., Ph.D. - Biochemistry
University of Düsseldorf, Germany

Dr. Helmut Sies conducted significant research on flavonoids, particularly their role in oxidative stress, cellular protection, and human health. He contributed to understanding how flavonoids function as antioxidants, their impact on cellular signaling, and their potential in disease prevention.

Sies was a pioneer in defining oxidative stress as an imbalance between free radicals and the body's antioxidant defenses. He studied how flavonoids help counteract oxidative damage by neutralizing reactive oxygen species and protecting lipids, proteins, and DNA from oxidative injury. His work provided evidence that flavonoids contribute to the body's overall redox balance, helping to maintain cellular health.

One of Sies's notable contributions was his research on flavonoids and nitric oxide (NO) metabolism. He explored how flavonoids influence NO production and bioavailability, which is critical for vascular function. His findings suggested that flavonoids improve endothelial function by enhancing NO signaling, leading to better circulation and cardiovascular health.

Sies investigated the potential of flavonoids in preventing chronic diseases, particularly cardiovascular diseases and cancer. He studied their anti-inflammatory properties, their role in protecting blood vessels, and their ability to modulate enzymes involved in detoxification and immune response. His research contributes to our understanding of how dietary flavonoids support long-term health.

Bruce A. Bohm, Ph.D. - Botany
University of British Columbia, Canada

Dr. Bruce A. Bohm's research on flavonoids focused on their classification, distribution, and evolutionary significance in plants. His work contributed to understanding how flavonoids serve as key chemical markers for plant systematics and taxonomy. Bohm extensively studied flavonoids as chemotaxonomic markers, helping to distinguish plant species and trace evolutionary relationships. By analyzing flavonoid profiles across different plant families, he provided insights into how these compounds vary among species and contribute to plant diversity.

Bohm's research also explored how flavonoids influence plant physiology, particularly in pigmentation, defense mechanisms, and environmental adaptations. He investigated the role of flavonoids in protecting plants from UV radiation, herbivores, and pathogens.

Bohm authored several works detailing flavonoid structures, biosynthesis, and their role in plant ecology. His contributions helped bridge the gap between phytochemistry and plant taxonomy, advancing the understanding of flavonoid diversity and function.

Jeffrey Bland, Ph.D. - Nutritional Biochemistry
University of Puget Sound, USA

Dr. Jeffrey S. Bland's work with flavonoids focused on their role in functional medicine, nutrition, and human health. He explored how flavonoids contribute to disease prevention and overall wellness by acting as antioxidants, modulating inflammation, and supporting detoxification processes.

Bland emphasized the importance of flavonoids in personalized nutrition and functional medicine, a field that integrates diet and lifestyle factors to optimize health. He studied how flavonoids influence metabolic pathways, immune function, and cellular signaling, highlighting their potential to prevent chronic diseases. His research examined how flavonoids help regulate inflammatory responses and support the body's natural detoxification systems and explored their ability to modulate enzyme activity in the liver, enhance gut health, and reduce oxidative stress, linking flavonoid-rich diets to improved metabolic function.

Bland's work contributed to our understanding of the role of flavonoids in preventing cardiovascular disease, neurodegenerative conditions, and metabolic disorders.

Michael H. Gordon, Ph.D. - Food Chemistry
University of Reading, England

Dr. Michael H. Gordon's research on flavonoids focused on their antioxidant properties, health benefits, and potential therapeutic applications. His work primarily explored how flavonoids protect against oxidative stress and contribute to reducing the risk of chronic diseases.

Gordon studied the ability of flavonoids to neutralize free radicals and reduce oxidative damage in cells. He investigated how flavonoids act as antioxidants, preventing damage to lipids, proteins, and DNA. His research also extended to the potential health benefits of flavonoids, particularly in relation to cardiovascular health. Gordon examined how flavonoids can improve blood vessel function, reduce inflammation, and protect against conditions like atherosclerosis. He also investigated their role in protecting against neurodegenerative diseases by reducing oxidative damage in the brain.

Gordon's work emphasized the importance of dietary flavonoids - particularly from fruits, vegetables, and beverages like tea - in promoting health. He studied the effects of flavonoid-rich diets on various biomarkers of health, linking higher flavonoid intake to improved overall well-being and reduced disease risk.

Brenda S.J. Winkel, Ph.D. - Biology
Virginia Polytechnic Institute and University, USA

Dr. Brenda S.J. Winkel's research on flavonoids focused on their biosynthesis, genetic regulation, and role in plant development and stress responses. Her work provided important insights into how plants produce flavonoids and how these compounds contribute to plant health and survival.

Winkel studied the genes and enzymes involved in flavonoid biosynthesis, particularly in the model plant Arabidopsis thaliana. She identified key regulatory pathways controlling the production of flavonoids, including anthocyanins and flavonols, which influence plant pigmentation, reproduction, and defense. Her research explored how flavonoids help plants respond to environmental stresses, such as UV radiation, oxidative stress, and pathogen attacks. She demonstrated that flavonoids act as protective compounds, shielding plants from damage and enhancing their adaptability to changing conditions.

Winkel investigated how flavonoids influence plant growth, seed development, and fertility. Her findings helped explain how these compounds affect cellular processes, such as hormone signaling and gene expression, contributing to overall plant fitness.

Erich Grotewold, Ph.D. - Plant Molecular Biology
Ohio State University, USA; Michigan State University, USA

Dr. Erich Grotewold's research on flavonoids primarily focused on their biosynthesis, genetic regulation, and role in plant development. His work helped uncover how plants produce flavonoids and how these compounds contribute to pigmentation, stress responses, and interactions with the environment.

Grotewold made significant contributions to understanding the genes and transcription factors that control flavonoid biosynthesis. He identified key regulatory proteins, such as MYB transcription factors, that influence the expression of flavonoid-related genes. His research demonstrated how these genetic regulators control the production of anthocyanins and other flavonoids in plants.

Grotewold's studies highlighted how flavonoids help plants respond to environmental stressors, such as UV radiation, pathogens, and herbivores. He also investigated how flavonoids act as protective compounds, aiding plant defense and signaling, and their role in seed development and plant fertility.

Catherine A. Rice-Evans, Ph.D - Biochemistry
King's College London, England

Dr. Catherine A. Rice-Evans's research focused on the antioxidant properties of flavonoids and their role in human health, particularly in disease prevention and cellular protection. Her work significantly contributed to understanding how flavonoids function at the molecular level and their impact on oxidative stress-related diseases. Rice-Evans studied the antioxidant capacity of flavonoids, investigating their ability to neutralize free radicals and protect biological molecules from oxidative damage. Her research provided a quantitative assessment of antioxidant activity in flavonoids, helping establish their role in reducing oxidative stress and preventing cellular damage. She also explored the potential cardiovascular benefits of flavonoids, demonstrating their ability to improve endothelial function, enhance nitric oxide availability, and reduce the oxidation of LDL cholesterol.

Rice-Evans's research extended to the potential role of flavonoids in protecting against neurodegenerative diseases and certain cancers. By studying their effects on oxidative stress and inflammation, she provided insights into how dietary flavonoids could help reduce the risk of chronic diseases. Rice-Evans also investigated how flavonoids are absorbed, metabolized, and utilized in the body, addressing the challenges of bioavailability and their implications for dietary recommendations. Her work emphasized the importance of consuming flavonoid-rich foods, such as fruits, vegetables, tea, and red wine, for optimal health benefits.

Rice-Evans also investigated how the structural variations of flavonoids influence their bioactivity, including their antioxidant efficiency and interactions with other biomolecules. Her work contributed to understanding which flavonoid subclasses, such as flavonols and flavan-3-ols, have the most potent biological effects.

Paul Kroon, Ph.D. - Nutritional Biochemistry
Quadram Institute, England

Dr. Paul Kroon's research focused on bioactivity, metabolism, and health benefits of flavonoids, particularly their effects on human health and their potential role in disease prevention. Kroon studied the bioavailability and metabolism of flavonoids, investigating how they are absorbed in the gut, processed by the liver, and metabolized by gut microbiota. His work highlighted the influence of these metabolic processes on the bioactive properties of flavonoids and their effectiveness in the human body.

Kroon explored the cardiovascular benefits of flavonoids, particularly their role in improving vascular function, reducing blood pressure, and protecting against heart disease. His research demonstrated that flavonoids could enhance endothelial function and reduce oxidative stress and inflammation, key factors in cardiovascular health. His work also extended to the anti-inflammatory and anticancer effects of flavonoids, investigating how flavonoids modulate inflammatory pathways and inhibit cancer cell growth, particularly by affecting gene expression, cell signaling, and apoptosis.

Kroon's research emphasized the importance of consuming flavonoid-rich foods, such as fruits, vegetables, and tea, to maintain health and prevent chronic diseases. He explored how diet influences flavonoid intake and the subsequent health benefits linked to regular consumption of these compounds.

Alan Crozier, Ph.D. - Plant Biochemistry
University of Glasgow, Scotland; University of California Davis, USA

Dr. Alan Crozier's research focused on the role of flavonoids in human nutrition, their bioavailability, and their potential health benefits, particularly in relation to antioxidant activity and disease prevention. Crozier studied how flavonoids are absorbed, metabolized, and distributed in the human body. He explored the challenges related to their bioavailability, identifying factors that influence their absorption in the digestive system and their bioactive effects in tissues.

Crozier's research also investigated the role of flavonoids in preventing chronic diseases, particularly cardiovascular disease, cancer, and neurodegenerative conditions. Crozier focused on how flavonoids, through their antioxidant and anti-inflammatory properties, can protect against oxidative damage and modulate pathways involved in inflammation and cell proliferation.

Like Kroon, Crozier emphasized the importance of dietary flavonoids, particularly those found in fruits, vegetables, and beverages like tea, and studied how a flavonoid-rich diet can contribute to health promotion and disease risk reduction, demonstrating the connection between flavonoid consumption and improved health outcomes.

Additionally, Crozier examined the metabolic fate of flavonoids in the human body, particularly how they are processed by gut microbiota and the liver, influencing their biological activity and health effects.

Vincenzo Fogliano, Ph.D. - Food Chemistry
University of Naples Federico II, Italy; Wageningen University, The Netherlands

Dr. Vincenzo Fogliano's research focused on the role of flavonoids in food science, health benefits, and bioactivity, with particular attention to their antioxidant properties and potential for disease prevention. Fogliano investigated how these compounds can scavenge free radicals and reduce oxidative stress in the body, highlighting the potential of flavonoids to protect cells and tissues from oxidative damage, which is linked to aging and various chronic diseases.

Fogliano's research also explored the incorporation of flavonoids into food products. He studied how flavonoids contribute to the nutritional quality of foods, focusing on their stability, bioavailability, and the effect of food processing on flavonoid content. His work aimed to optimize food formulations to retain and enhance the health benefits of flavonoids.

Fogliano's work extended to understanding the broader health benefits of flavonoids, particularly in relation to heart disease, cancer prevention, and metabolic health. He examined the role of flavonoids in regulating inflammation, reducing the risk of cardiovascular diseases, and supporting metabolic function.

Fogliano emphasized the importance of consuming flavonoid-rich foods, such as fruits, vegetables, and beverages like tea and wine, to support overall health, and explored the development of functional foods and supplements that maximize flavonoid content for improved public health outcomes.

Edward Giovannucci, M.D., M.P.H., Sc.D. - Nutritional Science
Harvard University, USA

Dr. Edward Giovannucci's research on flavonoids primarily focused on their relationship with human health, particularly in relation to chronic disease prevention, cancer, and cardiovascular health. Giovannucci conducted studies to examine how flavonoids, through their antioxidant and anti-inflammatory properties, may help reduce the risk of certain cancers. His work explored the potential of flavonoid-rich foods, such as fruits, vegetables, and tea, in lowering the incidence of cancers, particularly colorectal, prostate, and breast cancer. His research also looked into the cardiovascular benefits of flavonoids. Giovannucci focused on how dietary flavonoids can improve heart health by reducing inflammation, lowering blood pressure, and improving blood vessel function, ultimately helping to prevent cardiovascular diseases.

Giovannucci explored the broader implications of flavonoid intake on overall chronic disease prevention, linking higher consumption of flavonoid-rich foods to reduced risks of diabetes, obesity, and neurodegenerative conditions. His research provided evidence for the role of flavonoids in promoting long-term health and reducing disease burden. Giovannucci's work also emphasized the importance of dietary patterns rich in flavonoids and other phytonutrients for overall health, particularly in terms of reducing the risk of lifestyle-related diseases and improving quality of life.

Birgit Waltenberger, Ph.D. - Pharmacognosy
University of Innsbruck, Austria

Dr. Birgit Waltenberger's research focused on the bioactivity and therapeutic potential of flavonoids, particularly their anti-inflammatory properties. Waltenberger studied how flavonoids contribute to vascular health, focusing on their ability to improve endothelial function and reduce the risk of cardiovascular diseases. Her research highlighted how flavonoids help prevent the oxidation of LDL cholesterol, which plays a key role in the development of atherosclerosis, and also contributes to vasodilation by enhancing nitric oxide production.

Waltenberger investigated the anti-inflammatory effects of flavonoids, demonstrating their ability to modulate inflammatory pathways and reduce oxidative stress. Her work showed how flavonoids can inhibit the activation of inflammatory enzymes, such as cyclooxygenase (COX), and reduce the production of pro-inflammatory cytokines. She also explored the potential of flavonoids in preventing chronic conditions such as cancer, diabetes, and neurodegenerative diseases. Her studies examined how flavonoids can influence cellular signaling pathways involved in inflammation, apoptosis, and cell survival, thereby supporting their role in disease prevention.

Mark J.S. Miller, Ph.D. - Nutritional Science and Pharmacology
Louisiana State University, USA; Albany Medical College, USA

Dr. Mark J.S. Miller's research focused on the biological activities and health benefits of flavonoids, particularly their roles in inflammation, metabolism, and chronic disease prevention. Miller studied the anti-inflammatory effects of flavonoids, exploring how these compounds modulate inflammatory pathways and help reduce chronic inflammation. His work showed that flavonoids can inhibit pro-inflammatory enzymes and cytokines, which are implicated in diseases like arthritis, cardiovascular disease, and neurodegenerative conditions.

Miller's research also investigated the impact of flavonoids on metabolic health, particularly their role in regulating glucose metabolism and improving insulin sensitivity. Miller examined how flavonoids may help prevent or manage metabolic disorders such as obesity, type 2 diabetes, and metabolic syndrome.

Miller also explored the cardiovascular benefits of flavonoids, focusing on their potential to improve endothelial function, reduce blood pressure, and lower the risk of atherosclerosis. His studies highlighted how flavonoids may protect the cardiovascular system by reducing oxidative stress, improving lipid profiles, and enhancing vascular health.

Miller's work investigated the bioavailability of flavonoids, exploring how the body absorbs, metabolizes, and utilizes flavonoids, and how their bioactivity is influenced by factors such as food preparation and gut microbiota.

Li Tian, Ph.D. - Physiology and Plant Biochemistry
University of California Davis, USA

Dr. Li Tian's research primarily focused on the biological activities, health benefits, and disease-preventive properties of flavonoids, particularly in the context of cancer, cardiovascular health, inflammation, and metabolic diseases. Additionally, she has explored the role of flavonoids in cannabis, a unique area of her research.

Dr. Tian has studied the anticancer properties of flavonoids, investigating how these compounds influence cancer cell growth, apoptosis, and metastasis. Her work focuses on the molecular mechanisms by which flavonoids regulate key signaling pathways involved in cancer development and progression, supporting their potential use as chemopreventive agents.

Tian's research has also delved into the cardiovascular benefits of flavonoids. She explored how flavonoids can improve endothelial function, reduce oxidative stress, and prevent the development of atherosclerosis. Her studies show that flavonoids may help reduce inflammation, improve blood flow, and contribute to the prevention of heart disease and other cardiovascular conditions.

Dr. Tian has explored the anti-inflammatory effects of flavonoids, highlighting how these compounds can reduce inflammation by modulating key inflammatory mediators. Her research has provided evidence for the potential of flavonoids in managing chronic inflammatory conditions, such as arthritis and inflammatory bowel disease, by regulating inflammatory pathways.

In addition to their role in cancer and cardiovascular health, Dr. Tian's research has focused on flavonoids in metabolic diseases like diabetes and obesity. She has investigated how flavonoids can improve insulin sensitivity, regulate blood glucose levels, and assist in managing metabolic syndrome. Dr. Tian has also explored the impact of flavonoids on gut health. She studied how flavonoids can influence gut microbiota composition and function, contributing to better digestion, nutrient absorption, and immune function.

A key area of Dr. Tian's research involves the study of flavonoids in cannabis. She has investigated the unique flavonoid compounds found in cannabis plants – cannflavin A, B, and C – examining their potential therapeutic effects, including their anti-inflammatory, antioxidant, and anticancer properties. This research adds a new dimension to understanding the medicinal properties of cannabis and its potential benefits for various health conditions.

FLAVONOIDS 101

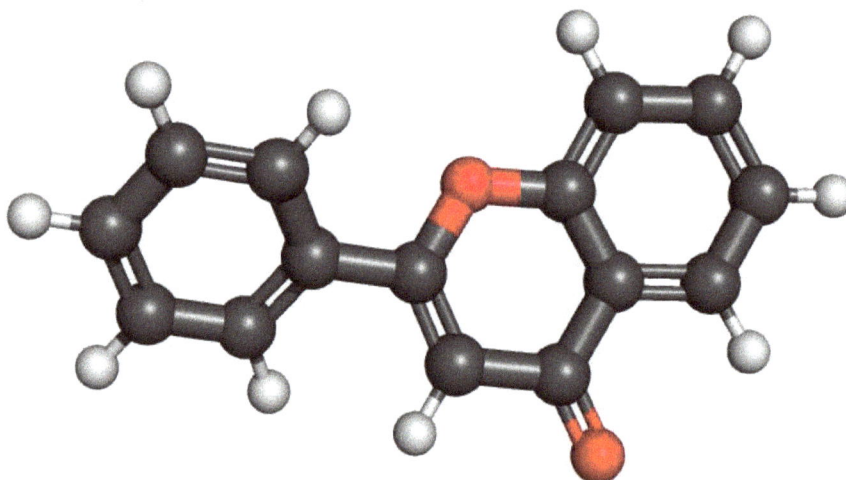

Figure 1: This is the basic structure of most flavonoids present in cannabis.

FLAVONOIDS IN THE PLANT KINGDOM

Flavonoids are phytochemicals made by plants, some bacteria[3], and some fungi[4]. Most edible plants produce flavonoids, and the human diet usually contains high amounts of these compounds. For the purpose of this book, we will discuss primarily plant-derived flavonoids, particularly those that are found in cannabis.

Flavonoids are generally thought to serve chiefly to attract pollinators by producing the yellow hues of some flowers and plant pollen. In fact, the name alone suggests this, as the word is derived from the Latin 'flavus,' or 'yellow.' But flavonoids are like terpenes in that they serve a wide variety of functions, with significant implications for human health and nutrition. Medical and nutritional value of flavonoids, of which the scientific literature is extensive, will be covered in each individual flavonoid section. This chapter discusses the uses of flavonoids by plants, and the chemical nature of these compounds.

HOW DO PLANTS USE FLAVONOIDS?

Although flavonoids and terpenes are different chemically, the way that plants use these phytochemicals is often similar. For instance, both terpenes and flavonoids possess antibacterial properties. Terpenes and flavonoids can attract pollinators, and both substances are proven antifungal agents. But diphenylpropane-based flavonoids diverge from isoprene-based terpenes in many ways. This is especially true of the UV light protection, antioxidant, and seed germination properties of flavonoids, as discussed below.

The most striking difference, however, is the visual signaling of flavonoids. These compounds are responsible for the yellow, orange, purple, red, and blue colors of many flowers, berries, fruits, and leaves, including in cannabis.

---

[3] Wang Y, Chen S, Yu 0. Metabolic engineering of flavonoids in plants and microorganisms. Applied Microbiology and Biotechnology. 2011;91(4):949-956.

[4] Du F, Zhang F, Chen F, et al. Advances in microbial heterologous production of flavonoids. African Journal of Microbiology Research. 2011;5(18):2566-2574.

So, while terpenes fill our senses with diverse aromas and tastes, flavonoids fill our world with splashes of vibrant color. But don't let these bright displays deceive you - these are functional colors. Just as terpenes are much more than a collection of smells with no meaning, flavonoids are far more than a palette of attractive colors devoid of purpose.

*Antioxidants

To understand how important antioxidants are, we need only look at the huge markets for products with antioxidant properties including vitamins, skin creams, beverages and sports-drinks, serums, lotions, shampoos, etcetera. In fact, many of these products contain flavonoids, precisely because they are effective natural antioxidants. Plants produce and use flavonoids as antioxidants, and humans in turn get their flavonoids from plants - the ones we eat, and the antioxidant flavonoids that go into skin cream, vitamins, and other products.

Oxidation in plants occurs when unstable oxygen molecules damage cells. This can be caused by biotic stress: that caused by living things like bacteria, fungi, or insects, or by abiotic stress: that caused by sun, wind, temperature, rain, etc. Oxidation can interfere with a plant's ability to photosynthesize, so evolution has provided at least one way to fight this - antioxidants like flavonoids that inhibit oxidation.

Flavonoids have been shown to exhibit antioxidant activity when plants are exposed to stress[5], with flavonoid production often focused on the site of reactive oxygen species[6], or the site of oxidation damage to plant tissue. In fact, the spectacular colors that we witness each autumn in deciduous trees and other plants is caused at least in part by flavonoids responding to oxidation, which offers protection to leaf cells against damage from photosynthetic activity or ultraviolet radiation[7]. In this case, flavonoids (together with carotenoids) can produce colors that aren't meant to attract pollinators; if anything, the colors have come to serve as a warning of impending winter.

*Pollinator Attraction

Perhaps the most visible role of flavonoids in plants is to attract pollinators. While flavonoids are well-known for producing yellow pigments in flowers and pollen that can be detected by insects and some animals, a variation of flavonoids called anthocyanidins produce a wide range of other colors in flowers and berries[8,9]. The health-foods industry has capitalized on this information, touting foods like dark-colored berries, eggplant, pomegranate, and beets as potent sources of antioxidants.

But because cannabis doesn't rely on pollinators for propagation, it is not likely that it uses flavonoids in this manner. However, some varieties of cannabis do feature flowers that are deep purple, red, and sometimes even blue hues, and this is caused by the anthocyanidin group of flavonoids.

So why exactly would cannabis need to display bright colors?

The answer is that the lively shades of anthocyanidins in some varieties of cannabis are likely related to UV protection or another response to abiotic stress, much in the same way as mentioned in the previous section, where

[5] Kumar, Shashank, and Abhay K. Pandey. Chemistry and Biological Activities of Flavonoids: An Overview. The Scientific World JOURNAL 2013.2013 (2013).

[6] Agati G, Azzarello E, Pollastri S, Tattini M. Flavonoids as antioxidants in plants: location and functional significance. *Plant Science*. 2012; 196:67-76.

[7] Feild T. S., Lee D. W., Holbrook N. M. Why leaves turn red in autumn. The role of anthocyanins in senescing leaves of red-osier dogwood. Plant Physiol. 127, 566-574. (2001).

[8] Falcone Ferreyra, Maria L., Sebastian P. Rius, and Paula Casati. Flavonoids: biosynthesis, biological functions, and biotechnological applications. Frontiers in Plant Science 3 (2012).

[9] Dudek B., Warskulat A.-C., Schneider B. The occurrence of flavonoids and related compounds in flower sections of Papaver nudicaule. Plants. 2016;5:28.

the colors of autumn can be attributed partly to the protective and restorative effects of flavonoids. Considering that the most common pollinators such as bees, flies, wasps, butterflies, moths, and beetles do not show a preference for the color purple[10] anyway, it seems even less likely that the bright purples of cannabis varieties like Lavender, Purple Urkle, or Mendocino Kush were meant to attract pollinators.

It's also likely that the prevalence of purple, red, and blue cannabis varieties is a result of years of deliberate clandestine breeding for visually appealing strains[11], without regard for developing a specific chemical profile.

*Symbiotic Relationships with Fungi & Bacteria

One of the most interesting uses of flavonoids by plants is to develop symbiotic relationships with certain microorganisms. For instance, a number of plants - including cannabis - manufacture and release flavonoids via the root system to encourage beneficial soil bacteria to attach to the roots[12]. Other flavonoids - such as those found in legume roots - stimulate the germination of beneficial fungi spores[13]. But in an odd sort of twist, flavonoids can also be used by plants as antibacterial and antifungal agents.

*Antibacterial Agents

While some flavonoids are made by plants to stimulate symbiotic relationships with bacteria, other flavonoids can be used as effective antibacterial agents when plants come under attack by harmful microorganisms. Several flavonoids have been found to possess robust antibacterial properties, including apigenin[14], a flavonoid often found in cannabis. Another flavonoid that commonly occurs in cannabis is quercetin, which has been attributed to the inhibition of the DNA gyrase of bacteria[15] (an essential bacterial enzyme).

The action of flavonoids on bacteria is potent, with some studies concluding that these compounds may inhibit the ability of bacteria to attach to other cells[16], turn off microbial enzymes, and alter microbial membranes[17], among other antibacterial functions. Impressively, some flavonoids may be able to overcome antibiotic tolerance of at least some bacteria[18]. And as is the case with terpenes, it is likely that plants use flavonoids synergistically. One Algerian study appears to have confirmed this, concluding that while a variation of the flavonoid quercetin had significant antibacterial properties, a synergistic blend of variations of quercetin, apigenin, and kaempferol was more effective against strains of clinical bacteria[19].

---

[10] Reverte, Sara, Javier Retana, Jose M. Gomez, and Jordi Bosch. Pollinators show flower colour preferences but flowers with similar colours do not attract similar pollinators. Annals of Botany 118.2 (2016): 249-257.

[11] Gagalova KK, Yan Y, Wang S, Matzat T, Castellarin SD, Birol I, Edwards D, Schuetz M. Leaf pigmentation in Cannabis sativa: Characterization of anthocyanin biosynthesis in colorful Cannabis varieties. Plant Direct. 2024 Nov 25;8(11):e70016.

[12] Dixon RA, Steele CL. Flavonoids and isoflavonoids a gold mine for metabolic engineering. Trends Plant Sci. 1999; 4(10):394-400.

[13] Bagga S, Straney D. Modulation of cAMP and phosphodiesterase activity by flavonoids which induce spore germination of Nectriahaematococca MP VI (Fusarium solani).Physiol. Mol. Plant Path. 2000; 56(2):51-61.

[14] Cushnie TPT, Lamb AJ. Antimicrobial activity of flavonoids. International Journal of Antimicrobial Agents. 2005;26(5):343-356.

[15] Cushnie TP, Lamb AJ. Antimicrobial activity of flavonoids. Int J Antimicrob Agents. 2006 Feb;27(2):181.

[16] Cowan MM. Plant products as antimicrobial agents. Clinical Microbiology Reviews. 1999;12(4):564-582.

[17] Xie Y, Yang W, Tang F, Chen X, Ren L. Antibacterial activities of flavonoids: structure-activity relationship and mechanism. Curr Med Chem. 2015;22(1):132-49.

[18] Nobakht, Motahareh, Stephen J. Trueman, Helen M. Wallace, Peter R. Brooks, Klrissa J. Streeter, and Mohammad Katouli. Antibacterial Properties of Flavonoids from Kino of the Eucalypt Tree, Corymbia torelliana. Plants 6.3 (2017).

[19] Akroum, Souad & Bendjeddou, Dalila & Satta, Dalila & Lalaoui, Korrichi. (2009). Antibacterial Activity And Acute Toxicity Effect of Flavonoids Extracted From Mentha longifolia. Am. Eurasian J. Sci. Res. 4.

Antifungal Agents

Flavonoids can protect plants from infestations of harmful fungi. Variations of kaempferol have exhibited antifungal activity against fusarium soil pathogens[20], which sometimes attack both drug varieties and fiber varieties of cannabis. Fortunately, cannabis has at least some native protection from these potentially dangerous fungi, as many strains have been found to contain kaempferol.

Quercetin also exhibits antifungal properties, with recent research suggesting that the flavonoid might be able to modulate (and thereby inhibit) the fatty acid synthase of certain types of fungi that commonly infect humans[21], while other studies have shown variations of quercetin and other flavonoids to suppress fungal growth of certain fungal species by up to 99%[22]. In fact, antifungal activities of six types of flavonoids were markedly more potent than Fluconazole[23], one of the most common antifungal medications in the United States.

UV Protection

Flavonoids provide plants with significant protection from harmful ultraviolet and other types of radiation. This includes both preventative and restorative functions, as evidenced in a study that took place on the international space station, which concluded that flavonoids absorbed cellular damage from UV radiation[24]. Other research has confirmed this, showing that flavonoids can repair damage from UV-B and UV-A radiation by inhibiting ROS (reactive oxygen species), and repairing damaged tissue[25].

Seed Germination & Rooting

Most plant seeds contain flavonoids that serve multiple purposes, including regulating critical physiological functions throughout the seed life stages, from dormancy period to maturation[26], followed by supporting seedling and root development[27]. Additionally, higher seed flavonoid content has been associated with darker seed hull color in some plants[28], corresponding with other studies that found a positive relationship between antioxidant activity and darker seed hull color[29,30]. This should be of particular interest to the cannabis industry, where urban myths -

[20] Francesco Galeotti, Elisa Barile, Paolo Curir, Marcello Dolci, Virginia Lanzotti. Flavonoids from carnation (Dianthus caryophyllus) and their antifungal activity, Phytochemistry Letters, Volume 1, Issue 1, 2008, Pages 44-48.

[21] Bitencourt, Tamires Aparecida, Tatiana Takahasi Komoto, Mozart Marins, and Ana Lucia Fachin. Antifungal activity of flavonoids and modulation of expression of genes of fatty acid synthesis in the dermatophyte Trichophyton rubrum. BMC Proceedings 8.Suppl. 4 (2014): P53-P53.

[22] Kanwal Q, Hussain I, Latif Siddiqui H, Javaid A. Antifungal activity of flavonoids isolated from mango (Mangifera indica L.) leaves. Nat Prod Res. 2010 Dec;24(20):1907-14.

[23] Orhan DD, Ozelik B, Ozgen S, Ergun F. Antibacterial, antifungal, and antiviral activities of some flavonoids. Microbiol Res. 2010 Aug 20;165(6):496-504.

[24] Takahashi A, Ohnishi T. The significance of the study about the biological effects of solar ultraviolet radiation using the exposed facility on the internal space station. Biol. Sci. Space2004; 18(4): 255-260.

[25] Agati G, Azzarello E, Pollastri S, Tattini M. Flavonoids as antioxidants in plants: location and functional significance. Plant Science. 2012; 196:67-76.

[26] Shirley BW, Flavonoids in seeds and grains: physiological function, agronomic importance, and the genetics of biosynthesis. Seed Sci. Res. 1998; 8: 415-422.

[27] Nandakumar, Lakshmi, and S. N.Rangaswamy. Effect of some Flavonoids and Phenolic Acids on Seed Germination and Rooting. Journal of Experimental Botany 36.8 (1985).

[28] O.S. Salawu, I.E. Oluwafemi, D. Oladipupo, B.O.S. Bukola. Effect of Callosobruchus maculatus infestation on the nutrient-antinutrient composition, phenolic composition and antioxidant activities of some varieties of cowpeas (Vigna unguiculata) Adv. J. Food Sci. Technol., 6 (2014), pp. 322-332.

[29] B.D. Oomah, F. Caspar, L.J. Malcolmson, A.S. Bellido. Phenolics and antioxidant activity of lentil and pea hulls Food Res. Int., 44 (2011), pp. 436-441.

[30] P. Polthum, A. Ahromrit GABA content and antioxidant activity of Thai waxy corn seeds germinated by hypoxia method. Songklanakarin J. Sci. Technol., 36 (2014), pp. 309-316.

now apparently confirmed - have circulated for decades that darker seeds contain better genetics. What makes those seeds darker? Higher antioxidant flavonoid content.

*Detoxification

Flavonoids are excellent detoxification agents, especially of metals. For instance, in Ginko biloba, flavonoids were shown to increase plant tissue production by up to twelve times the normal level in response to heavy metal stress[31]. Many of the flavonoids that are found in a large number of today's cannabis varieties, such as quercetin, kaempferol, luteolin, and apigenin, have been shown to be effective at detoxifying both copper and iron[32]. However, in a true testament to the versatility of flavonoids, these phytochemicals can exhibit both antioxidant activity and prooxidant activity, depending on the combination of flavonoids and concentration levels of metal present[33].

*Regulation of Cells

Flavonoids play multiple roles in the regulation of plant cells. One such role is the modulation of the movement of the plant hormone auxin[34], which elongates cells and regulates plant growth. This is an important consideration for cannabis growers because the act of tipping (removing the topmost flowers of cannabis or bending the top flowers to be level or lower than buds farther down the plant) is a method of manipulating auxin flow, whereby the topmost flowers inhibit the growth of flowers below them with a downward flow of auxins.

Plant cells contain further phytochemicals that function as flavonoid membrane transporters[35], while antioxidant flavonoids regulate parts of cell growth and differentiation, and thus help control the development of individual plant parts, and the whole plant[36]. Flavonoids play these roles in the natural state, but they also respond similarly under manipulation in the laboratory setting, reacting to altered genes and suppressing growth of cultured cells[37].

While much research is ongoing, and more is extremely needed, the already prolific roles of flavonoids in the regulation of plant cells has led one of the world's leading flavonoid researchers, Brenda Winkel, to conclude that one can "consider flavonoid biosynthesis, not as an assemblage of independent components, but as part of a large, complex, and tightly orchestrated metabolic network[38]." Other scientists have echoed this complex nature of flavonoids, even pointing out that it can be difficult to determine whether a particular flavonoid or group of flavonoids are acting primarily as cell regulators, or as defense compounds against herbivores, as antioxidants, or other roles.

[31] Samanta Amalesh, Das Gouranga, Das, Sanjoy. (2011). Roles of flavonoids in Plants. International Journal of pharmaceutical science and technology. 6. 12-35.

[32] Fernandez MT, Mira ML, Florencio MH, Jennings KR. Iron and copper chelation by flavonoids: an electrospray mass spectrometry study. J Inorg Biochem. 2002 Nov 11;92(2):105-11.

[33] Cherrak, Sabri Ahmed, Nassima Mokhtari-Soulimane, Farid Berroukeche, Bachir Bensenane, Angeline Cherbonnel, Hafida Merzouk, and Mourad Elhabiri. In Vitro Antioxidant versus Metal Ion Chelating Properties of Flavonoids: A Structure-Activity Investigation. PLoS ONE 11.10 (2016).

[34] Wendy Ann Peer, Angus S. Murphy, Flavonoids and auxin transport: modulators or regulators? Trends in Plant Science, Volume 12, Issue 12, 2007, Pages 556-563.

[35] Passamonti S, Terdoslavich M, Franca R, Vanzo A, Tramer F, Braidot E, Petrussa E, Vianello A. Bioavailability of flavonoids: a review of their membrane transport and the function of bilitranslocase in animal and plant organisms. Current Drug Metabolism. 2009 May;10(4):369-94.

[36] Agati G, Azzarello E, Pollastri S, Tattini M. Flavonoids as antioxidants in plants: location and functional significance. Plant Sci. 2012 Nov;196:67-76.

[37] Woo, Ho-Hyung, Byeong Jeong, and Martha Hawes. Flavonoids: from cell cycle regulation to biotechnology. Biotechnology Letters 27.6 (2005): 365-374.

[38] Winkel-Shirley, Brenda. Flavonoid Biosynthesis. A Colorful Model for Genetics, Biochemistry, Cell Biology, and Biotechnology. 126.2 (2001)485.

Allelopathic Properties

Like terpenes, flavonoids possess allelopathic functions[39]; flavonoids engage in chemical warfare with competing plants. For instance, when released from the roots of buckwheat, the flavonoid quercetin strongly inhibited the growth of nearby lettuce seedlings[40]. Other examples include legumes that exude the flavonoids quercetin and kaempferol - beneficial for seed germination at low concentrations, but inhibitory to the growth of seedlings at high concentrations[41]; the flavonoid luteolin, which, as isolated from dogwood, inhibited both seed germination and seedling growth of nearby plants[42]. In fact, the success of the world's most tenacious weeds is largely attributed to allelopathic properties of flavonoids[43]. One study has even concluded that a crude extract of flavonoids had a 100% inhibitory effect on the seed germination and plant growth for M. pigra[44] (the Giant Sensitive Plant), a staggering testament to the allelopathic properties of flavonoids.

CHEMICAL NATURE OF FLAVONOIDS

Like terpenes, flavonoids are formed in the cytosol of plant cells from the derivatives of acetic acids. However, while this occurs in the methyl erythritol phosphate pathway for monoterpenes and the mevalonate pathway for sesquiterpenes, the biosynthesis process for most flavonoids occurs initially in the shikimic acid pathway[45], and is completed in the phenylpropanoid pathway. Found in the nucleus of mesophyll cells (plant leaf tissue specialized for photosynthesis) and within centers of reactive oxygen species generation[46], flavonoids are one of the most widely occurring plant phenolic compounds, with more than 5,000 distinct flavonoids reported by 2018.

As mentioned, plants produce flavonoids in the phenylpropanoid metabolic pathway[47] (which is the origination point for multiple key plant compounds), where an amino acid creates several substances leading to the development of a compound containing two benzene (phenyl) rings, referred to as Ring A and Ring B. These two rings are joined in the center by a 3-carbon ring of pyran containing oxygen, called Ring C. This forms the classic fifteen carbon skeleton structure of flavonoids, which can be abbreviated as $C_6C_3C_6$ (see image of flavonoid skeleton at the beginning of this chapter).

Compounds consisting of this basic chemical structure can be divided into six groups; the first being the **chalcones**, which consist only of the two benzene rings, ring A and ring B. In this way, chalcones are the base structure of higher flavonoids. Other compounds in this family include **flavones, flavonols, flavandiols, anthocyanins,** and

---

[39] Weston, Leslie, and Ulrike Mathesius. Flavonoids: Their Structure, Biosynthesis and Role in the Rhizosphere, Including Allelopathy. Journal of Chemical Ecology 39.2 (2013): 283-297.

[40] Zahida Iqbal, Syuntaro Hiradate and Yoshiharu Fujii. Allelopathic flavonoids from buckwheat (Fagopyrum tataricum Gaertn.) (NIAES) J. Weed Sci. Tech. Vol. 48 (Sup.)

[41] Mariana Palma-Tenango, Marcos Soto-Hernandez and Eva Aguirre- Hernandez. Flavonoids in Agriculture. InTechOpen Published: August 23rd 2017.

[42] Zhang, H.Y. & Qi, Shan-Shan & Dai, Zhi-Cong & Zhang, M. & Sun, J.F. & Du, D.L.. (2017). Allelopathic potential of flavonoids identified from invasive plant Conyza canadensis on Agrostis stolonifera and Lactuca sativa. Allelopathy Journal. 41. 223-237.

[43] Alford, Elan R., Jorge M. Vivanco, and Mark W. Paschke. The Effects of Flavonoid Allelochemicals from Knapweeds on Legume-Rhizobia Candidates for Restoration. Restoration Ecology 17.4 (2009): 506-514.

[44] Tikamporn Yongvanich, Waraporn Juntarajumnong, Sittichoke Nakapong. Allelopathic effect of flavonoids from Typha angustifolia on seed growth of Mimosa pigra. Thai Journal of Agricultural Science (Thailand) 2002.

[45] Wang TY, Li Q, Bi KS. Bioactive flavonoids in medicinal plants: Structure, activity and biological fate. Asian J Pharm Sci. 2018 Jan;13(1):12-23.

[46] Kumar, Shashank, and Abhay K. Pandey. Chemistry and Biological Activities of Flavonoids: An Overview. The Scientific World Journal 2013.2013 (2013).

[47] Falcone Ferreyra Maria Lorena, Rius Sebastian, Casati Paula. Flavonoids: biosynthesis, biological functions, and biotechnological applications. Frontiers in Plant Science Volume 3 2020.

tannins – with the latter two being responsible for the bright purple, red, and blue hues of many plants, including cannabis. Flavones and flavonols, on the other hand, are often responsible for yellow and white coloration. When ring B is attached at position 2 on ring C, this formation comprises the flavones, flavonols, flavnones, flavanonols, catechins, and anthocyanins. However, the rings can be attached in other positions; when ring B is linked at position 3 on ring C, these compounds are called isoflavones; when ring B is linked at position 4 on ring C, the resulting compounds are called neoflavonoids.

Flavonoid Variations

The activity or function of flavonoids depends on the chemical structure of the compound[48], and there are seemingly endless possible modifications or variations. Flavonoid variations occur through several biochemical modifications that influence their structure, solubility, and biological activity:

Hydroxylation – the addition of hydroxyl (-OH) groups (an oxygen atom bonded to a hydrogen atom) at different positions on the flavonoid skeleton, affecting antioxidant activity and solubility.

Methylation – the transfer of methyl (-CH$_3$) groups (one carbon atom bonded to three hydrogen atoms), usually to hydroxyl groups, increasing lipophilicity and stability, which can enhance bioavailability and membrane permeability.

Glycosylation – the attachment of sugar molecules (glucose, rhamnose, etc.) to hydroxyl groups, improving flavonoid solubility, stability, and transport in plants and humans.

Acylation – the addition of acyl groups (malonyl, acetyl, etc.) to flavonoids, modifying their stability, solubility, and interactions with cellular components.

Prenylation – the attachment of prenyl (isoprene-derived) groups, enhancing lipid solubility and often increasing biological activity, including antibacterial and anticancer properties.

Sulfonation – the addition of sulfate (-SO$_3$H) groups (a sulfur atom bonded to three oxygen atoms and a hydroxyl (-OH) group), influencing water solubility, detoxification, and metabolic excretion in organisms.

Halogenation – the incorporation of halogen atoms (fluorine, chlorine, bromine, iodine) into the flavonoid structure, which can enhance biological activity, increase stability, and modify interactions with enzymes and receptors.

C-ring Modifications – structural changes to the central flavonoid ring, including oxidation or reduction, leading to different flavonoid subclasses (for example, flavones vs. flavanones).

These modifications contribute to the vast diversity of flavonoids, affecting their function, bioavailability, and role in plant and human health. Like terpenes, flavonoids can be built chemically in many ways, with each modification serving one or more purposes. We have seen this in the preceding sections, where flavonoids have been shown to possess both antioxidant and prooxidant, antibacterial and pro-bacterial, antifungal and pro-fungal, and other conflicting properties. Slight variations in flavonoid structure can result in entirely distinct functions, making flavonoids among the most versatile of phytochemicals.

---

[48] Kelly EH, Anthony RT, Dennis JB. Flavonoid antioxidants: chemistry, metabolism and structure-activity relationships. Journal of Nutritional Biochemistry. 2002;13(10):572-584.

Flavonoids Biosynthesis

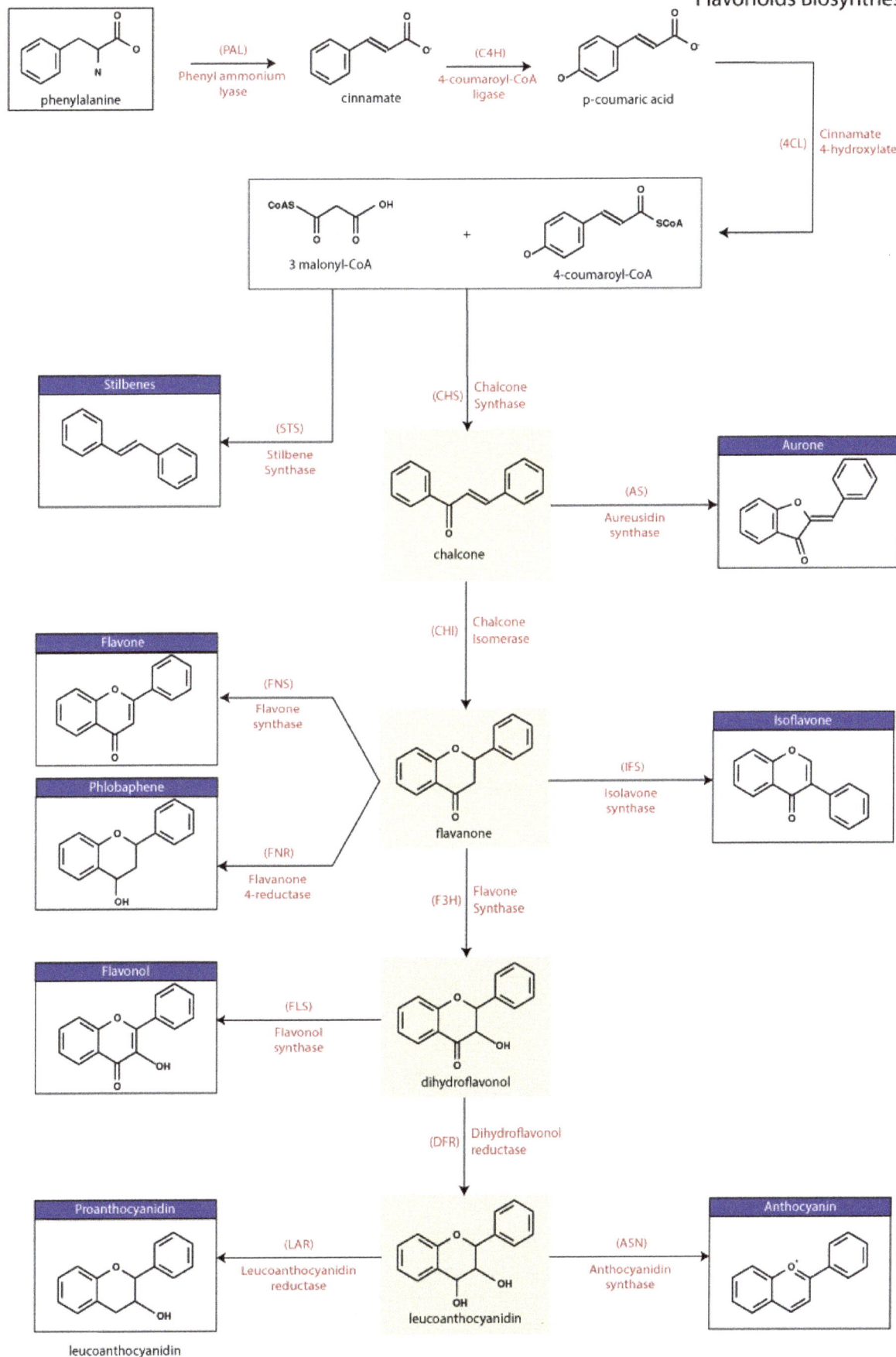

However, not all flavonoids nor all variations occur in cannabis. Laboratory testing of cannabis in the United States typically shows that most strains contain some or all of just 14 flavonoids as listed below. Unfortunately, just because these are the flavonoids most cannabis testing laboratories are testing for does not mean that these are the most prevalent flavonoids in cannabis. Laboratories can only test for compounds that they have known standards for, coupled with detailed SOPs on how to quantify the compound using the standard. Meaning, a cultivator may have a variety where the primary flavonoid constituent is chrysin, but if the laboratory where they send their samples doesn't test for that compound, then the results will instead show some other compound, or that there is an unknown compound, where further testing and analysis is required. Additionally, it must be noted that most laboratories don't test for all of the below flavonoids – in general, cannabis labs test for 4-8 of the most common flavonoids.

Top Flavonoids in Typical US Cannabis Laboratory Flavonoid Testing Panels

Cannflavin A
Cannflavin B
Orientin
Rutin
Apigenin
Quercetin
Cannflavin C
Silymarin
Kaempferol
Vitexin
Isovitexin
Luteolin
Wogonin
Beta-sitosterol

Interestingly, despite the fact that the beta-sitosterol compound is NOT a flavonoid and is in fact a plant sterol, many laboratories erroneously consider this compound a flavonoid and test for it in cannabis strains accordingly. Because the compound is so often included in flavonoids testing panels, the author chose to include it in this list and allocated the compound a separate chapter. Additionally, silymarin is mistakenly listed and tested for as an individual molecule, when it is actually a flavonoid complex consisting of several compounds. The author has also included a short chapter explaining this situation.

Most of these flavonoids have been mentioned in the preceding sections, and their proven functions in other plants are likely similar to their functions in cannabis. Nevertheless, the exceptional ability of flavonoids to organize and act in thousands of different ways has lent cannabis some unique compounds in the form of cannflavin A, B, and C, which are generally thought to only occur in cannabis[49], although cannflavin A has also been found in Mimulus bigelovii[50].

Remarkably, the flavonoid constituent profile of most types of cannabis, including 'indica,' 'sativa,' and 'ruderalis,' are similar enough that a 1971 study seeking to determine if cannabis is monotypic (a genus containing only one species) or polytypic (containing more than one species) concluded that cannabis has only one species, cannabis sativa L.[51], and further mused that all variations are phenotypes that can be bred into or out of existence in a single season. Most cannabis breeders know this to be true from experience, but these conclusions were based on the

---

[49] Brenneisen R. (2007) Chemistry and Analysis of Phytocannabinoids and Other Cannabis Constituents. In: ElSohly M.A. (eds) Marijuana and the Cannabinoids. Forensic Science and Medicine. Humana Press.

[50] Salem MM, Capers J, Rito S, Werbovetz KA. Antiparasitic activity of C-geranyl flavonoids from Mimulus bigelovii. Phytother Res. 2011 Aug;25(8):1246-9.

[51] Murray Nelson Clark. A Study of InfraSpecific Flavonoid Variation of Cannabis Sativa L. (Cannabaceae) Brandon University, Department of Botany, 1971.

analysis of cannabis-extracted flavonoids originating from different regions of the world, using both feral and cultivated varieties. The overall flavonoid profiles were not dissimilar enough to warrant the classification of three species of cannabis, though the colloquial debate of the subject continues.

This monotypic theory is also evidenced by the voracious interbreeding between varieties of cannabis; there are virtually no propagation barriers between strains or phenotypes, and resultant hybrid offspring do not experience reduced reproductive capacity. This further cements the theory that there is only one cannabis plant, but that the wide variety of distinct chemotypes and phenotypes is based on the chemical profile of each plant, including the ratios of cannabinoids, terpenes, flavonoids, esters, aldehydes, sulfurs, and other compounds.

CONCENTRATION OF FLAVONOIDS IN PLANTS

The concentration or total dry weight of flavonoids in plants varies widely, depending on the species. Among ninety-one different edible plant species, flavonoid content ranged from 0-254 mg per 100g of fresh plant[52]. Leafy vegetables showed a higher flavonoid content than fruit crops, with parsley containing 14.4 mg of quercetin per gram of dry weight[53], the highest for the leafy vegetables examined in the study. A separate study found that onion leaves had the highest flavonoid content out of sixty-two edible tropical plants, containing 1497.5 mg/kg quercetin, 391.0 mg/kg luteolin, and 832.0 mg/kg kaempferol[54].

In the cannabis industry, it is generally thought that flavonoids comprise around 2.5% of dry weight of cannabis sativa inflorescences, although data concerning the total and specific flavonoid content in trichomes is scarce as of the publishing of this text. Recent work has shown that flavonoid content is higher in cannabis leaves than in flowers, with some research showing between 0.07–0.14% flavonoids in flowers and 0.34–0.44% in leaves[55], and similar work showing between 0.028–0.284% in flowers and 0.051–0.470% in leaves[56].

Interestingly, research conducted in 2020 found that the Kompolti variety of hemp accumulated significantly more of the prenylated flavone cannflavin A at high altitude in the mountains, versus at low altitude in the plains[57], an important consideration for cannabis cultivators.

HUMAN CONSUMPTION AND METABOLISM OF FLAVONOIDS

Flavonoids are common in the human diet, and are largely responsible for the color, prevention of fat oxidation, and protection of vitamins and enzymes in foods[58]. These compounds are so ubiquitous in the human diet that some researchers have suggested that flavonoids in human urine can be measured and used as biomarkers for intake of

[52] Yang RY, Lin S, Kuo G. Content and distribution of flavonoids among 91 edible plant species. Asia Pac J Clin Nutr. 2008;17 Suppl 1:275-9.

[53] Chandra, Suman, Shabana Khan, Bharathi Avula, Hemant Lata, Min Hye Yang, Mahmoud A. ElSohly, and Ikhlas A. Khan. Assessment of Total Phenolic and Flavonoid Content, Antioxidant Properties, and Yield of Aeroponically and Conventionally Grown Leafy Vegetables and Fruit Crops: A Comparative Study. Evidence-based Complementary and Alternative Medicine: eCAM 2014 (2014).

[54] Miean KH, Mohamed S. Flavonoid (myricetin, quercetin, kaempferol, luteolin, and apigenin) content of edible tropical plants. J Agric Food Chem. 2001 Jun;49(6):3106-12.

[55] Jin D, Dai K, Xie Z, Chen J. Secondary Metabolites Profiled in Cannabis Inflorescences, Leaves, Stem Barks, and Roots for Medicinal Purposes. Sci Rep. 2020 Feb 24;10(1):3309.

[56] Jin D, Henry P, Shan J, Chen J. Identification of Chemotypic Markers in Three Chemotype Categories of Cannabis Using Secondary Metabolites Profiled in Inflorescences, Leaves, Stem Bark, and Roots. Front Plant Sci. 2021 Jul 1;12:699530.

[57] Giupponi L, Leoni V, Pavlovic R, Giorgi A. Influence of Altitude on Phytochemical Composition of Hemp Inflorescence: A Metabolomic Approach. Molecules. 2020 Mar 18;25(6):1381.

[58] Yao LH, Jiang YM, Shi J, et al. Flavonoids in food and their health benefits. Plant Foods for Human Nutrition. 2004;59(3):113-122.

fruits and vegetables[59]. Flavonoids are also extremely pharmacologically active, and there is a substantial amount of literature that has established the potential medical value of flavonoids, which we will discuss in detail for each individual flavonoid chapter.

Flavonoids undergo extensive metabolic transformations in the human body, which influences their bioavailability and biological effects. Understanding the metabolic fate of flavonoids is crucial for elucidating their roles in human health.

Absorption and Initial Metabolism

Upon ingestion, flavonoid glycosides—flavonoids bound to sugar moieties—are subjected to enzymatic hydrolysis in the small intestine. Enzymes such as lactase-phlorizin hydrolase (LPH) located on the brush border membrane, and cytosolic β-glucosidase within enterocytes, cleave these glycosidic bonds, releasing the aglycone forms. These aglycones can be absorbed directly into enterocytes. Once inside, they undergo phase II metabolism, primarily involving O-methylation, sulfation, and glucuronidation, facilitated by enzymes like catechol-O-methyltransferase, sulfotransferases, and uridine 5'-diphospho-glucuronosyltransferases. These conjugation reactions enhance the hydrophilicity of flavonoids, promoting their transport into the bloodstream and subsequent distribution to various tissues.

Role of the Colonic Microbiota

A significant portion of dietary flavonoids escape absorption in the small intestine and reaches the colon, where they encounter the gut microbiota. Intestinal bacteria play a pivotal role in the further metabolism of these compounds. One notable microbial transformation is the cleavage of the central C-ring of flavonoids. Studies have identified specific human intestinal bacteria capable of performing this C-ring cleavage, leading to the production of phenolic acids such as 3,4-dihydroxyphenylacetic acid[60]. This process not only alters the structure and function of flavonoids but also affects their bioavailability and biological activity.

Systemic Circulation and Excretion

The metabolites resulting from both host and microbial metabolism can be absorbed into the systemic circulation. These metabolites may exert various biological effects, including antioxidant, anti-inflammatory, and cardioprotective activities. Ultimately, flavonoid metabolites are excreted from the body primarily through urine and bile. The efficiency of excretion depends on factors such as the degree of conjugation and the molecular size of the metabolites.

Factors Influencing Flavonoid Metabolism

Several factors can influence the metabolism and bioavailability of flavonoids, including the specific structure of the flavonoid, the composition of the individual's gut microbiota, genetic variations affecting metabolic enzymes, and interactions with other dietary components. Understanding these factors is essential for assessing the health implications of flavonoid consumption and for developing dietary recommendations.

Potential Toxicity of Flavonoids

While flavonoids are widely regarded for their general safety and beneficial effects on human health, including their antioxidant, anti-inflammatory, and anticancer properties, these compounds could be toxic under certain conditions.

---

[59] Nielsen SE, Freese R, Kleemola P, Mutanen M. Flavonoids in human urine as biomarkers for intake of fruits and vegetables. Cancer Epidemiol Biomarkers Prev. 2002 May;11(5):459-66.

[60] Winter J, Moore LH, Dowell VR Jr, Bokkenheuser VD. C-ring cleavage of flavonoids by human intestinal bacteria. Appl Environ Microbiol. 1989 May;55(5):1203-8.

The bioactivity and safety of flavonoids depends on various factors, including their dose, chemical structure, and the metabolic pathways they undergo in the body.

Flavonoids interact with a wide range of enzymes, including those involved in drug metabolism, such as cytochrome P450 enzymes. These interactions can lead to both positive and negative effects. For example, some flavonoids can inhibit cytochrome P450 enzymes, potentially interfering with the metabolism of other drugs and leading to adverse drug interactions[61]. Additionally, high doses of certain flavonoids, particularly those with potent estrogenic effects, may alter hormonal balance and lead to adverse health outcomes, including reproductive toxicity.

As mentioned previously, flavonoids are subject to metabolism by gut microbiota, which can produce both beneficial and potentially harmful metabolites. For example, while some microbial metabolites have demonstrated health-promoting effects, others may contribute to toxicity, depending on the individual's gut microbiome composition and the flavonoid's structure.

Additionally, excessive consumption of flavonoid-rich supplements may increase the risk of liver toxicity due to the accumulation of conjugated metabolites. Flavonoids such as quercetin, for example, have been shown to exhibit toxic effects at high doses, including hepatotoxicity and renal toxicity, primarily due to their impact on metabolic enzymes.

Overall, while moderate intake of flavonoid-rich foods is generally safe and beneficial, caution is warranted with high-dose flavonoid supplements, particularly in individuals with compromised liver function or those on medications that interact with flavonoids. More research is required to fully understand the potential toxicity of flavonoids, as well as their safe dosage ranges for various populations.

Flavonoid Metabolism - Do we get flavonoids from cannabis?

The metabolism and fate of flavonoids in the human body is complex and variable. Many readers interested in cannabis probably want to know if these compounds are bioavailable exactly as they occur in various cannabis products. The most direct answer is that consuming raw cannabis leaves and flowers is likely to result in the dietary acquisition of flavonoids, including cannflavins A, B, and C. However, the metabolism and fate of flavonoids – if any – in cannabis products that are smoked, vaped, inhaled, or those that are found in extracts, concentrates, or other products is unknown and/or unreported.

However, we can make some assumptions. For instance, if flavonoids are present in a vaped cannabis product, and the product is vaporized at a temperature lower than the point of combustion, it is likely that some of the flavonoids will be inhaled. The fate of those flavonoids in the human oral cavities, throat, esophagus, lungs, and blood is unknown.

If they contain flavonoids, products like hemp or cannabis skin creams may be efficacious, depending on the desired effect. As an example, cannabis-based creams that contain wogonin or baicalin are likely to permit absorption through the skin, where the molecules can act as tightening or whitening/brightening agents. The effects of a particular flavonoid will probably be similar, regardless of whether the compound is sourced from cannabis, made in the lab, or extracted from another plant source.

Finally, cannabis edibles, when formulated correctly, may allow the acquisition of flavonoids during consumption.

---

[61]Skibola CF, Smith MT. Potential health impacts of excessive flavonoid intake. Free Radic Biol Med. 2000 Aug;29(3-4):375-83.

Flavonoids 101 Review

Answer the following questions to test your knowledge of this section:

Question #1: Flavonoids serve what primary biological role in plants:

     a. Attract pollinators by flavor or taste
     b. Defend against predators with warning colors
     c. Deter herbivores with bitter flavorings
     d. Attract pollinators with colors

Question #2: Name three ways that plants use flavonoids:

_____, _____, _____

Question #3: How many variations of flavonoids have been discovered as of 2018?

     a. 15,000
     b. 25,000
     c. 7,500
     d. 5,000

Question #4: The basic flavonoid molecule contains how many rings?

     a. Four
     b. Three
     c. Two
     d. Five

Question #5: Name the three flavonoids that (as of 2025) have mostly only been found in cannabis:

_____, _____, _____

Question #6: Flavonoid biosynthesis occurs in which pathway?

     a. Skikmich ludentine pathway
     b. Mevalonate pathway
     c. Isoprenic pathway
     d. Phenylpropanoid pathway

Question #7: Name two potential flavonoid modifications or variations:

_____, _____

Question #8: Name one primary difference between terpenes and flavonoids:

_____

Question #9: What is the primary way that flavonoid metabolites are excreted through the body?

_____

Question #10: Name 3 subclasses of flavonoids:

_____, _____, _____

For the answer key to Flavonoids 101, please visit www.cannabischemistry.org

SYNERGY

Oxford Language Definition of Synergy:

"The interaction or cooperation of two or more organizations, substances, or other agents to produce a combined effect greater than the sum of their separate effects."

[Editor's note: The Introduction and Synergy 101 sections below are adapted from Edition 2 of The Big Book of Terps]

INTRODUCTION

Synergy as we define it today (particularly in the cannabis industry) was first proposed as a concept by Shimon Ben-Shabat in 1998[62], who soon thereafter refined the theory together with Raphael Mechoulam[63] and mutual colleagues. At the time, Ben-Shabat referred to synergistic phenomena observed between compounds in the endocannabinoid system as the entourage effect, although it is only quite recently that this concept has been trending in the cannabis industry. However, evidence for synergy in plants has been known for at least a half century, and this is particularly true of the evidence for synergy among phytochemicals like terpenes, flavonoids, and other compounds.

Initially, Ben-Shabat and Mechoulam referred primarily to synergy between endogenous and exogenous cannabinoids, and later Dr. Ethan Russo extended the concept to include terpenes, terpenoids, and potentially other compounds including flavonoids. The principal idea behind the entourage effect is that pure compounds in cannabis - particularly the primary psychoactive compounds like THC and CBD - are modulated or supplemented by other compounds, mostly in a beneficial and additive way, resulting in therapeutic effects that are more potent than those elicited by single molecules like D9-THC.

In recent years, the cannabis industry has seized on the concept of the entourage effect, but with strikingly little understanding of the mechanisms behind it. Cannabis products and marketing materials in the United States, Canada, Europe, Israel, South America, and other places around the world tout the entourage effect as a proven, well-established, and unflinchingly positive aspect of their much-venerated plant.

Unfortunately, many of the benefits claimed by these products, groups, and marketing materials are untested and disseminated without sound science - or even basic knowledge of chemistry - to support them. Perhaps worse is that potential negative or detrimental synergies are ignored.

Although there is little research into synergy between flavonoids and other compounds in cannabis, there is ample and well-established (albeit neglected) evidence for synergy in non-cannabis plants. We can make some logical assumptions about how these synergies might work in cannabis by studying the literature about synergy in other plants. Understanding synergy and how it could apply to cannabis is critical for the nascent medical and recreational markets, and especially for thousands of ill people and the healthcare workers who treat them, the latter of which often are not aware that cannabinoid monotherapies may not be as efficacious as blended preparations or formulations[64].

---

[62] Ben-Shabat S., Fride E., Sheskin T., Tamiri T., Rhee M. H., Vogel Z., et al. (1998). An entourage effect: inactive endogenous fatty acid glycerol esters enhance 2-arachidonoyl-glycerol cannabinoid activity. Eur. J. Pharmacol. 353 23-31.

[63] Mechoulam R., Ben-Shabat S. (1999). From gan-zi-gun-nu to anandamide and 2-arachidonoylglycerol: the ongoing story of cannabis. Nat. Prod. Rep. 16 131-143.

[64] Sanchez-Ramos, Juan. The entourage effect of the phytocannabinoids. Annals of Neurology 77.6 (2015): 1083- 1083.

SYNERGY 101

To begin, we must define three similar yet distinct concepts:

1. *Polypharmacology*

In polypharmacology, individual molecules are applied therapeutically to affect multiple targets or pathways, or in some cases, multiple drugs to affect a specific target[65], but without the concept of implied synergy.

2. *The Entourage Effect*

The entourage effect describes multiple components in cannabis all acting together, with one or two primary active compounds being supported by multiple inactive compounds, all taking place within the endocannabinoid system.

3. *Synergy*

Synergy describes the enhanced effects caused by interactions between two or more compounds, with most compounds involved being considered active. This synergy can exist outside of the endocannabinoid system.

While the general concept of synergy in phytochemicals can be explained simply by the wisdom that it is nutritionally better to eat whole fruit than it is to take vitamins or supplements of singular compounds[66], the difference between this and the entourage effect needs clearer delineation. Apart from existing outside of the endocannabinoid system, the most significant difference is the active nature of the compounds involved in synergy, versus the inactive view of the compounds in the entourage effect.

Synergy Implications

The viewpoint that synergy is always positive or beneficial is relative, and often leaves equally important negative synergies ignored. Terpenes, flavonoids, cannabinoids, and other compounds may interact synergistically, but whether this action is beneficial or detrimental depends on the individual situation and desired outcome.

Synergy Characters

Flavonoids and cannabinoids can potentially act synergistically with a virtually unlimited number of other compounds. These interactions can include activity and engagement between flavonoids and chemotherapeutics, flavonoids and cannabinoids, flavonoids and antimalarial drugs, flavonoids and terpenes, flavonoids and neuroprotective agents, and many other combinations.

Mechanisms and Types of Synergy

As documented often in the individual compound chapters of this text, plants can produce specific blends of flavonoids and other compounds in response to a variety of stimuli. These blends appear to act synergistically and can serve an astonishing number of complex functions.

Cannabinoid, flavonoid, and terpene/terpenoid formulations created in the pharmacological setting can also serve many functions, with synergy occurring via one or more of several different mechanisms including additive and

---

[65] Reddy, A Srinivas, and Shuxing Zhang. Polypharmacology: drug discovery for the future. Expert review of clinical pharmacology vol. 6,1 (2013): 41-7.

[66] Liu, R. H. (2013). Health-promoting components of fruits and vegetables in the diet. Adv. Nutr. Int. Rev. J. 4, 384S-392S.

competitive agonism or antagonism, regulation of cells, allosteric effects, multi-target effects, pharmacokinetic effects, modulation of adverse events[67], and other effects.

DIRECT EVIDENCE FOR NON-CANNABIS FLAVONOID SYNERGY

Many studies have found evidence of synergy between flavonoids and other compounds, including synthetic drugs. The following 55 quotes taken directly from peer-reviewed studies or research papers offer a distinct and fascinating glimpse into the synergistic nature of flavonoids. The author encourages readers to review all of these, as well as the underlying studies, to gain an understanding of how often we encounter synergy between compounds. Note that in the studies listed below, the authors didn't always set out to specifically study synergy, and in many cases discovered it unintentionally.

"The combination of ascorbic acid and rutin had higher antioxidant properties compared to the activity of the single compound alone, and showed a stronger effect against UV-induced reactive oxygen species generation[68]."

"Subcutaneous and orthotopic xenograft studies also showed that tumor volumes were significantly lower in mice receiving combined TMZ/Rutin treatment as compared to TMZ or rutin alone treatment. Moreover, immunoblotting analysis showed that TMZ activated JNK activity to induce protective response autophagy, which was blocked by rutin, resulting in decreased autophagy and increased apoptosis, suggesting that rutin enhances TMZ efficacy both in vitro and in vivo[69]."

"Compared with Quercetin and Rutin, FPL (an extract of persimmon leaves where these two compounds are the primary constituents) showed higher cytotoxicity at 12.5 and 25 µg/ml concentrations and also presented lower IC50 in PC-3 cells[70]."

"Furthermore, the CD26 effect was enhanced when apigenin was paired with chemotherapeutic agents utilized in the treatment of advanced colorectal cancer including irinotecan, 5-fluorouracil and oxaliplatin. For irinotecan, apigenin caused a 4-fold increase in the potency of the drug[71]."

"Apoptosis assay revealed that TRAIL or apigenin alone induced a marked apoptosis in RAFLS and their combination yielded a synergistic increase in RAFLS apoptosis[72]."

"When quercetin was combined with quercitrin, enhancement of anti-DENV-2 activity and reduced cytotoxicity were observed. However, the synergistic efficacy of the flavonoid combination was still less than that of the EA fraction[73]."

---

[67] Wagner H, Ulrich-Merzenich G. Synergy research: approaching a new generation of phytopharmaceuticals. Phytomedicine. 2009 Mar;16(2-3):97-110.

[68] Gegotek, Agnieszka, Ewa Ambrozewicz, Anna Jastrzab, Iwona Jarocka-Karpowicz, and Elzbieta Skrzydlewska. Rutin and ascorbic acid cooperation in antioxidant and antiapoptotic effect on human skin keratinocytes and fibroblasts exposed to UVA and UVB radiation. Archives of Dermatological Research 311.3 (2019): 203-219.

[69] Zhang, P., Sun, S., Li, N. et al. Rutin increases the cytotoxicity of temozolomide in glioblastoma via autophagy inhibition. J Neurooncol 132, 393-400 (2017).

[70] Ding, Yan, Kai Ren, Huanhuan Dong, Fei Song, Jing Chen, Youtian Guo, Yanshan Liu, Weijie Tao, and Yali Zhang. Flavonoids from persimmon (Diospyros kaki L.) leaves inhibit proliferation and induce apoptosis in PC-3 cells by activation of oxidative stress and mitochondrial apoptosis. Chemico-Biological Interactions 275 (2017): 210-217.

[71] Lefort, Emilie, and Jonathan Blay. The dietary flavonoid apigenin enhances the activities of the anti-metastatic protein CD26 on human colon carcinoma cells. Clinical & Experimental Metastasis 28.4 (2011): 337-349.

[72] Sun, Q., Jiang, S., Yang, K. et al. Apigenin enhances the cytotoxic effects of tumor necrosis factor-related apoptosis-inducing ligand in human rheumatoid arthritis fibroblast-like synoviocytes. Mol Biol Rep 39, 5529-5535 (2012).

[73] Chiow, K.H., M.C. Phoon, Thomas Putti, Benny K.H. Tan, and Vincent T. Chow. Evaluation of antiviral activities of Houttuynia cordata Thunb. extract, quercetin, quercetrin and cinanserin on murine coronavirus and dengue virus infection. Asian Pacific Journal of Tropical Medicine 9.1 (2015): 1-7.

"The combination of cisplatin with quercetin synergistically inhibits cell growth and triggers apoptosis in HepG2 cells. Our data revealed that the combination of quercetin and cisplatin was significantly ($P<?0.05$) effective in inducing growth suppression and apoptosis in HepG2 cells, when compared with single agent treatment[74]."

"The experiment results suggest that quercetin can increase chemosensitivity of DDP and VC on the two cell lines (A549 human lung adenocarcinoma)[75]."

"We found that Quercetin synergistically enhanced rituximab-induced growth inhibition and apoptosis in DLBCL cell lines[76]."

"Co-administration of luteolin and paclitaxel resulted in an increase in apoptosis compared with the treatment of paclitaxel alone...immunoblotting analysis also showed that the co-administration of luteolin and paclitaxel activated caspase-8 and caspase-3 and increased the expression of Fas[77]."

"Combined administration of luteolin and 5-FU in Solid Ehrlich Carcinoma model increased levels of p53, p21, caspase-3, DRAM and survivability...current results proved the antitumor therapeutic effects of luteolin alone or combined with 5-FU as a novel strategy for cancer therapy[78]."

"In particular, combined celecoxib and luteolin treatment significantly decreased the growth of MDA-MB-231 cancer cells in vivo compared with either agent alone[79]."

"A combinational treatment of cisplatin and luteolin induced more effectively cell growth inhibition, compared to cisplatin treatment alone[80]."

"In the majority of cases, luteolin, when combined with IFN-1, had additive effects in modulating cell proliferation, IL-11, TNF-a, MMP-9 and TIMP-1[81]."

"Luteolin enhanced anti-proliferation effect of cisplatin on cisplatin-resistant ovarian cancer CAOV3/DDP cells. Flow cytometry revealed that luteolin enhanced cell apoptosis in combination with cisplatin. Western blotting and qRT-PCR assay revealed that luteolin increased cisplatin-induced downregulation of Bcl-2 expression. In addition, wound-healing assay and Matrigel invasion assay showed that luteolin and cisplatin synergistically inhibited

[74] Zhao, Ji-ling, Jing Zhao, and Hong-jun Jiao. Synergistic Growth-Suppressive Effects of Quercetin and Cisplatin on HepG2 Human Hepatocellular Carcinoma Cells. Applied Biochemistry and Biotechnology 172.2 (2013):784-791.

[75] Zhan, Xuejun, Runxiang Zhang, Yanping Xu, Shuhua Yang, Daze Xie, and Liwei Tan. Empirical studies about quercetin increasing chemosensitivity on human lung adenocarcinoma cell line A549. The Chinese-German Journal of Clinical Oncology 11.7 (2012): 380-383.

[76] Li, Xin, Xinhua Wang, Mingzhi Zhang, Aimin Li, Zhenchang Sun, and Qi Yu. Quercetin Potentiates the Antitumor Activity of Rituximab in Diffuse Large B-Cell Lymphoma by Inhibiting STAT3 Pathway. Cell Biochemistry and Biophysics 70.2 (2014): 1357-1362.

[77] Yang, Mon-Yuan, Chau-Jong Wang, Nai-Fang Chen, Wen-Hsin Ho, Fung-Jou Lu, and Tsui-Hwa Tseng. Luteolin enhances paclitaxel-induced apoptosis in human breast cancer MDA-MB-231 cells by blocking STAT3. Chemico-biological interactions 213 (2014): 60-68.

[78] Soliman, Nema A, Rania N Abd-Ellatif, Amira A ELSaadany, Shahinaz M Shalaby, and Asmaa E Bedeer. Luteolin and 5-flurouracil act synergistically to induce cellular weapons in experimentally induced Solid Ehrlich Carcinoma: Realistic role of P53; a guardian fights in a cellular battle. Chemico-biological interactions 310 (2019).

[79] Jeon, Ye, Young Ahn, Won Chung, Hyun Choi, and Young Suh. Synergistic effect between celecoxib and luteolin is dependent on estrogen receptor in human breast cancer cells. Tumor Biology 36.8 (2015): 6349-6359.

[80] Wu, Bin, Qiang Zhang, Weiming Shen, and Jun Zhu. Anti-proliferative and chemosensitizing effects of luteolin on human gastric cancer AGS cell line. Molecular and Cellular Biochemistry 313.2 (2008): 125-132.

[81] Sternberg, Zohara, Kailash Chadha, Alicia Lieberman, Allison Drake, David Hojnacki, Bianca Weinstock-Guttman, and Frederick Munschauer. Immunomodulatory responses of peripheral blood mononuclear cells from multiple sclerosis patients upon in vitro incubation with the flavonoid luteolin: additive effects of IFN-1. Journal of Neuroinflammation 6 (2009): 28-28.

migration and invasion of CAOV3/DDP cells. Moreover, in vivo, luteolin enhanced cisplatin-induced reduction of tumor growth as well as induction of apoptosis[82]."

"Vitexin can cooperate with HBO to sensitize the glioma radiotherapy...The present results showed that the combination of HBO and vitexin could synergistically sensitize the glioma radiotherapy[83]."

"Furthermore, administration of N-acetylcysteine (NAC) and kaempferol significantly rescued more mice than a low dose of NAC only did when a lethal dose of propacetamol injected and therapized at a delayed time point[84]."

"After the combined application of both phenols, a synergistic effect of kaempferol plus low but not high doses of (-)-epicatechin was observed[85]."

"Of particular interest is that in combination, the two PACs [oridonin and wogonin] were synergistic in their cytotoxicity to five of six of the primary cultures and to both the cell lines[86]."

"Wogonin significantly sensitized resistant HNC cells to cisplatin both in vitro and in vivo. Our study revealed that wogonin could act synergistically with cisplatin and thereby circumvent the resistance to cisplatin in HNC cells[87]."

"Treatment with rutin and metformin in combination significantly reduced PE-induced contraction and increased ACh-induced and SNP-induced relaxation in diabetes when compared to rutin or metformin alone...Significant histological improvements were seen with combination therapy[88]."

"Additionally, it was discovered that rutin and hesperidin combined therapy was the most effective at restoring liver function and histological integrity in paclitaxel-administered rat models[89]."

"While apigenin alone did not significantly affect the biofilm biomass, apigenin + RGO reduced the biomass in an apigenin concentration-dependent manner[90]."

"Cell viability was determined after incubating with different concentrations of Api, Nar, or the combination of Api and Nar (CoAN) for 24h. Analysis using the CompuSyn software revealed that the CI value of each combined dose

[82] Wang, Haixia, Youjun Luo, Tiankui Qiao, Zhaoxia Wu, and Zhonghua Huang. Luteolin sensitizes the antitumor effect of cisplatin in drug-resistant ovarian cancer via induction of apoptosis and inhibition of cell migration and invasion. Journal of Ovarian Research 11.1 (2018): 1-12.

[83] Xie, T., J.-R. Wang, C.-G. Dai, X.-A. Fu, J. Dong, and Q. Huang. Vitexin, an inhibitor of hypoxia-inducible factor-1a, enhances the radiotherapy sensitization of hyperbaric oxygen on glioma. Clinical and Translational Oncology 22.7 (2020): 1086-1093.

[84] Tsai, Ming-Shiun, Ying-Han Wang, Yan-Yun Lai, Hsi-Kai Tsou, Gan-Guang Liou, Jiunn-Liang Ko, and Sue-Hong Wang. Kaempferol protects against propacetamol-induced acute liver injury through CYP2E1 inactivation, UGT1A1 activation, and attenuation of oxidative stress, inflammation and apoptosis in mice. Toxicology letters 290 (2018): 97-109.

[85] Escandon, R.A., del Campo, M., Lopez-Solis, R. et al. Antibacterial effect of kaempferol and (-)-epicatechin on Helicobacter pylori . Eur Food Res Technol 242, 1495-1502 (2016).

[86] Chen, Sophie, Matt Cooper, Matt Jones, Thumuluru Madhuri, Julie Wade, Ashleigh Bachelor, and Simon Butler Manuel. Combined activity of oridonin and wogonin in advanced-stage ovarian cancer cells. Cell Biology and Toxicology 27.2 (2010): 133-147.

[87] Kim, Eun, Hyejin Jang, Daiha Shin, Seung Baek, and Jong-Lyel Roh. Targeting Nrf2 with wogonin overcomes cisplatin resistance in head and neck cancer. Apoptosis 21.11 (2016): 1265-1278.

[88] David SR, Lai PPN, Chellian J, Chakravarthi S, Rajabalaya R. Influence of rutin and its combination with metformin on vascular functions in type 1 diabetes. Sci Rep. 2023 Aug 1;13(1):12423.

[89] Ali YA, Soliman HA, Abdel-Gabbar M, Ahmed NA, Attia KAA, Shalaby FM, El-Nahass ES, Ahmed OM. Rutin and Hesperidin Revoke the Hepatotoxicity Induced by Paclitaxel in Male Wistar Rats via Their Antioxidant, Anti-Inflammatory, and Antiapoptotic Activities. Evid Based Complement Alternat Med. 2023 May 26;2023:2738351.

[90] Kim MA, Min KS. Combined effect of apigenin and reduced graphene oxide against Enterococcus faecalis biofilms. J Oral Sci. 2023 Jul 1;65(3):163-167.

was < 1, depicting that the two drugs had a synergistic inhibitory effect. The results established that CoAN treatment caused significant cytotoxicity with cell cycle arrest at G2/M phases. Furthermore, CoAN significantly enhanced mitochondria dysfunction, elevated oxidative stress, and activated the apoptotic pathway versus Api or Nar alone groups[91]."

"In summary, our research observed that in vitro experiments using Que[recetin] combined with 5-FU on 5-FU-resistant HCT-116 cells showed a good synergistic effect[92]."

"Combination testing of QUE[rcetin] and AZ[ithromycin] in a ratio of 2:1 (QUE:AZ) showed an IC50 value of 0.081 μM. Interestingly, a fractional inhibitory index value of 0.28 was observed, indicating a strong synergy. Overall, the results indicate that QUE[rcetin] is a novel lead capable of synergizing with AZ[ithromycin] for inhibiting T. gondii growth[93]."

"In this study, we investigated the synergistic effects of IVX and cisplatin (DDP) in non-small cell lung cancer (NSCLC) A549 and H1975 cells. The results showed that the combined treatment with IVT and DDP markedly inhibited proliferation and induced apoptosis of the two NSCLC cells[94]."

"This synergism was, at least partially, ascribed to the induction of mitotic catastrophe. The switch from the cytoprotective autophagy to the autophagic cell death was also implicated in the mechanism of the synergistic action of fisetin and paclitaxel in the A549 cells. In addition, we revealed that the synergism between fisetin and paclitaxel was cell line-specific as well as that fisetin synergizes with arsenic trioxide[95]."

"A synergism effect of quercetin/fisetin and naringenin (CI < 1) was observed for both cell lines. Combination therapies were significantly more effective in cell growth reduction, migration suppression and apoptosis induction than single therapies. Quercetin/fisetin enhances the anti-proliferative and anti-migratory activities in combination with naringenin[96]."

"Importantly, the results of our research indicate that substances such as peimine, fisetin, bardoxolone methyl and especially astaxanthin increase the effectiveness of morphine, buprenorphine and/or oxycodone, which is an exceptionally important outcome from a clinical point of view[97]."

"A significant synergism effect [of treatment with quercetin and fisetin] was observed for all cell lines. Combination therapy was significantly more effective in colony formation and wound healing assays compared to single therapies. The expression level of potential effectors was also showed a greater change. In vivo study confirmed the

[91] Liu X, Zhao T, Shi Z, Hu C, Li Q, Sun C. Synergism Antiproliferative Effects of Apigenin and Naringenin in NSCLC Cells. Molecules. 2023 Jun 23;28(13):4947.

[92] Tang Z, Wang L, Chen Y, Zheng X, Wang R, Liu B, Zhang S, Wang H. Quercetin reverses 5-fluorouracil resistance in colon cancer cells by modulating the NRF2/HO-1 pathway. Eur J Histochem. 2023 Aug 7;67(3):3719.

[93] Abugri DA, Wijerathne SVT, Sharma HN, Ayariga JA, Napier A, Robertson BK. Quercetin inhibits Toxoplasma gondii tachyzoite proliferation and acts synergically with azithromycin. Parasit Vectors. 2023 Aug 3;16(1):261.

[94] Chen RL, Wang Z, Huang P, Sun CH, Yu WY, Zhang HH, Yu CH, He JQ. Isovitexin potentiated the antitumor activity of cisplatin by inhibiting the glucose metabolism of lung cancer cells and reduced cisplatin-induced immunotoxicity in mice. Int Immunopharmacol. 2021 May;94:107357.

[95] Klimaszewska-Wisniewska A, Halas-Wisniewska M, Tadrowski T, Gagat M, Grzanka D, Grzanka A. Paclitaxel and the dietary flavonoid fisetin: a synergistic combination that induces mitotic catastrophe and autophagic cell death in A549 non-small cell lung cancer cells. Cancer Cell Int. 2016 Feb 16;16:10.

[96] Jalalpour Choupanan M, Shahbazi S, Reiisi S. Naringenin in combination with quercetin/fisetin shows synergistic anti-proliferative and migration reduction effects in breast cancer cell lines. Mol Biol Rep. 2023 Sep;50(9):7489-7500.

[97] Ciapała K, Rojewska E, Pawlik K, Ciechanowska A, Mika J. Analgesic Effects of Fisetin, Peimine, Astaxanthin, Artemisinin, Bardoxolone Methyl and 740 Y-P and Their Influence on Opioid Analgesia in a Mouse Model of Neuropathic Pain. Int J Mol Sci. 2023 May 19;24(10):9000.

in vitro results and showed how significantly their synergism promotes their singular function in inhibiting cancer progression[98]."

"The results showed that PANC-1 cell viability decreased significantly after co-treatment with 20-μM gemcitabine and fisetin, which showed that fisetin might have a potential to be used as a complementary therapy in cases of chemotherapy resistance by enhancing the chemosensitivity of pancreatic cancer cells[99]."

"Treatment of PIK3CA-mutant cells with fisetin and 5-FU reduced the expression of PI3K, phosphorylation of AKT, mTOR, its target proteins, constituents of mTOR signaling complex and this treatment increased the phosphorylation of AMPKα. In addition, the combination of fisetin and 5-FU also reduced the total number of intestinal tumors. Fisetin could be used as a preventive agent plus an adjuvant with 5-FU for the treatment of PIK3CA-mutant colorectal cancer[100]."

"…investigate the antifungal activities of baicalin and/or sodium bicarbonate (SB) against 29 C. albicans isolates including 27 clinical ones. By using broth microdilution method and checkerboard assay, it was observed that the minimum inhibitory concentrations (MICs) of baicalin and SB alone were > 2048 μg/mL, and those of baicalin and SB in combination decreased 16-32 folds with fractional inhibitory concentration index (FICI) in a range of 0.094-0.375. The results presented the strong synergism between SB and baicalin in 27 clinical C. albicans isolates and provided an alternative choice against C. albicans[101]."

"Baicalin increased the sensitivity of MESO924 to the chemotherapeutic drugs doxorubicin, cisplatin, and pemetrexed[102]."

"It was found that even with the same content of baicalin, the corresponding functions of the linen fabric treated by SBE (Scutellaria baicalensis (chinese skullcap) extracts) are obviously better than treated by baicalin. The advantages of the SBE over just baicalin on the functionalization of linen fabric was analyzed and verified by experiments[103]."

"The results showed that baicalin combined with emodin at a lower dose had the same effect as the two drugs alone...The combined treatment decreased the expression of CD14/TLR4/NF-κB pathway proteins and increased the expression of PPAR-γ protein in the colon of colitis mice. Further study in vitro has shown that baicalin decreased the expression of CD14, whereas emodin increased the expression of PPAR-γ, both of which inhibited the activity of NF-κB and exerted antiinflammatory effects. Furthermore, compared to the treatment using the two drugs individually, baicalin combined with emodin had more significant effects on the expression of CD14 and PPAR-γ[104]."

[98] Hosseini SS, Ebrahimi SO, Haji Ghasem Kashani M, Reiisi S. Study of quercetin and fisetin synergistic effect on breast cancer and potentially involved signaling pathways. Cell Biol Int. 2023 Jan;47(1):98-109.

[99] Ding G, Xu X, Li D, Chen Y, Wang W, Ping D, Jia S, Cao L. Fisetin inhibits proliferation of pancreatic adenocarcinoma by inducing DNA damage via RFXAP/KDM4A-dependent histone H3K36 demethylation. Cell Death Dis. 2020 Oct 22;11(10):893.

[100] Khan N, Jajeh F, Eberhardt EL, Miller DD, Albrecht DM, Van Doorn R, Hruby MD, Maresh ME, Clipson L, Mukhtar H, Halberg RB. Fisetin and 5-fluorouracil: Effective combination for PIK3CA-mutant colorectal cancer. Int J Cancer. 2019 Dec 1;145(11):3022-3032.

[101] Shao J, Xiong L, Zhang MX, Wang TM, Wang CZ. Report: Synergism of sodium bicarbonate and baicalin against clinical Candida albicans isolates via broth microdilution method and checkerboard assay. Pak J Pharm Sci. 2019 May;32(3):1103-1105.

[102] Xu WF, Liu F, Ma YC, Qian ZR, Shi L, Mu H, Ding F, Fu XQ, Li XH. Baicalin Regulates Proliferation, Apoptosis, Migration, and Invasion in Mesothelioma. Med Sci Monit. 2019 Oct 31;25:8172-8180.

[103] Li, H., Li, Z., Liu, Y. et al. Advantages of Scutellaria baicalensis extracts over just baicalin in the ultrasonically assisted multi-functional treatment of linen fabrics. Cellulose 27, 4831–4846 (2020).

[104] Xu B, Huang S, Chen Y, Wang Q, Luo S, Li Y, Wang X, Chen J, Luo X, Zhou L. Synergistic effect of combined treatment with baicalin and emodin on DSS-induced colitis in mouse. Phytother Res. 2021 Oct;35(10):5708-5719.

"BA is a promising preventive or adjuvant therapy in breast cancer treatment with 5-FU mainly via cooperative inhibition of inflammation, angiogenesis, and triggering apoptotic cell death[105]."

"The results showed that baicalin could decrease the expression level of GSK3β, while upregulate the expression level of DCX, BDNF, Cyclin D1-cyclin dependent kinase 4/6 (CDK4/6), thus promoted cell proliferation and survival in corticosterone (CORT) induced PC-12 cells. Moreover, this effect was enhanced when baicalin and lithium chloride were coadministration[106]."

"The combination of baicalin and 5-FU demonstrated synergistic activity against 5-FU-resistant RKO-R10 cells. The combination significantly inhibited in vivo tumor growth greater than each treatment alone[107]."

"While both baicalin and E. coli DH5α-lux/βG could inhibit tumor growth as monotherapy, an enhanced inhibition was observed when animals were subjected to combination therapy[108]."

"Baicalin was efficient in destroying the biofilm and exerted a synergistic bactericidal effect when combined with linezolid. Based on these findings, baicalin combined with linezolid may be efficacious in controlling S. aureus biofilm-related infections[109]."

"Taken together, our results demonstrate that both flavonoids inhibited cells growth in a dose-dependent manner and the association of quercetin improved chrysin's toxic effect over the cell lines. Both quercetin and chrysin presented a cytotoxic effect over the studied cancer cell lines, with chrysin being the compound with greater isolated effect. However, the association between them showed a more satisfactory result, since the preincubation with quercetin potentiated the effect observed by chrysin in the induction of cell death, presenting greater toxicity in the tumor cell lines[110]."

"Findings revealed that chrysin notably increased the cytotoxicity of NK-92 and T cells towards MCF-7 and MDA-MB-231 (breast cancer) cells[111]."

"We determined that pyrotinib combined with chrysin yielded a potent synergistic effect to induce more evident cell cycle arrest, inhibit the proliferation of BT-474 and SK-BR-3 BC cells, and repress in vivo tumor growth in xenograft mice models[112]."

[105] Shehatta NH, Okda TM, Omran GA, Abd-Alhaseeb MM. Baicalin; a promising chemopreventive agent, enhances the antitumor effect of 5-FU against breast cancer and inhibits tumor growth and angiogenesis in Ehrlich solid tumor. Biomed Pharmacother. 2022 Feb;146:112599.

[106] Wang Z, Cheng Y, Lu Y, Sun G, Pei L. Baicalin Coadministration with Lithium Chloride Enhanced Neurogenesis via GSK3β Pathway in Corticosterone Induced PC-12 Cells. Biol Pharm Bull. 2022 May 1;45(5):605-613.

[107] Liu H, Liu H, Zhou Z, Chung J, Zhang G, Chang J, Parise RA, Chu E, Schmitz JC. Scutellaria baicalensis enhances 5-fluorouracil-based chemotherapy via inhibition of proliferative signaling pathways. Cell Commun Signal. 2023 Jun 19;21(1):147.

[108] Jafari B, Bahrami AR, Matin MM. Targeted bacteria-mediated therapy of mouse colorectal cancer using baicalin, a natural glucuronide compound, and E. coli overexpressing β-glucuronidase. Int J Pharm. 2023 Jul 25;642:123099.

[109] Du Z, Han J, Luo J, Bi G, Liu T, Kong J, Chen Y. Combination effects of baicalin with linezolid against Staphylococcus aureus biofilm-related infections: in vivo animal model. New Microbiol. 2023 Sep;46(3):258-263.

[110] Ramos PS, Ferreira C, Passos CLA, Silva JL, Fialho E. Effect of quercetin and chrysin and its association on viability and cell cycle progression in MDA-MB-231 and MCF-7 human breast cancer cells. Biomed Pharmacother. 2024 Oct;179:117276.

[111] Durmus E, Ozman Z, Ceyran IH, Pasin O, Kocyigit A. Chrysin Enhances Anti-Cancer Activity of Jurkat T Cell and NK-92 Cells Against Human Breast Cancer Cell Lines. Chem Biodivers. 2024 Oct;21(10):e202400806.

[112] Liu X, Zhang X, Shao Z, Zhong X, Ding X, Wu L, Chen J, He P, Cheng Y, Zhu K, Zheng D, Jing J, Luo T. Pyrotinib and chrysin synergistically potentiate autophagy in HER2-positive breast cancer. Signal Transduct Target Ther. 2023 Dec 18;8(1):463.

"CurChr (curcumin-chrysin)-loaded NPs (nanoencapsulated) had a considerable synergistic cytotoxicity against MDA-MB-231 (breast cancer) cells with more cell accumulation in G2/M phase compared to the other groups[113]."

"In vitro experiments revealed that the combination of diosmetin and chrysin could induce apoptosis, enhance autophagy, reduce inflammatory mediator production, and improve the tumor cell microenvironment by inhibiting the PI3K/AKT/mTOR/NF-κB signaling pathway. Notably, the synergy score for the combination of diosmetin (25 μM) and chrysin (10 μM) was 16. Thus, the diosmetin-chrysin combination shows promise as an effective therapeutic approach for hepatocellular carcinoma due to its strong synergistic effect[114]."

"SW480 and HCT-116 cells were treated with either apigenin or chrysin alone or two-drug combination...Apigenin (25 μM) combined with chrysin (25 μM) were determined to be optimal. Treatment with the combination of apigenin (25 μM) and chrysin (25 μM) significantly reduced cell clone numbers, migration, and invasion ability, while increased the cell apoptosis in both CRC cell lines. The combined effect was higher than chrysin or apigenin alone[115]."

"Curcumin (Cur) and Chrysin (Chr), were co-encapsulated in PEGylated PLGA NPs and investigated their synergistic inhibitory effect against Caco-2 cancer cells. The results showed that free drugs and nano-formulations exhibited a dose-dependent cytotoxicity against Caco-2 cells and especially, Cur-Chr-PLGA/PEG NPs had more synergistic antiproliferative effect and significantly arrested the growth of cancer cells than the other groups[116]."

"We examined the effects of chrysin alone and in combination with radiation on ICD induction in B16-F10 cells. Combination therapy exhibited a synergistic effect, with an optimum combination index of 0.66. The synergistic anti-cancer effect correlated with increased cell apoptosis in cancer cells. Our findings revealed that chrysin could induce ICD and intensify the RT-induced immunogenicity...In conclusion, chrysin synergistically potentiated the anti-cancer effect of RT in B16-F10 cells with hyperactive STAT3 by inducing ICD and elevating the levels of DAMPs molecules, such as CRT, HMGB1, HSP70, and ATP[117]."

"We examined the effect of RT in combination with chrysin as a possible radiosensitizing agent in an MDA-MB-231 cell line as a model of a TNBC...Treatment of MDA-MB-231 cells with chrysin in combination with RT caused synergistic antitumor effects, with an optimum combination index (CI) of 0.495. Our results indicated that chrysin synergistically potentiated RT-induced apoptosis in MDA-MB-231 compared with monotherapies[118]."

"The obtained micelle was used for co-delivery of the anticancer drug docetaxel (DTX) and Chrysin (CHS) as an adjuvant on the CSCs originated from Human colon adenocarcinoma cell line...Data demonstrated that the micelles harbouring DTX@CHS had potential to reduce cancer stem cell viability compared to free DTX@CHS, single-drug

[113] Javan N, Khadem Ansari MH, Dadashpour M, Khojastehfard M, Bastami M, Rahmati-Yamchi M, Zarghami N. Synergistic Antiproliferative Effects of Co-nanoencapsulated Curcumin and Chrysin on MDA-MB-231 Breast Cancer Cells Through Upregulating miR-132 and miR-502c. Nutr Cancer. 2019;71(7):1201-1213.

[114] Yu X, Zhang D, Hu C, Yu Z, Li Y, Fang C, Qiu Y, Mei Z, Xu L. Combination of Diosmetin With Chrysin Against Hepatocellular Carcinoma Through Inhibiting PI3K/AKT/mTOR/NF-κB Signaling Pathway: TCGA Analysis, Molecular Docking, Molecular Dynamics, In Vitro Experiment. Chem Biol Drug Des. 2024 Oct;104(4):e70003.

[115] Zhang X, Zhang W, Chen F, Lu Z. Combined effect of chrysin and apigenin on inhibiting the development and progression of colorectal cancer by suppressing the activity of P38-MAPK/AKT pathway. IUBMB Life. 2021 May;73(5):774-783.

[116] Lotfi-Attari J, Pilehvar-Soltanahmadi Y, Dadashpour M, Alipour S, Farajzadeh R, Javidfar S, Zarghami N. Co-Delivery of Curcumin and Chrysin by Polymeric Nanoparticles Inhibit Synergistically Growth and hTERT Gene Expression in Human Colorectal Cancer Cells. Nutr Cancer. 2017 Nov-Dec;69(8):1290-1299.

[117] Jafari S, Ardakan AK, Aghdam EM, Mesbahi A, Montazersaheb S, Molavi O. Induction of immunogenic cell death and enhancement of the radiation-induced immunogenicity by chrysin in melanoma cancer cells. Sci Rep. 2024 Oct 5;14(1):23231.

[118] Jafari, S., Dabiri, S., Mehdizadeh Aghdam, E. et al. Synergistic effect of chrysin and radiotherapy against triple-negative breast cancer (TNBC) cell lines. Clin Transl Oncol 25, 2559–2568 (2023).

formulations and the control group (p < 0.05). The combination effect of DTX and CHS formulated in micelle was synergistic in CSCs (CI < 1)[119]."

"To compare the antioxidant, anticholinesterase, and behavioral effects of CHR with its glycosylated form (CHR bonded to β-d-glucose tetraacetate, denoted as LQFM280), we employed an integrated approach using both in vitro (SH-SY5Y cells) and in vivo (aluminum-induced neurotoxicity in Swiss mice) models...LQFM280 demonstrated higher antioxidant activity than CHR in both models. Remarkably, LQFM280 proved more effective than CHR in recovering memory loss and counteracting neuronal death in the aluminum chloride mice model, suggesting its increased bioavailability at the brain level[120]."

"Based on the results, it can be concluded that resveratrol and naringenin can decrease cell viability in retinoblastoma cells in an in vitro dose/time-dependent manner. Albeit more studies are needed to shed the light on the mechanism of action, our data reveal a potential synergistic cytotoxic effect of naringenin and resveratrol on Y79 cells in 48 hours[121]."

"This study investigated the latent synergistic antiproliferative functions of Api and Nar in A549 and H1299 NSCLC cells...combination of Api and Nar (CoAN)...The results established that CoAN treatment caused significant cytotoxicity with cell cycle arrest at G2/M phases. Furthermore, CoAN significantly enhanced mitochondria dysfunction, elevated oxidative stress, and activated the apoptotic pathway versus Api or Nar alone groups[122]."

[119] Ghamkhari A, Pouyafar A, Salehi R, Rahbarghazi R. Chrysin and Docetaxel Loaded Biodegradable Micelle for Combination Chemotherapy of Cancer Stem Cell. Pharm Res. 2019 Oct 23;36(12):165.

[120] Okoh VI, Campos HM, Yasmin de Oliveira Ferreira P, Pereira RM, Souza Silva Y, Arruda EL, Pagliarani B, de Almeida Ribeiro Oliveira G, Lião LM, Franco Dos Santos G, Vaz BG, Sabino JR, Alcantara Dos Santos FC, Costa EA, Tarozzi A, Menegatti R, Ghedini PC. Chrysin bonded to β-d-glucose tetraacetate enhances its protective effects against the neurotoxicity induced by aluminum in Swiss mice. J Pharm Pharmacol. 2024 Apr 3;76(4):368-380.

[121] Rakhshan R, Atashi HA, Hoseinian M, Jafari A, Haghighi A, Ziyadloo F, Razizadeh N, Ghasemian H, Nia MMK, Sefidi AB, Arani HZ. The Synergistic Cytotoxic and Apoptotic Effect of Resveratrol and Naringenin on Y79 Retinoblastoma Cell Line. Anticancer Agents Med Chem. 2021 Oct 28;21(16):2243-2249.

[122] Liu X, Zhao T, Shi Z, Hu C, Li Q, Sun C. Synergism Antiproliferative Effects of Apigenin and Naringenin in NSCLC Cells. Molecules. 2023 Jun 23;28(13):4947.

Synergy Review

Answer the following questions to test your knowledge of this section:

Question #1: Match the term with the correct loose definition:

a.	Synergy	1. One compound affecting multiple targets
b.	Polypharmacology	2. A nonsensical term
c.	Synergistic bonding	3. Synergy occurring within the ECS
d.	Entourage Effect	4. Two or more compounds acting together

Question #2: Synergy characters can include:

    a.   Flavonoids and flavonoids
    b.   Flavonoids and cannabinoids
    c.   Flavonoids, terpenoids, and anticancer agents
    d.   Antibacterial agents, flavonoids, and cannabinoids
    e.   Flavonols, analgesics, terpenes
    f.   All of the above

Question #3: Name the researcher that originally coined the term 'entourage effect':

    a.   Ben Simon
    b.   Ethan Russo
    c.   Raphael Mechnoulam
    d.   Shimon Ben-Shabat

Question #4: Define the term "polypharmacology:"

_____

Question #5: Define the term "entourage effect:"

_____

Question #6: Define the term "synergy:"

_____

Question #7: Is synergy between compounds in cannabis therapeutically beneficial?

    a.   Yes
    b.   No
    c.   Depends on the desire outcome

For the answer key to Synergy, please visit www.cannabischemistry.org

CANNFLAVIN A

Type: Flavone
Chemical Formula: $C_{26}H_{28}O_6$
Molecular Weight: 436.5 g/mol
Boiling Point: 182 °C (360 °F)
Flash Point: 222 °C (431.6 °F)
Melting Point: >145°C[123]
Solubility: Ethanol, DMSO, slightly soluble in acetone, ethyl acetate, methanol
Oral LD50: Unknown
Biological Role: Unknown
Therapeutic Role: Anti-inflammatory, neurological agent, anticancer agent
Commercial Use: Research chemical
Occurrence in Cannabis: #1 Flavonoid by concentration

Occurs in Cannabis Strains: Cheese, Critical Dream, Fruit Punch, Huckleberry, Northern Lights, Purple Moose, Purple Punch, Skywalker, Space Candy, Strawberry Cough, Strawberry Banana, Super Lemon Haze, White Cookies. Several varieties of hemp have also been found to contain cannflavin A, including the Kompoti, Tiborszallasi, Antal, and Carmagnola cultivars[124]. Likely occurs as a major compound in most cannabis strains.

INTRODUCTION

Although most of the flavonoids in cannabis also occur in many other plants, cannflavin A is mostly unique to cannabis, although the compound has also been found in Mimulus bigelovii[125]. Discovered and named in 1985 by Barrett et al, early work has shown that this compound exhibits significant anti-inflammatory properties, which can be useful in the prevention and treatment of numerous diseases and conditions.

Cannflavin A belongs to the flavonoid subclass of flavones and typically occurs together with cannflavin B and cannflavin C, both of which are also believed to be mostly unique to cannabis, although the latter was only discovered in 2008, and lacks research.

The author sent fifteen samples of unique strains for flavonoid testing to ACS laboratory in Florida, and in all but one sample, cannflavin A was the top flavonoid by concentration. Where it didn't occur as the number one constituent, it was a close second. Outside work has confirmed this compound's position as the primary flavonoid constituent in cannabis; a 2018 study of multiple samples of hemp found that cannflavin A "was observed to be the main compound in almost all the samples[126]."

Interestingly, research has shown that accumulation of cannflavin A in cannabis is determined primarily by genetics, but that accumulation can be increased by a low air temperature[127].

[123] Toronto Research Chemicals, Data Sheet for Compound C175398, from: https://www.trc-canada.com/product-detail/?C175398. Accessed October 31, 2023.

[124] Izzo L, Castaldo L, Narváez A, Graziani G, Gaspari A, Rodríguez-Carrasco Y, Ritieni A. Analysis of Phenolic Compounds in Commercial Cannabis sativa L. Inflorescences Using UHPLC-Q-Orbitrap HRMS. Molecules. 2020 Jan 31;25(3):631.

[125] Salem MM, Capers J, Rito S, Werbovetz KA. Antiparasitic activity of C-geranyl flavonoids from Mimulus bigelovii. Phytother Res. 2011 Aug;25(8):1246-9.

[126] Pellati, Federica et al. New Methods for the Comprehensive Analysis of Bioactive Compounds in Cannabis sativa L. (hemp). Molecules (Basel, Switzerland) vol. 23,10 2639. 14 Oct. 2018.

[127] D. Calzolari, G. Magagnini, L. Lucini, G. Grassi, G.B. Appendino, S. Amaducci, High added-value compounds from Cannabis threshing residues, Industrial Crops and Products, Volume 108, 2017, Pages 558-563, ISSN 0926-6690.

CHEMICAL STRUCTURE

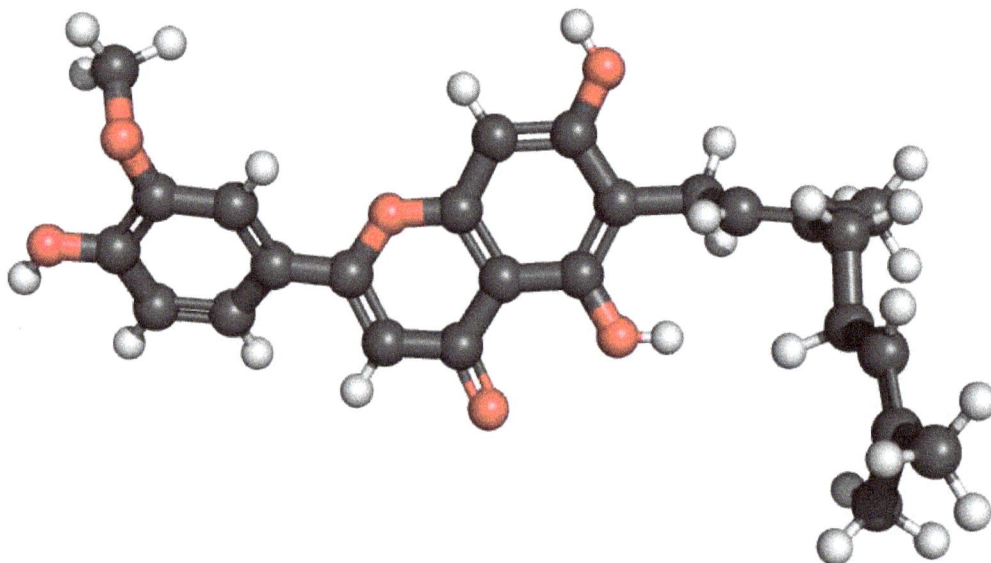

Cannflavin A is formed in the phenylpropanoid metabolic pathway from geranyl diphosphate, which in turn forms luteolin. The methylated flavone chrysoeriol is further derived from luteolin to produce cannflavin A[128]. The molecular structure of cannflavin A consists of three rings, all containing oxygen atoms, with all three featuring endocyclic double-bonds. Overall, there are nine sets of double-bonds in this molecule, along with twenty-six carbon atoms, twenty-eight hydrogen atoms, and six oxygen atoms, scientifically notated as $C_{26}H_{28}O_6$.

OCCURRENCE IN PLANTS

While cannflavin A is largely thought of as being produced solely by cannabis, it has also been found in Mimulus bigelovii[129], a plant in the Phrymaceae family (lopseed family).

BIOLOGICAL ACTIVITY IN PLANTS

At the time of publication of this book, there were no studies or research papers available regarding the biological roles that cannflavin A plays in cannabis or other plants. More research is needed, though we can assume some of these roles based on the general activities of flavonoids in plants, which include UV protection, antioxidant activity, and allelopathic activity, among other potential functions.

USES IN INDUSTRY

When purified, cannflavin A appears as a pale yellow or off-white solid possessing a waxy texture. There are currently no industrial or commercial uses of this flavonoid; this compound is currently available for research purposes only.

---

[128] Rea, Kevin A, Jose A Casaretto, M Sameer Al-Abdul-Wahid, Arjun Sukumaran, Jennifer Geddes-McAlister, Steven J Rothstein, and Tariq A Akhtar. Biosynthesis of cannflavins A and B from Cannabis sativa L. Phytochemistry 164 (2019): 162-171.

[129] Bautista JL, Yu S, Tian L. Flavonoids in Cannabis sativa: Biosynthesis, Bioactivities, and Biotechnology. ACS Omega. 2021 Feb 18;6(8):5119-5123.

POTENTIAL USES IN MEDICINE

While minimal research into the medical value of cannflavin A has been published as of early 2025, this flavonoid has been found to possess anti-inflammatory properties and may one day make an excellent treatment for Alzheimer's disease, among many other potential medical uses based on the pharmacological capacities of other flavonoids common to both cannabis and other plants.

Anti-Inflammatory Agent: The anti-inflammatory properties of cannflavin A have been found to be approximately thirty times more potent than aspirin[130] as an inhibitor of prostaglandin production[131]. Researchers have proposed cannflavin A for the treatment of inflammation-related diseases – particularly rheumatoid arthritis – after finding that the compound modulates inflammatory signaling pathways as a TAK1 (an enzyme encoded by the MAP3K7 gene that controls a variety of cell functions including transcription regulation and apoptosis) inhibitor[132].

Treatment for Alzheimer's Disease: Recently, cannflavin A has been shown to increase PC12 (rat adrenal medulla) cell viability by 40% via hormetic and neuroprotective activity against amyloid beta-mediated neurotoxicity[133], which may be useful in the treatment of neurodegenerative diseases like Alzheimer's.

Treatment for Bladder Cancer: As an isolated compound, cannflavin A was shown to activate apoptosis (cell death) via caspase-3 cleavage and reduce invasion of bladder cancer cells by 50%, while also exhibiting antagonistic, additive, and synergistic effects with chemotherapeutic agents gemcitabine and cisplatin[134].

Anti-nociceptive Agent: Canadian researchers found that cannflavin A interacts with and desensitizes vanilloid receptors (involved in pain and heat sensation) in Caenorhabditis elegans (a nematode), inducing remarkable antinociceptive activity in the worms, with effects comparable to capsaicin – a well-known vanilloid receptor ligand[135]. However, the researchers warned that the duration of the effect is likely short.

Treatment for Neurodegenerative/Neuroinflammatory Disease: As derived from hemp, cannflavin A inhibited 99% of the kynurenine-3-monooxygenase (KMO) protein (a mitochondrial enzyme), with effects comparable to Ro 61-8048, a known KMO (hydroxylase) inhibitor[136]. This could offer potential in the treatment of neuroinflammatory and neurodegenerative diseases.

IMPLICATIONS FOR HUMAN HEALTH & NUTRITION

Because cannflavin A is found primarily in cannabis, consumption of cannabis products is likely the only viable method of increasing dietary intake of this flavonoid.

[130] M.L. Barrett, A.M. Scott, and F.J. Evans Cannflavin A and B, prenylated flavones from Cannabis Sativa L. 28 May 1985.

[131] Barrett, M L, D Gordon, and F J Evans. Isolation from Cannabis sativa L. of cannflavin--a novel inhibitor of prostaglandin production. Biochemical pharmacology 34.11 (1985): 2019-24.

[132] Chuanphongpanich S, Racha S, Saengsitthisak B, Pirakitikulr P, Racha K. Computational Assessment of Cannflavin A as a TAK1 Inhibitor: Implication as a Potential Therapeutic Target for Anti-Inflammation. Scientia Pharmaceutica. 2023; 91(3):36.

[133] Eggers, Carly, Masaya Fujitani, Ryuji Kato, and Scott Smid. Novel cannabis flavonoid, cannflavin A displays both a hormetic and neuroprotective profile against amyloid B-mediated neurotoxicity in PC12 cells: Comparison with geranylated flavonoids, mimulone and diplacone. Biochemical pharmacology 169 (2019): 113609-113609.

[134] Tomko AM, Whynot EG, Dupré DJ. Anti-cancer properties of cannflavin A and potential synergistic effects with gemcitabine, cisplatin, and cannabinoids in bladder cancer. J Cannabis Res. 2022 Jul 22;4(1):41.

[135] Lahaise M, Boujenoui F, Beaudry F. Cannflavins isolated from Cannabis sativa impede Caenorhabditis elegans response to noxious heat. Naunyn Schmiedebergs Arch Pharmacol. 2023 Jul 22.

[136] Puopolo T, Chang T, Liu C, Li H, Liu X, Wu X, Ma H, Seeram NP. Gram-Scale Preparation of Cannflavin A from Hemp (Cannabis sativa L.) and Its Inhibitory Effect on Tryptophan Catabolism Enzyme Kynurenine-3-Monooxygenase. Biology (Basel). 2022 Sep 28;11(10):1416.

Cannflavin A Review

Answer the following questions to test your knowledge of this flavonoid:

Question #1: What type of flavonoid is cannflavin A?

      a.   Flavonol
      b.   Flavone
      c.   Chalcone
      d.   Flavanones

Question #2: The base flavonoid in the biosynthesis of cannflavin A is:

      a.   Cannabisin
      b.   Rutin
      c.   Quercetin
      d.   Luteolin

Question #3: How common is cannflavin A in cannabis?

      a.   Top three
      b.   Top five
      c.   Top ten
      d.   Secondary

Question #4: What is the chemical formula for cannflavin A?

      a. $C_{24}H_{26}O$
      b. $C_{15}H_{24}O_6$
      c. $C_{26}H_{28}O_6$
      d. $N_{15}H_{26}$

Question #5: Which researcher discovered and coined the name for the molecule cannflavin A?

      a.   ElSohly
      b.   Barrett
      c.   Russo
      d.   Rosenthal

Question #6: Name two potential medical uses of cannflavin A:

1 _____ 2 _____

For the answer key to Cannflavin A, please visit www.cannabischemistry.org

CANNFLAVIN B

Type: Flavone
Chemical Formula: $C_{21}H_{20}O_6$
Molecular Weight: 368.4 g/mol
Boiling Point: 616.7 °C at 760 mmHg (estimated)[137]
Flash Point: 219.4 °C (estimated by GuideChem)
Melting Point: >154°C[138]
Solubility: Slightly soluble in water[139], slightly soluble in acetone, methanol
Oral LD50: Unknown
Biological Role: Unknown
Therapeutic Role: Antileishmanial, anticancer, anti-inflammatory agent
Commercial Use: Research chemical
Occurrence in Cannabis: Top three

Occurs in Cannabis Strains: Cheese, Critical Dream, Fruit Punch, Huckleberry, Northern Lights, Purple Moose, Purple Munch, Skywalker, Space Candy, Strawberry Cough, Strawberry Banana, Super Lemon Haze, White Cookies. Cannflavin B has also been found in a methanolic extract of industrial hemp cultivar Futura 75[140], and in hexane extracts of high potency cannabis grown from Mexican seeds of an unknown variety[141]. Additionally, it is likely that this compound occurs in most cannabis varieties.

INTRODUCTION

Cannflavin B is one of several flavonoids thought to be mostly unique to cannabis. Belonging to the flavone subclass of flavonoids, cannflavin B was discovered and named together with cannflavin A in 1985 by Barrett et al. Research has shown that this compound may be effective in the treatment of leishmaniasis, and that a synthetic derivative of cannflavin B can be used in the treatment of pancreatic cancer.

---

[137] GuideChem Chemical Trading Guide Data Sheet for Cannflavin B, from; https://www.guidechem.com/dictionary/en/1083197-70-9.html. Accessed November 1, 2023.

[138] Toronto Research Chemicals Data Sheet for Cannflavin B or Compound C175405, from: https://www.trc-canada.com/product-detail/?C175405. Accessed October 31, 2023.

[139] BenchChem Data Sheet for Cannflavin B, from; https://www.benchchem.com/product/b1205605. Accessed November 1, 2023.

[140] De Vita S, Finamore C, Chini MG, Saviano G, De Felice V, De Marino S, Lauro G, Casapullo A, Fantasma F, Trombetta F, Bifulco G, Iorizzi M. Phytochemical Analysis of the Methanolic Extract and Essential Oil from Leaves of Industrial Hemp Futura 75 Cultivar: Isolation of a New Cannabinoid Derivative and Biological Profile Using Computational Approaches. Plants (Basel). 2022 Jun 24;11(13):1671.

[141] Radwan MM, Ross SA, Slade D, Ahmed SA, Zulfiqar F, Elsohly MA. Isolation and characterization of new Cannabis constituents from a high potency variety. Planta Med. 2008 Feb;74(3):267-72.

CHEMICAL STRUCTURE

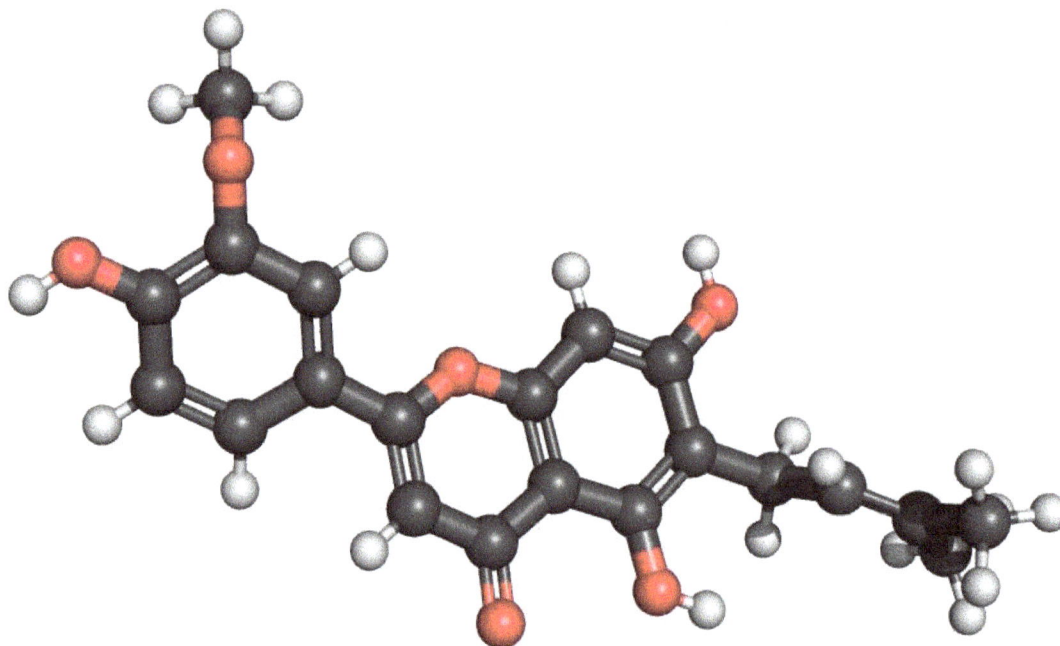

Cannflavin B is formed in the phenylpropanoid metabolic pathway from geranyl diphosphate, which then forms luteolin, from which the methylated flavone chrysoeriol is derived to ultimately produce both cannflavin A and B[142]. Cannflavin B differs from cannflavin A in that it does not contain a 5-carbon alkyl unit at the C-4 position[143]. This molecule features three rings, seven endocyclic double bonds, two exocyclic double bonds, twenty-one carbon atoms, twenty hydrogen atoms, and six oxygen atoms, notated as $C_{21}H_{20}O_6$.

Cannflavin B also exists as an isomer called isocannflavin B, where the prenyl moiety normally attached to position C-6 is instead attached at C-8[144].

OCCURRENCE IN PLANTS

As of publication, cannflavin B has only been found in cannabis and Dorstenia mannii (a small broadleaf evergreen perennial). While this compound is clearly found in drug-type cannabis chemovars, which generally express cannflavin B as a top three flavonoid constituent in cannabis, it has also been found recently in samples of industrial hemp[145].

[142] Rea, Kevin A, Jose A Casaretto, M Sameer Al-Abdul-Wahid, Arjun Sukumaran, Jennifer Geddes-McAlister, Steven J Rothstein, and Tariq A Akhtar. Biosynthesis of cannflavins A and B from Cannabis sativa L. Phytochemistry 164 (2019): 162-171.

[143] M.L. Barrett, A.M. Scott, and F.J. Evans Cannflavin A and B, prenylated flavones from Cannabis Sativa L. 28 May 1985.

[144] Moreau M, Ibeh U, Decosmo K, et al. Flavonoid derivative of cannabis demonstrates therapeutic potential in preclinical models of metastatic pancreatic cancer. Front Oncol 2019;9:660.

[145] Izzo, Luana, Luigi Castaldo, Alfonso Narvaez, Giulia Graziani, Anna Gaspari, Yelko Rodriguez-Carrasco, and Alberto Ritieni. Analysis of Phenolic Compounds in Commercial Cannabis sativa L. Inflorescences Using UHPLC- Q-Orbitrap HRMS. Molecules (Basel, Switzerland) 25.3 (2020)1.

BIOLOGICAL ACTIVITY IN PLANTS

As of early 2025, no studies or other work were available regarding the biological functions of cannflavin B in cannabis. Based on the general activities of other flavonoids, we can assume that this compound provides UV protection, antioxidant activity, and allelopathic activity, among other potential functions.

USES IN INDUSTRY

Cannflavin B appears as a pale yellow or off-white solid with a waxy texture. There are currently no industrial or commercial uses of this flavonoid; this compound is currently available for research purposes only, as it is difficult to isolate and purify via crystallization between its isomers[146].

Interestingly, cannflavin B has been found in a variety of hemp that has been proposed for mass food production. Researchers found significant levels of cannflavin B in hemp sprouts from the Ermo variety of cannabis sativa L.[147], proposing hemp sprouts as a new anti-inflammatory hemp food product for potential development and mass production.

Finally, cannflavin B derivatives (primarily dihydroxycannflavins) have been produced via microbial transformation in Beauveria bassiana[148] (a soil-dwelling fungus), however, researchers found no significant antifungal, antibacterial, antileishmanial, or antimalarial activity in the cannflavin metabolites produced by the fungi.

POTENTIAL USES IN MEDICINE

Cannflavin B offers therapeutic and medical potential in the treatment of cancer and inflammatory diseases, but it should be noted that the compound possibly interferes with some types of neuronal signaling. Researchers found that cannflavin A and cannflavin B had an inhibitory effect on brain-derived neurotrophic factor (BDNF)-induced Arc expression through disruption of receptor signaling[149]. BDNF and Arc are related to learning and memory.

Antileishmanial Agent: As a primary constituent in a variety of cannabis sativa L., cannflavin B exhibited moderate antileishmanial activity[150] as one of only two compounds in the strain to exhibit these effects.

Treatment for Pancreatic Cancer: A synthetic derivative of cannflavin B called FBL-03G has been demonstrated to significantly increase apoptosis in vitro in two pancreatic cancer models and showed therapeutic efficacy in delaying local and metastatic tumor progression in vivo in animal models of pancreatic cancer[151].

[146] Russo, Ethan, and Marcu, Jahan. Cannabis Pharmacology: The Usual Suspects and a Few Promising Leads. Advances in pharmacology (San Diego, Calif.) 80 (2017): 67-134.

[147] Oliver Werz, Julia Seegers, Anja Maria Schaible, Christina Weinigel, Dagmar Barz, Andreas Koeberle, Gianna Allegrone, Federica Pollastro, Lorenzo Zampieri, Gianpaolo Grassi, Giovanni Appendino. Cannflavins from hemp sprouts, a novel cannabinoid-free hemp food product, target microsomal prostaglandin E2 synthase-1 and 5- lipoxygenase PharmaNutrition Volume 2, Issue 3, July 2014, Pages 53-60.

[148] Ibrahim AK, Radwan MM, Ahmed SA, Slade D, Ross SA, ElSohly MA, Khan IA. Microbial metabolism of cannflavin A and B isolated from Cannabis sativa. Phytochemistry. 2010 Jun;71(8-9):1014-9.

[149] Jennifer Holborn, Alicyia Walczyk-Mooradally, Colby Perrin, Begüm Alural, Cara Aitchison, Adina Borenstein, Nina Jones, Jibran Y. Khokhar, Tariq A. Akhtar, Jasmin Lalonde, Interference of neuronal TrkB signaling by the cannabis-derived flavonoids cannflavins A and B, Phytomedicine Plus, Volume 3, Issue 1, 2023, 100410, ISSN 2667-0313.

[150] Radwan, Mohamed M et al. Isolation and characterization of new Cannabis constituents from a high potency variety. Planta medica vol. 74,3 (2008): 267-72. doi:10.1055/s-2008-1034311.

[151] Moreau, Michele, Udoka Ibeh, Kaylie Decosmo, Noella Bih, Sayeda Yasmin-Karim, Ngeh Toyang, Henry Lowe, and Wilfred Ngwa. Flavonoid Derivative of Cannabis Demonstrates Therapeutic Potential in Preclinical Models of Metastatic Pancreatic Cancer. Frontiers in oncology 9 (2019): 660.

IMPLICATIONS FOR HUMAN HEALTH & NUTRITION

Because cannflavin B is found primarily in cannabis, consumption of cannabis products is the only viable method of increasing dietary intake of this flavonoid.

Cannflavin B Review

Answer the following questions to test your knowledge of this flavonoid:

Question #1: What type of flavonoid is cannflavin B?

    a.   Chalcone
    b.   Flavanones
    c.   Flavonol
    d.   Flavone

Question #2: The base flavonoid in the biosynthesis of cannflavin B is:

    a.   Cannabisin
    b.   Rutin
    c.   Quercetin
    d.   Luteolin

Question #3: How common is cannflavin B in cannabis?

    a.   Top three
    b.   Top five
    c.   Top ten
    d.   Secondary

Question #4: What is the chemical formula for cannflavin B?

    a. $C_{24}H_{26}O$
    b. $C_{21}H_{20}O_6$
    c. $C_{26}H_{28}O_6$
    d. $N_{15}H_{26}$

Question #5: Which researcher discovered and coined the name for the molecule cannflavin B?

    a.   Elsohly
    b.   Barrett
    c.   Russo
    d.   Rosenthal

Question #6: Name two potential medical uses of cannflavin B:

1 _____ 2 _____

For the answer key to Cannflavin B, please visit www.cannabischemistry.org

ORIENTIN

Type: Flavone
Chemical Formula: $C_{21}H_{20}O_{11}$
Molecular Weight: 448.38 g/mol
Boiling Point: 816.10 °C @ 760.00 mm Hg (estimated)[152]
Flash Point: 289.10 °C (estimated by TGSC)
Melting Point: 265-267 °C (decomposes)[153]
Solubility: Soluble in water
Oral LD50: Unreported/unknown
Biological Role: Allelopathic agent, abiotic stress reducer
Therapeutic Role: Anticancer, neuroprotective agent, anti-diabetes agent
Commercial Use: Research chemical
Occurrence in Cannabis: Top three

Occurs in Cannabis Strains: Cheese, Critical Dream, Fruit Punch, Huckleberry, Purple Moose, Strawberry Banana, Strawberry Cough, Super Lemon Haze. Orientin has also been found in several hemp and CBD varieties, including Skunk, Fourway, Kompolti, Fasamo[154], and CBD Mango Haze[155]. This flavone likely occurs in many cannabis varieties in addition to those listed here, however, flavonoid testing is rarely carried out for cannabis and cannabis products.

INTRODUCTION

Orientin is well-studied from a medical perspective, offering dozens of potential therapeutic and medical uses, especially in the prevention and treatment of cancer and diabetes, and as a potent antioxidant.

Orientin is common in cannabis, generally occurring within the top three flavonoids by concentration in many strains as detailed above. Interestingly, the only flavonoids that are more prevalent in cannabis are two that are thought to be mostly unique to the plant; cannflavin A and cannflavin B, while a third, cannflavin C, is much less common. However, orientin also occurs in many other plants.

---

[152] The Good Scents Company Data Sheet for Orientin, from: https://www.thegoodscentscompany.com/data/rw1699391.html. Accessed November 1, 2023.

[153] FoodB Data Sheet for Orientin, from: https://foodb.ca/compounds/FDB002510. Accessed November 1, 2023.

[154] Isidore E, Karim H, Ioannou I. Extraction of Phenolic Compounds and Terpenes from Cannabis sativa L. By-Products: From Conventional to Intensified Processes. Antioxidants (Basel). 2021 Jun 10;10(6):942.

[155] Jin, D., Dai, K., Xie, Z. et al. Secondary Metabolites Profiled in Cannabis Inflorescences, Leaves, Stem Barks, and Roots for Medicinal Purposes. Sci Rep 10, 3309 (2020).

CHEMICAL STRUCTURE

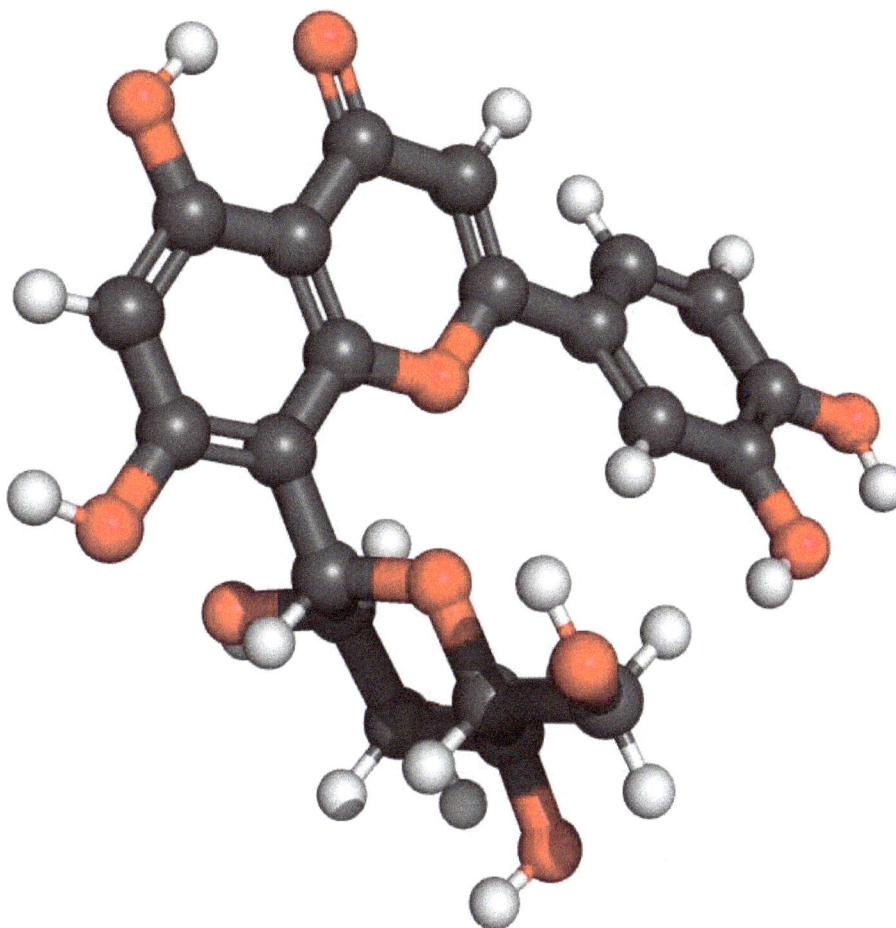

Orientin is formed from amino acid derivatives in the phenylpropanoid metabolic pathway of plants, where two rings of benzene are joined together by one ring of pyrane to form the base compound for many flavonoids, luteolin. Luteolin is then modified by glycosylation at position 8 of the C ring to become orientin. The molecular skeleton of orientin consists of four rings and eight sets of double bonds, all but one of which are endocyclic, with a total of twenty-one carbon atoms, twenty hydrogen atoms, and eleven oxygen atoms scientifically notated as $C_{21}H_{20}O_{11}$. Other variations of orientin include:

Orientin 7,3'-dimethyl ether: has methyl groups (CH3-) attached to the 7th and 3' positions.

Orientin 7-O-sulfate: has a sulfate group (-OSO3H) attached to the 7th position.

Orientin 3'-O-glucoside: has a glucoside moiety (-O-glucose) attached to the 3'-position.

Orientin 2"-O-xyloside-6"-ferulate: contains a xyloside group (-O-xylose) at the 2" position, and a ferulate moiety (-O-ferulic acid) at the 6" position

Isoorientin: a structural isomer of orientin.

OCCURRENCE IN PLANTS

Orientin occurs in many plants, with production of this compound in plants likely varying according to light and dark conditions[156], along with the time of year.

Acai	Apricot	Bamboo
Barley	Basil	Blueberry
Buckwheat	Cannabis	Corn silk
Cucumber	Fenugreek	Flax
Marjoram	Oregano	Passion flower
Pheasant's eye	Salsify	Tamarind
Yopo		

BIOLOGICAL ACTIVITY IN PLANTS

The specific biological functions of orientin in plants is not well-studied, however, limited research has shown that this flavone is effective at reducing some types of abiotic stress, and can serve as an allelopathic agent:

Abiotic Stress Reliever: Buckwheat was shown to increase production of orientin, rutin, and vitexin in response to salinity stress by as much as 153%, while also showing an increase in antioxidant activity and overall levels of phenolic compounds[157].

Allelopathic Agent: As a primary constituent in the aqueous and ethanolic extracts made from psychotria viridis (a perennial shrub in the coffee family) leaves, orientin contributed to the significant inhibition of the germination of seeds and the growth of seedlings of lactuca sativa[158] (lettuce). As the primary constituent of the aromatic shrub tinospora tuberculate, orientin contributed to the inhibition of seed germination and seedling growth of rice plants[159].

USES IN INDUSTRY

Commercial samples of orientin are for experimental or research use only, and appear as a clear, yellow, or yellow-orange powder.

POTENTIAL USES IN MEDICINE

Orientin offers a surprising number of potential medical uses, especially in cancer prevention and treatment, diabetes treatment, and as a potent antioxidant, which can be useful in the treatment of many diseases and conditions.

---

[156] Li, Xiaohua, Aye Aye Thwe, Nam Il Park, Tatsuro Suzuki, Sun Ju Kim, and Sang Un Park. Accumulation of phenylpropanoids and correlated gene expression during the development of tartary buckwheat sprouts. Journal of agricultural and food chemistry 60.22 (2012): 5629-35.

[157] Lim, Jeong-Ho, Kee-Jai Park, Bum-Keun Kim, Jin-Woong Jeong, and Hyun-Jin Kim. Effect of salinity stress on phenolic compounds and carotenoids in buckwheat (Fagopyrum esculentum M.) sprout. Food Chemistry 135.3 (2012): 1065-1070.

[158] Amanda O. Andrade, et al. Allelopathic Effects of Psychotria viridis Ruiz & Pavon on the Germination and Initial Growth of Lactuca sativa L. Journal of Agricultural Science; Vol. 9, No. 1; 2017 ISSN 1916-9760.

[159] Aslani, Farzad (2015) Allelopathic suppression of weeds in rice field by Tinospora tuberculata beumee. Doctoral thesis, Universiti Putra Malaysia.

Treatment for Bladder Cancer: Orientin has been shown to inhibit the proliferation of and cause apoptosis in T24 bladder carcinoma cells, while also causing cell cycle arrest, reduced cell viability, and inhibiting the expression of inflammatory mediators[160].

Treatment for Colorectal Cancer: As isolated from rooibos and tulsi leaves, orientin exhibited remarkable cytotoxic and antiproliferative effects against HT29 colorectal carcinoma cells, with researchers proposing that this flavonoid "could be a potent chemotherapeutic agent against colorectal cancer[161]."

Treatment for Esophageal Cancer: As a primary constituent in an isolate made from the dried flowers of Trollius chinensis, orientin markedly inhibited the proliferation of and induced apoptosis in EC-109 esophageal cancer cells[162].

Prevention of Cancer Metastasis: Orientin has been shown to exhibit significant antimigratory and anti-invasive effects in MCF-7 breast cancer cells[163] and has been suggested as a possible therapeutic agent for the treatment of cancer metastasis.

Radioprotective Agent in Cancer Treatment: As a primary constituent in holy basil (ocimum sanctum), orientin has been shown to exhibit radioprotective effects[164], and may be used to attenuate damage to normal cells caused by radiation treatment.

Neuroprotective Agent: As isolated from the leaves of cyperus esculentus, orientin lowered lipid peroxidation and reactive oxygen species formation, decreased protein oxidation, and attenuated brain water content and cerebral infarct volume in rats with induced ischemia/reperfusion injury[165].

Treatment for Neuropathic Pain: Orientin has been shown to alleviate allodynia (a neurological condition where normal stimuli become painful) associated with temperature and mechanical stimuli in rats with spinal nerve ligation by suppressing pro-inflammatory cytokines and increasing levels of anti-inflammatory cytokines[166], among other mechanisms of action.

Cardioprotective Agent: Researchers have demonstrated that orientin can promote the induction of autophagy (a process of cellular cleanup and recycling) during myocardial ischemia reperfusion injury, and increases the formation of autophagosomes, among numerous other effects associated with enhanced cell viability and decreased apoptosis[167]. Additionally, as a primary constituent in a methanolic extract of lagenaria siceraria (the calabash or

---

[160] Tian, Fenghao et al. The Effects of Orientin on Proliferation and Apoptosis of T24 Human Bladder Carcinoma Cells Occurs Through the Inhibition of Nuclear Factor-kappaB and the Hedgehog Signaling Pathway. Medical science monitor - international medical journal of experimental and clinical research vol. 25 9547-9554. 14 Dec. 2019.

[161] Thangaraj, Kalaiyarasu et al. Orientin Induces G0/G1 Cell Cycle Arrest and Mitochondria Mediated Intrinsic Apoptosis in Human Colorectal Carcinoma HT29 Cells. Biomolecules vol. 9,9 418. 27 Aug. 2019.

[162] An, Fang et al. Effects of orientin and vitexin from Trollius chinensis on the growth and apoptosis of esophageal cancer EC-109 cells. Oncology letters vol. 10,4 (2015): 2627-2633.

[163] Kim, Soo-Jin, Thu-Huyen Pham, Yesol Bak, Hyung-Won Ryu, Sei-Ryang Oh, and Do-Young Yoon. Orientin inhibits invasion by suppressing MMP-9 and IL-8 expression via the PKCa/ ERK/AP-1/STAT3-mediated signaling pathways in TPA-treated MCF-7 breast cancer cells. Phytomedicine 50 (2018): 35-42.

[164] Suresh Rao, et al. Radioprotective Effects of the Ocimum Flavonoids Orientin and Vicenin: Observations from Preclinical Studies. Polyphenols in Human Health and Disease Volume 2, 2014, Pages 1367-1371.

[165] Jing, Si-Qun et al. Neuroprotection of Cyperus esculentus L. orientin against cerebral ischemia/reperfusion induced brain injury. Neural regeneration research vol. 15,3 (2020): 548-556.

[166] Guo, Dongdong, Xinyi Hu, Haojie Zhang, Chenghua Lu, Guangwei Cui, and Xingjing Luo. Orientin and neuropathic pain in rats with spinal nerve ligation. International Immunopharmacology 58 (2018): 72-79.

[167] Liu, Liya, Youxi Wu, and Xiulan Huang. Orientin protects myocardial cells against hypoxia-reoxygenation injury through induction of autophagy. European Journal of Pharmacology 776 (2016): 90-98.

bottle gourd) fruit powder, orientin has been shown to reduce hypertension in rats by reducing necrosis, heart inflammation, and serum cholesterol, while also exhibiting antioxidant activity[168].

Post Heart Attack Treatment: As isolated from bamboo leaves, orientin has been shown to inhibit reactive oxygen species (ROS) generation, repolarize mitochondrial membrane potential, suppress mitochondrial cytochrome C release, enhance B-cell lymphoma two levels, inhibit BAX (apoptosis regulator) levels[169], and, like the two studies referred to above, showed protective effects against ischemia/reperfusion injury.

Treatment for Atherosclerosis: Orientin has been shown to inhibit ROS generation and increase endothelial nitric oxide synthase expression, significantly reverse the effects of oxidized low-density lipoproteins (ox-LDL), and downregulate scavenger receptors of ox-LDL[170], among other effects that may be useful in the treatment of atherosclerosis.

Anti-inflammatory Agent: Orientin has been shown to significantly decrease lipopolysaccharide-stimulated production of proinflammatory mediators by dramatically inhibiting tumor necrosis factor (TNF)-a, IL-6 (interleukin), IL-18, and IL-1B, prostaglandin E2, and nitric oxide[171].

Prevention & Treatment of Osteoporosis: As isolated from rooibos tea, orientin stimulated the significant increase of mineral content in human osteoblast cells and was associated with increased alkaline phosphatase and mitochondrial activity[172].

Treatment for Diabetes: As the primary constituent in a methanolic extract made from the leaves of ambay pumpwood (cecropia pachystachya), orientin exhibited significant blood-glucose lowering activity in alloxan-induced diabetic rats, with overall effects comparable to the standard diabetes drugs metformin and glibenclamide[173]. As the primary constituent in a hydroethanolic extract made from the aerial parts of parkinsonia (cercidium), orientin significantly reduced serum and urinary glucose, urinary urea, and triglyceride levels[174], also in alloxan-induced diabetic rats. Another study in rats found that: "acai can modulate reactive oxygen species production by neutrophils and that it has a significant favorable effect on the liver antioxidant defense system under physiological conditions of oxidative stress and partially revert deleterious effects of diabetes in the liver[175]." Acai berries are known to produce high levels of orientin. Finally, in a cultured skeletal muscle cell model of mitochondrial dysfunction, administration of orientin reduced production of intracellular reactive oxygen species and enhanced the expression of genes involved in mitochondrial function, with the study authors concluding that

[168] Mali, Vishal R, V Mohan, and Subhash L Bodhankar. Antihypertensive and cardioprotective effects of the Lagenaria siceraria fruit in NG-nitro-L-arginine methyl ester (L-NAME) induced hypertensive rats. Pharmaceutical biology 50.11 (2013): 1428-35.

[169] Lu, Na, Yiguo Sun, and Xiaoxiang Zheng. Orientin-induced cardioprotection against reperfusion is associated with attenuation of mitochondrial permeability transition. Planta medica 77.10 (2011): 984-991.

[170] Li, Chunmeng, Chanchun Cai, Xiangjian Zheng, Jun Sun, and Liou Ye. Orientin suppresses oxidized low-density lipoproteins induced inflammation and oxidative stress of macrophages in atherosclerosis. Bioscience, biotechnology, and biochemistry 84.4 (2020): 774-779.

[171] Xiao, Qingfei et al. Orientin Ameliorates LPS-Induced Inflammatory Responses through the Inhibitory of the NF-KB Pathway and NLRP3 Inflammasome. Evidence-based complementary and alternative medicine: eCAM vol. 2017 (2017): 2495496.

[172] Nash, Leslie A., Philip J. Sullivan, Sandra J. Peters, and Wendy E. Ward. Rooibos flavonoids, orientin and luteolin, stimulate mineralization in human osteoblasts through the Wnt pathway. Molecular Nutrition & Food Research 59.3 (2015): 443-453.

[173] Aragao, Danielle M O, Lyvia Guarize, Juliana Lanini, Juliana C da Costa, Raul M G Garcia, and Elita Scio. Hypoglycemic effects of Cecropia pachystachya in normal and alloxan-induced diabetic rats. Journal of ethnopharmacology 128.3 (2010): 629-633.

[174] Leite, Ana Catarina Rezende et al. Characterization of the Antidiabetic Role of Parkinsonia aculeata (Caesalpineaceae). Evidence-based complementary and alternative medicine: eCAM vol. 2011 (2011): 692378.

[175] Guerra, Joyce Ferreira da Costa et al. Dietary a9ai modulates ROS production by neutrophils and gene expression of liver antioxidant enzymes in rats. Journal of clinical biochemistry and nutrition vol. 49,3 (2011): 188-94.

orientin and several related flavonoids "have the potential to be as effective as established pharmacological drugs such as metformin and insulin in protecting against mitochondrial dysfunction[176]."

Treatment for Retinal Disease: As the primary constituent isolated from black bamboo (phyllostachys nigra), orientin was shown to contribute to the reduction of reactive oxygen species and replenished reduced glutathione levels in retinal ganglion cells[177], indicating significant antioxidant effects.

Prevention of Obesity: As a primary component in an ethanolic extract made from duckweed (spirodela polyrhiza), orientin contributed to potent anti-adipogenesis activity by decreasing expression of adipogenic transcription factors[178], which is essential in the formation of fat cells from stem cells.

Analgesic Agent: As a primary constituent in an extract made from the leaves of piper solmsianum, and as a pure compound, orientin was found to exhibit potent dose-dependent effects against acetic acid-induced writhing and capsaicin- and glutamate-induced nociception (pain reception) in mice, and overall was approximately twenty times more potent than aspirin, and three and a half times more active than indomethacin[179].

Genoprotective Agent Against Insecticides: As a primary constituent in an extract made from the leaves of holy basil (ocimum sanctum), orientin contributed to the scavenging of reactive intermediates that are capable of binding to proteins and DNA; it also showed an overall genoprotective effect in rats exposed to the insecticide chlorpyrifos[180].

Antioxidant: In addition to the several studies above that refer to the antioxidant effects of orientin, many other studies have also confirmed these activities. For instance, in a study published by the *Journal of Agriculture and Food Chemistry*, researchers found that the antioxidant properties of various edible blueberry species were attributed mostly to flavonoids, including orientin and vitexin[181]. Similar antioxidant activities have also been attributed to orientin as found in an n-butanol-soluble fraction of celtis africana[182], and to a methanolic crude extract of launaea procumbens[183] (creeping launaea, a perennial herb). Finally, as the primary flavonoid constituent in the fruits of livistona chinensis (Chinese fan palm), the free radical scavenging activity of orientin was found to be stronger than those of two known antioxidants: vitamin C and baicalin[184].

---

[176] Mthembu SXH, Muller CJF, Dludla PV, Madoroba E, Kappo AP, Mazibuko-Mbeje SE. Rooibos Flavonoids, Aspalathin, Isoorientin, and Orientin Ameliorate Antimycin A-Induced Mitochondrial Dysfunction by Improving Mitochondrial Bioenergetics in Cultured Skeletal Muscle Cells. Molecules. 2021 Oct 18;26(20):6289.

[177] Lee, Hee Ju, Kyung-A Kim, Kui Dong Kang, Eun Ha Lee, Chul Young Kim, Byung Hun Um, and Sang Hoon Jung. The compound isolated from the leaves of Phyllostachys nigra protects oxidative stress-induced retinal ganglion cells death. Food and chemical toxicology: an international journal published for the British Industrial Biological Research Association 48.6 (2010): 1721-7.

[178] Kim, JinPyo, IkSoo Lee, JeongJu Seo, MunYhung Jung, YoungHee Kim, NamHui Yim, and KiHwan Bae. Vitexin, orientin and other flavonoids from Spirodela polyrhiza inhibit adipogenesis in 3T3-L1 cells. Phytotherapy Research 24.10 (2010): 1543-1548.

[179] Silva, Rosi, Rosendo Yunes, Marcia Souza, Franco Monache, and Valdir Cechinel-Filho. Antinociceptive properties of conocarpan and orientin obtained from Piper solmsianum C. DC. var. solmsianum (Piperaceae). Journal of Natural Medicines 64.4 (2010): 402-408.

[180] Khanna, Asha et al. Role of Ocimum sanctum as a Genoprotective Agent on Chlorpyrifos-Induced Genotoxicity. Toxicology international vol. 18,1 (2011): 9-13.

[181] Dastmalchi, Keyvan et al. Edible neotropical blueberries: antioxidant and compositional fingerprint analysis. Journal of agricultural and food chemistry vol. 59,7 (2011): 3020-6.

[182] Perveen, Shagufta, Azza Muhammed El-Shafae, Areej Al-Taweel, Ghada Ahmed Fawzy, Abdul Malik, Nighat Afza, Mehreen Latif, and Lubna Iqbal. Antioxidant and urease inhibitory C-glycosylflavonoids from Celtis africana. Journal of Asian natural products research 13.9 (2011): 799-804.

[183] Khan, Rahmat Ali et al. Assessment of flavonoids contents and in vitro antioxidant activity of Launaea procumbens. Chemistry Central journal vol. 6,1 43. 22 May. 2012.

[184] Yao, Hong, Yan Chen, Peiying Shi, Juan Hu, Shaoguang Li, Liying Huang, Jianhua Lin, and Xinhua Lin. Screening and quantitative analysis of antioxidants in the fruits of Livistona chinensis R. Br using HPLC-DAD- ESI/MS coupled with pre-column DPPH assay. Food Chemistry 135.4 (2012): 2802-2807.

Treatment for Intervertebral Disc Degeneration: In an intervertebral disc degeneration (IVDD) pathological model in rats, treatment with orientin upregulated AMPK (an enzyme that plays a role in cellular energy homeostasis) and SIRT1 (a protein that regulates DNA damage signaling) in nucleus pulposus (intervertebral disc fluid) cells, maintaining extracellular matrix and endoplasmic reticulum balance and decreasing the oxidative stress response[185].

Treatment for Rheumatoid Arthritis: Orientin may be useful in the treatment of rheumatoid arthritis after it was shown to inhibit cell viability, migration, and invasion, facilitate apoptosis and decrease the secretion of cytokines induced by tumor necrosis factor alpha (TNF-α), and inactivate the mitogen-activated protein kinase (MAPK)-related signaling pathway in human rheumatoid arthritis fibroblast-like synoviocytes[186].

Treatment for Liver Injury and Failure: In a model of d-galactosamine- and lipopolysaccharides (d-GalN/LPS)-induced liver injury in mice, administration of orientin improved the hepatic histological changes and reduced the levels of hepatic and serum alanine aminotransferase (an enzyme normally found in low levels) and aspartic acid aminotransferase (an enzyme that helps regulate amino acid metabolism; when levels of this are high, it can be an indicator of liver disease), significantly suppressing oxidative stress and ameliorating liver injury in the mice[187]. Orientin has also been shown to reverse acetaminophen-induced acute liver failure by inhibiting oxidative stress and mitochondrial dysfunction by inhibiting the expression of cytochrome P450 2E1 (an enzyme involved in ethanol metabolism), maintaining a normal liver structure, and reducing the levels of serum alanine transaminase and serum aspartate aminotransferase (AST), among other effects in C57/BL6 mice[188].

Treatment for Drug-resistant Pathogenic Staphylococcus Aureus Infection: In a murine model of drug-resistant pathogenic Staphylococcus aureus pneumonia infections in mice, treatment with orientin effectively inhibited the activity of Sortase A (an enzyme important in host colonization), inhibited the binding of S. aureus to fibrinogen, diminished biofilm formation and the attaching of Staphylococcal protein A to the cell wall in vitro, while also attenuating S. aureus virulence in vivo[189].

IMPLICATIONS FOR HUMAN HEALTH & NUTRITION

Dietary consumption of orientin can be accomplished by consuming plant-based foods or cannabis where this flavonoid is a primary constituent. Considering eating fruits like acai berries, apricots, blueberries, passion fruit, or tamarind. For an easy to prepare vegetable, cucumbers are high in orientin. When preparing, cooking, or serving food, consider working with herbs like basil, marjoram, and oregano. Products containing grains like barley, buckwheat, or flax are also likely to contain orientin.

While it is likely that most cannabis varieties contain some level of orientin, the following are strains where lab tests have confirmed this flavone in the top ten flavonoids by concentration: Cheese, Critical Dream, Fruit Punch, Huckleberry, Purple Moose, Strawberry Banana, Strawberry Cough, and Super Lemon Haze.

[185] Zhang Z, Wu J, Teng C, Wang J, Yu J, Jin C, Wang L, Wu L, Lin Z, Yu Z, Lin Z. Orientin downregulating oxidative stress-mediated endoplasmic reticulum stress and mitochondrial dysfunction through AMPK/SIRT1 pathway in rat nucleus pulposus cells in vitro and attenuated intervertebral disc degeneration in vivo. Apoptosis. 2022 Dec;27(11-12):1031-1048.

[186] Ji W, Xu W. Orientin inhibits the progression of fibroblast-like synovial cells in rheumatoid arthritis by regulating MAPK-signaling pathway. Allergol Immunopathol (Madr). 2022 Nov 1;50(6):154-162.

[187] Li F, Liao X, Jiang L, Zhao J, Wu S, Ming J. Orientin Attenuated d-GalN/LPS-Induced Liver Injury through the Inhibition of Oxidative Stress via Nrf2/Keap1 Pathway. J Agric Food Chem. 2022 Jul 6;70(26):7953-7967.

[188] Xiao Q, Zhao Y, Ma L, Piao R. Orientin reverses acetaminophen-induced acute liver failure by inhibiting oxidative stress and mitochondrial dysfunction. J Pharmacol Sci. 2022 May;149(1):11-19.

[189] Wang L, Jing S, Qu H, Wang K, Jin Y, Ding Y, Yang L, Yu H, Shi Y, Li Q, Wang D. Orientin mediates protection against MRSA-induced pneumonia by inhibiting Sortase A. Virulence. 2021 Dec;12(1):2149-2161.

Experimentation with purified orientin should be carried out with extreme caution, as these compounds could be toxic in some situations.

Orientin Review

Answer the following questions to test your knowledge of this flavonoid:

Question #1: What type of flavonoid is orientin?

  a. Flavonol
  b. Flavorall
  c. Chalcone
  d. Flavone

Question #2: The base flavonoid in the biosynthesis of orientin is:

  a. Cannabisin
  b. Rutin
  c. Quercetin
  d. Luteolin

Question #3: How common is orientin in cannabis?

  a. Top three
  b. Top five
  c. Top ten
  d. Secondary

Question #4: What is the chemical formula for orientin?

  a. $C_{24}H_{26}O$
  b. $C_{15}H_{24}O_6$
  c. $C_{26}H_{28}O_6$
  d. $C_{21}H_{20}O_{11}$

Question #5: Name two biological roles of orientin in plants:

1 _____   2 _____

Question #6: Name two potential medical uses of orientin:

1 _____   2 _____

For the answer key to Orientin, please visit www.cannabischemistry.org

RUTIN

Type: Flavonol
Chemical Formula: $C_{27}H_{30}O_{16}$
Molecular Weight: 610.5 g/mol
Boiling Point: 983.1 °C at 760 mmHg[190]
Flash Point: 325.4 °C[191]
Melting Point: 125.00 °C @ 760.00 mm Hg[192]
Solubility: Slightly soluble in water, soluble in ethanol, DMSO, dimethyl formamide[193]
Intraperitoneal LD50: 2,000 mg/kg (rat)
Biological Role: Allelopathic, antibacterial
Therapeutic Role: Anti-diabetes, neuroprotective, anticancer agent, chemotherapeutic adjuvant
Commercial Use: Antioxidant ingredient
Occurrence in Cannabis: Top five

Occurs in Cannabis Strains: Cheese, Critical Dream, Fruit Punch, Huckleberry, Northern Lights, Purple Punch, Purple Moose, Skywalker, Space Candy, Strawberry Banana, Strawberry Cough, Super Lemon Haze, White Cookies. Rutin has also been found in several varieties of industrial hemp including the four commercial cultivars Kompoti, Tiborszallasi, Antal, and Carmagnola[194], and likely occurs in a majority of cannabis chemovars.

INTRODUCTION

Rutin is one of the most common flavonoids in plants, including cannabis, where it generally occurs within the top five flavonoids by concentration in most cannabis varieties. A member of the flavonoid subclass called flavonols, rutin is widely studied and used in medicine, including as an anticancer agent, and in the treatment of neurological conditions. Rutin also serves antibacterial and allelopathic roles in a wide range of plants, including many common fruits, vegetables, herbs, and spices.

---

[190] ChemSpider Data Sheet for Rutin, from: https://www.chemspider.com/Chemical-Structure.4444362.html. Accessed November 2, 2023.

[191] ChemSrc Data Sheet for Rutin, from: https://www.chemsrc.com/en/cas/153-18-4_894794.html. Accessed November 2, 2023.

[192] The Good Scents Company Data Sheet for Rutin, from: https://www.thegoodscentscompany.com/data/rw1302941.html. Accessed November 2, 2023.

[193] Cayman Chemical Product Information Sheet for Rutin Hydrate, from: https://cdn.caymanchem.com/cdn/insert/19868.pdf. Accessed November 2, 2023.

[194] Izzo L, Castaldo L, Narváez A, Graziani G, Gaspari A, Rodríguez-Carrasco Y, Ritieni A. Analysis of Phenolic Compounds in Commercial *Cannabis sativa* L. Inflorescences Using UHPLC-Q-Orbitrap HRMS. Molecules. 2020 Jan 31;25(3):631.

CHEMICAL STRUCTURE

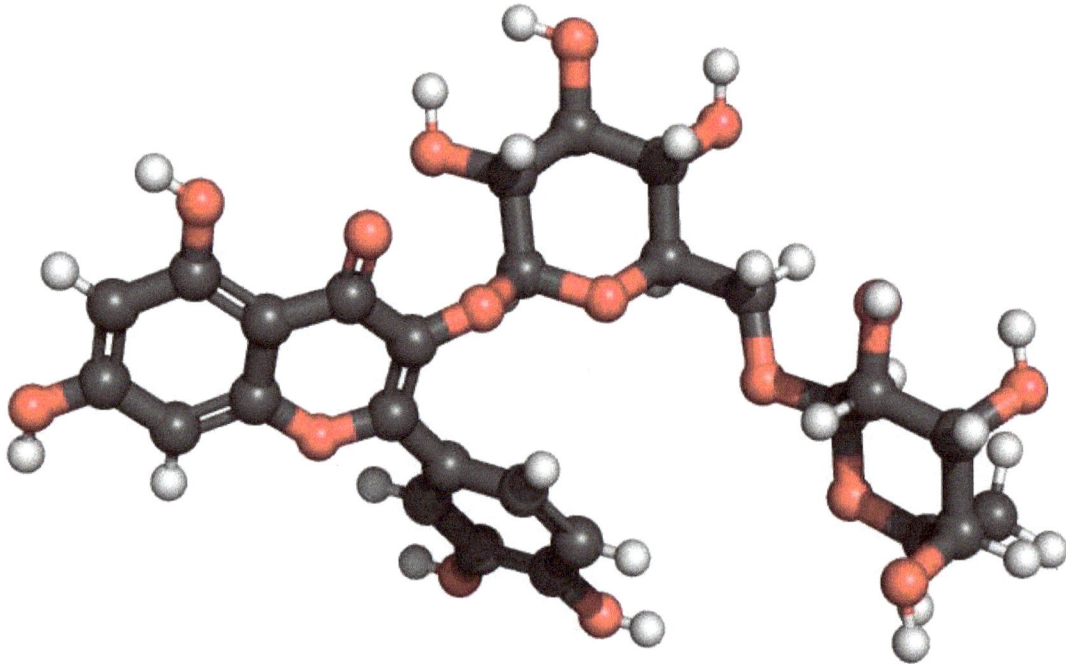

Biosynthesized in the dihydroquercetin pathway, rutin is known as a rutinoside or glycoside (referring to the addition of a sugar moiety) because the molecule is formed from quercetin and rutinose. Rutin contains twenty-seven carbon atoms, thirty hydrogen atoms, and sixteen oxygen atoms notated as $C_{27}H_{30}O_{16}$. This molecule belongs to the subclass of flavonoids called flavonols because of the way ring B is attached to ring C-2, with the hydroxy group at C-3 substituted for glucose and rhamnose (sugar) moieties. Other molecular skeleton features that make rutin unique include seven endocyclic double bonds, and one exocyclic double-bonded oxygen atom.

There are at least six known variations of rutin, with each variant possessing distinct characteristics and applications due to the added atoms, functional groups, and/or molecules, which can affect their solubility, bioavailability, and potential uses in the pharmaceutical, nutraceutical, or cosmetic industries:

Rutin Trihydrate - rutin combined with three water molecules.

Rutin Erucate - rutin combined with Erucic acid, a monounsaturated omega-9 fatty acid.

Rutin Hydrate - similar to rutin trihydrate; this form contains a variable number of water molecules.

Rutin Sulfate - rutin combined with sulfate ions, which likely changes solubility.

Rutin Linolenate - rutin combined with linolenic acid, an omega-3 fatty acid.

Rutin Sodium Sulfate - rutin combined with sodium and sulfate ions.

OCCURRENCE IN PLANTS

Rutin is found in many plants, particularly in fruits and vegetables that are easy to obtain at most supermarkets and farm stands. Often, rutin occurs in plants that are purple or red in hue, such as blackberries, blueberries, black beans, pomegranate, tomatoes, rhubarb, cherries, and plums.

Anise	Apple	Apricot
Asparagus	Banana	Basil
Bay laurel	Blackberry	Black bean
Black currant	Black elder	Black pepper
Black raspberry	Cannabis	Carpobrotus
Cayenne	Celery	Chamomile
Cherry	Chicory	Coriander
Cotton	Date palm	Dill
Fennel	Fenugreek	Horseradish
Hydrangea	Lemon	Lemongrass
Loquat	Lotus	Marjoram
Mugwort	Olive	Onion
Orange	Parsley	Parsnip
Rheum	Rhubarb	Ruta graveolens
Sorrel	Soybean	Spinach
Tarragon	Tobacco	Tomato
Watercress	Watermelon	Zucchini

BIOLOGICAL ACTIVITY IN PLANTS

Based on the body of currently available research, one of the most significant roles that rutin plays in plants is as an allelopathic agent against numerous species of weeds, flowers, and other plants. Interestingly, it has been shown that weather conditions can affect the production of rutin in the flowers, stems, and achenes of buckwheat, with different concentrations of rutin found in the grass depending on the overall weather for each production year[195]. Researchers have also shown that light quality, photoperiod, CO_2 concentration, and air temperature increase the production of rutin in Lactuca sativa L.[196] (lettuce).

Allelopathic Agent: As the primary constituent in an extract prepared from sweet chestnut (castanea sativa), rutin was shown to reduce the seed germination and root and epicotyl growth in radish (raphanus sativus) plants[197]. As a primary component in an extract made from medicago sativa (alfalfa), rutin contributed to the inhibition of growth of several plants, including digitaria ciliaris (southern crab grass), chenopodium album (pigweed), amaranthus lividus (purple amaranth), portulaca oleracea (common purslane), and commelina communis (asiatic dayflower)[198]. Interestingly, this same study showed that, by itself, rutin significantly inhibited alfalfa seed germination. As an individual component, rutin has also been shown to reduce the concentration of chlorophyll pigments, inhibit photosynthetic efficiency, decrease fluorescence quenching, and markedly decrease excitation energy fluxes in

[195] Kalinova J., Dadakova E. Varietal and year variation of rutin content in common buckwheat (Fagopyrum esculentum Moench) Cereal Research Communications Volume 34, Issue 4 March 28, 2007.

[196] Naoya Fukuda ME, Yoshida H, Kusano M. Effects of light quality, photoperiod, CO2 concentration, and air temperature on chlorogenic acid and rutin accumulation in young lettuce plants. Plant Physiol Biochem. 2022 Sep 1;186:290-298.

[197] Basile, A, S Sorbo, S Giordano, L Ricciardi, S Ferrara, D Montesano, R Castaldo Cobianchi, M L Vuotto, and L Ferrara. Antibacterial and allelopathic activity of extract from Castanea sativa leaves. Fitoterapia 71 Suppl 1 (2003) : S110-S6.

[198] Ghimire, Bimal Kumar, Balkrishna Ghimire, Chang Yeon Yu, and Ill-Min Chung. Allelopathic and Autotoxic Effects of Medicago sativa-Derived Allelochemicals. Plants 8.7 (2019).

Arabidopsis thaliana[199] (thale cress). Rutin has also been shown to inhibit the growth of rice seedlings (oryza sativa L.)[200]. Finally, rutin has been shown to activate defenses against Bemisia tabaci (silverleaf whitefly), reducing the developmental rate of nymphs, and the fecundity and feeding efficiency of adult females on plants grown from tomato seeds that were pre-treated with this flavonol[201].

Antibacterial Agent: In work published in the *European Journal of Plant Pathology*, rutin was shown to stimulate defense genes in tomato plants, leading to an induced resistance to xanthomonas perforans[202], an often-devastating bacterial disease in tomato crops.

USES IN INDUSTRY

Rutin appears as a pale yellow to amber powder and is primarily used in medical research and applications listed below, however, it can also be found as an added antioxidant ingredient in products like hair and skin conditioners. In fact, as of early 2025, there were more than 33,700 patents for products containing rutin listed in PubChem's patent database[203], however, according to The Good Scents Company, this compound is generally not for flavor or fragrance use.

POTENTIAL USES IN MEDICINE

Rutin has been used and studied in a wide variety of medical applications and settings. It has been proven as a potent treatment for several neurological diseases and conditions, and as a direct and supplemental treatment for brain, prostate, and other cancers. However, the low bioavailability of rutin as a result of poor absorption and rapid excretion[204] currently limits the widespread use of this flavonol in medicine.

Treatment for Chromosomal, DNA Damage: Rutin was shown to decrease levels of malondialdehyde (a marker of oxidative stress), micronuclei formations, and DNA fragmentation in 2,5-hexanedione (a toxic metabolite of hexane)-induced chromosomal and DNA damage in rats[205].

Neuroprotective Agent: In a cellular model of Parkinson's disease, rutin was shown to attenuate several genes in rat PC12 cells treated with 6-hydroxydopamine (a neurotoxic agent), while also upregulating the TH, ion transport, and antiapoptotic genes[206]. Rutin was also shown to protect against acrylamide cytotoxicity in rat PC12 cells due to inhibition of reactive oxygen species (ROS) production[207]. Other work investigating the neuroprotective effects of

[199] M. Iftikhar Hussain, Manuel Reigosa Roger Plant secondary metabolite rutin affects the photosynthesis and excitation energy flux responses in Arabidopsis thaliana Allelopathy Journal 38(2):215-228. January 2016.

[200] Li Pengcheng, Sun Junwei and ZhuCheng Effect of Rutin in Amaranthus spinosus L. on Antioxidative Metabolism for Rice (Oryza sativa L.) Advance Journal of Food Science and Technology 12(10): 562-567, 2016.

[201] Tang J, Shen H, Zhang R, Yang F, Hu J, Che J, Dai H, Tong H, Wu Q, Zhang Y, Su Q. Seed priming with rutin enhances tomato resistance against the whitefly Bemisia tabaci. Pestic Biochem Physiol. 2023 Aug;194:105470.

[202] Safaie Farahani, Ali, and S. Mohsen Taghavi. Rutin promoted resistance of tomato against Xanthomonas perforans. European Journal of Plant Pathology 151.2 (2017): 527-531.

[203] PubChem, National Library of Medicine, Data Sheet for Rutin, from: https://pubchem.ncbi.nlm.nih.gov/compound/Rutin#section=Patents. Accessed February 6, 2025.

[204] Enogieru, Adaze Bijou et al. Rutin as a Potent Antioxidant: Implications for Neurodegenerative Disorders. Oxidative medicine and cellular longevity vol. 2018 6241017. 27 Jun. 2018.

[205] Muhammad A, Arthur DE, Babangida S, Erukainure OL, Malami I, Sani H, Abdulhamid AW, Ajiboye IO, Hamza NM, Asema S, Ado ZM, Musa TI. Modulatory role of rutin on 2,5-hexanedione-induced chromosomaland DNA damage in rats: validation of computational predictions. Drug Chem Toxicol. 2020 Mar;43(2):113-126.

[206] Magalingam KB, Radhakrishnan A, Ramdas P, Haleagrahara N. Quercetin glycosides induced neuroprotection by changes in the gene expression in a cellular model of Parkinson's disease. J Mol Neurosci. 2015 Mar;55(3):609-17.

[207] Motamedshariaty, V.S., Amel Farzad, S., Nassiri-Asl, M. et al. Effects of rutin on acrylamide-induced neurotoxicity. DARU J Pharm Sci 22, 27 (2014).

rutin has found that the flavonol "protects against the neurodegenerative effects of prion accumulation by increasing production of neurotropic factors and inhibiting apoptotic pathway activation in neuronal cells[208]."

Treatment for Parkinson's Disease & CNS Diseases: As isolated from Japanese honeysuckle (lonicera japonica), rutin significantly decreased sodium nitroprusside-induced reactive oxygen species in rat PC12 (derived from a pheochromocytoma of the rat adrenal medulla) cells, reversing the declined GSH/GSSG (reduced glutathione and oxidized glutathione, respectively) ratio and mitochondrial membrane potential while also activating a key protein and kinase signaling pathways[209]. These findings indicate the potential to use rutin as a treatment for central nervous system diseases related to nitric oxide neurotoxicity. Additionally, in a model of aminochrome toxicity of Parkinson's Disease - which mimics dopaminergic neuronal loss – rutin prevented lysosomal dysfunction and aminochrome-induced cell death in SHSY-5Y cells (neuroblastoma cell line from a metastatic bone tumor), protected PCMC (primary culture of mesencephalic cells) against aminochrome cytotoxicity, and prevented in-vivo loss of dopaminergic neurons in substantia nigra pars compacta (SNPc), as well as microgliosis and astrogliosis[210], among other potentially anti-Parkinson's effects.

Protection Against Mitochondrial Damage: In rats subjected to isoproterenol-induced cardiotoxicity, rutin reversed increases in the levels of heart mitochondrial lipids, lipid peroxidation products, and calcium, while also attenuating a significant decrease in the activities/levels of mitochondrial antioxidants, enzymes, and adenosine triphosphate[211], thereby reducing the overall extent of mitochondrial damage induced by isoproterenol.

Immunomodulatory Agent: Rutin has been shown to increase immune activity by cellular and humoral mediated mechanisms, significantly restoring the functioning of leucocytes in cyclophosphamide (an immune suppressant) treated rats, and augmenting the phagocytic index (measures bacteria consumed by phagocytes) in the carbon clearance assay, while also increasing antibody titers and immunoglobulin levels[212], among other mechanisms.

Treatment for Radiation Injury: The combination of podophyllotoxin (an anti-tumor drug) and rutin protected intestinal stem cells from damage caused by acute radiation, exhibiting significant functional and structural intestine regeneration in irradiated subjects, enhancing crypt stem cells, upregulating catenin target genes, and minimizing acute inflammation in a murine model[213].

Cell Protective Agent Against Nephrotoxicity: Rutin has been shown to offer significant protection from vancomycin (a last-resort antibacterial drug)-induced nephrotoxicity (toxicity in the kidneys) by decreasing intracellular reactive oxygen species, increasing superoxide dismutase and catalase activities, while further protecting cells from vancomycin-induced caspase activation, mitochondrial membrane depolarization, and subsequent apoptosis in porcine (pig) renal cells[214].

[208] Na, J., Kim, S., Song, K. et al. Rutin Alleviates Prion Peptide-Induced Cell Death Through Inhibiting Apoptotic Pathway Activation in Dopaminergic Neuronal Cells. Cell Mol Neurobiol 34, 1071-1079 (2014).

[209] Wang, Rikang, Yongbing Sun, Hesong Huang, Lan Wang, Jinlong Chen, and Wei Shen. Rutin, A Natural Flavonoid Protects PC12 Cells Against Sodium Nitroprusside-Induced Neurotoxicity Through Activating PI3K/Akt/mTOR and ERK1/2 Pathway. Neurochemical Research 40.9 (2015): 1945-1953.

[210] De Araújo FM, Frota AF, de Jesus LB, Cuenca-Bermejo L, Ferreira KMS, Santos CC, Soares EN, Souza JT, Sanches FS, Costa ACS, Farias AA, de Fatima Dias Costa M, Munoz P, Menezes-Filho JA, Segura-Aguilar J, Costa SL, Herrero MT, Silva VDA. Protective Effects of Flavonoid Rutin Against Aminochrome Neurotoxicity. Neurotox Res. 2023 Jun;41(3):224-241.

[211] Punithavathi, V., K. Shanmugapriya, and P. Stanely Mainzen Prince. Protective Effects of Rutin on Mitochondrial Damage in Isoproterenol-Induced Cardiotoxic Rats: An In Vivo and In Vitro Study. Cardiovascular Toxicology 10.3 (2010): 181-189.

[212] Ganeshpurkar, Aditya, and Ajay K. Saluja. Protective effect of rutin on humoral and cell mediated immunity in rat model. Chemico-Biological Interactions 273 (2017): 154-159.

[213] Kalita, Bhargab, Rajiv Ranjan, and Manju Gupta. Combination treatment of podophyllotoxin and rutin promotes mouse Lgr5+ ve intestinal stem cells survival against lethal radiation injury through Wnt signaling. Apoptosis 24.4 (2019): 326-340.

[214] Qu, Shaoqi, Cunchun Dai, Hui Guo, Cuncai Wang, Zhihui Hao, Qihe Tang, Haixia Wang, and Yanping Zhang. Rutin attenuates vancomycin-induced renal tubular cell apoptosis via suppression of apoptosis, mitochondrial dysfunction, and oxidative stress. Phytotherapy Research 33.8 (2019): 2056-2063.

Cytoprotective Agent Against Skin Damage: The combination of rutin and ascorbic acid reduced increased levels of ROS generation on human keratinocytes and fibroblasts exposed to UVA and UVB radiation, while also increasing antioxidant enzyme activity and quelling UV-induced expression of pro-inflammatory factor NFKB and pro-apoptotic proteins[215], among other effects.

Treatment for Brain Cancer: Rutin has been shown to suppress the growth of brain cancer by decreasing the number of viable cancer cells and mitochondrial metabolism, damaging mitochondria and rough endoplasmic reticulum, inducing apoptosis, inducing a delay in cell migration and an increase in intra- and extracellular expression of fibronectin[216], and other effects in human GL-15 glioblastoma cells. Rutin has also been shown to enhance the efficacy of the glioblastoma chemotherapeutic agent temozolomide, reducing tumor volumes in mice, while also decreasing autophagy and increasing apoptosis of brain cancer cells[217].

Treatment for Prostate Cancer: In a Chinese study conducted in 2017, rutin was found to contribute to the significant reduction of proliferation and migration of PC-3 (human prostate cancer) cells, while also inducing apoptosis by activating ROS production and mitochondrial-related cell death[218]. In this study, rutin was a primary constituent in an extract made from diospyros kaki (persimmon) leaves.

Treatment for Chemotherapeutic Agent Toxicity: In Wistar rats, liver structural damage caused by cyclophosphamide (a powerful but toxic anti-cancer agent) has been shown to be reversed by rutin via the up regulation of antioxidant enzyme activities and the down regulation of serum toxicity markers[219], suggesting the potential for co-administration of cyclophosphamide and rutin for cancer patients. Rutin has also been shown to significantly reduce ROS expression levels, decrease p38 mitogen-activated protein kinase (helps manage cellular responses) and jun N-terminal kinase (an important cell regulator), and upregulate the ratio of p-AKT/AKT in cochlear hair cells exposed to cisplatin[220]. Cisplatin is a common chemotherapeutic agent for many types of cancer. Finally, researchers have demonstrated that oral administration of rutin, hesperidin, and their combination could counteract paclitaxel-induced liver damage and toxicity in rats by strengthening the antioxidant defense system and decreasing oxidative stress and apoptosis[221]. Paclitaxel is used to treat several types of cancer and is also known as Taxol.

Anticancer Agent: Rutin has been shown to be a potent general anticancer drug by researchers who observed that the flavonol exhibited antioxidant and cytotoxic effects against leukemia, multiple myeloma, and melanoma cell

---

[215] Gygotek, Agnieszka, Ewa Ambrozewicz, Anna Jastrz£b, Iwona Jarocka-Karpowicz, and Elzbieta Skrzydlewska. Rutin and ascorbic acid cooperation in antioxidant and antiapoptotic effect on human skin keratinocytes and fibroblasts exposed to UVA and UVB radiation. Archives of Dermatological Research 311.3 (2019): 203-219.

[216] Santos, Balbino L., Mona N. Oliveira, Paulo L.C. Coelho, Bruno P.S. Pitanga, Alessandra B. da Silva, Tais Adelita, Victor Diogenes A. Silva, Maria de F.D. Costa, Ramon S. El-Bacha, Marcienne Tardy, Herve Chneiweiss, Marie-Pierre Junier, Vivaldo Moura-Neto, and Silvia L. Costa. Flavonoids suppress human glioblastoma cell growth by inhibiting cell metabolism, migration, and by regulating extracellular matrix proteins and metalloproteinases expression. Chemico-Biological Interactions 242 (2015): 123-138.

[217] Zhang, P., Sun, S., Li, N. et al. Rutin increases the cytotoxicity of temozolomide in glioblastoma via autophagy inhibition. J Neurooncol 132, 393-400 (2017).

[218] Ding, Yan, Kai Ren, Huanhuan Dong, Fei Song, Jing Chen, Youtian Guo, Yanshan Liu, Weijie Tao, and Yali Zhang. Flavonoids from persimmon (Diospyros kaki L.) leaves inhibit proliferation and induce apoptosis in PC-3 cells by activation of oxidative stress and mitochondrial apoptosis. Chemico-Biological Interactions 275 (2017): 210-217.

[219] Nafees, Sana, Summya Rashid, Nemat Ali, Syed Kazim Hasan, and Sarwat Sultana. Rutin ameliorates cyclophosphamide induced oxidative stress and inflammation in Wistar rats: Role of NFKB/MAPK pathway. Chemico-Biological Interactions 231 (2015): 98-107.

[220] Zheng S, Liu C, Tang D, Zheng Z, Yan R, Wu C, Zuo N, Ma J, He Y, Liu S. The protective effect of rutin against the cisplatin-induced cochlear damage in vitro. Neurotoxicology. 2022 May;90:102-111.

[221] Ali YA, Soliman HA, Abdel-Gabbar M, Ahmed NA, Attia KAA, Shalaby FM, El-Nahass ES, Ahmed OM. Rutin and Hesperidin Revoke the Hepatotoxicity Induced by Paclitaxel in Male Wistar Rats via Their Antioxidant, Anti-Inflammatory, and Antiapoptotic Activities. Evid Based Complement Alternat Med. 2023 May 26;2023:2738351.

lines in vitro[222], with no toxicity to surrounding healthy cells. As a primary constituent in ethanolic extracts of Quercus mongolica Fisch leaves, rutin was found to induce apoptosis and inhibit cell proliferation in MCF-7 human breast cancer cell lines, SMMC-7721 human hepatocellular carcinoma cells, HeLa human cervical carcinoma cell lines, and SKOV3 human ovarian carcinoma cell lines[223].

Anti-Inflammatory Agent: In a study published in Poultry Science, rutin was demonstrated to attenuate the effects of lipopolysaccharide (LPS)-induced inflammatory responses in the muscle cells of mice, reducing production of ROS and blocking activation of NF-KB[224] (a critical cell protein that controls cell survival). Rutin has also been shown to significantly ameliorate the effects of inflammatory bowel disease (IBD) in experimental animals, while aiding in the delivery of another potential flavonoid-based treatment for IBD, quercetin[225]. Finally, rutin potently inhibited the release of the HMGB1 protein, a late mediator of severe vascular inflammatory conditions, while also down-regulating HMGB1-dependent inflammatory responses in human endothelial cells, and inhibited HMGB1-mediated hyperpermeability and leukocyte migration in mice[226], among several other effects against vascular inflammation.

Treatment for Liver Disease: In rats subjected to carbon tetrachloride hepatoxicity, rutin restored the alteration in expression of genes in the interleukin-6 pathway (responsible for producing pro-inflammatory cytokines) via anti-apoptotic, anti-inflammatory and antioxidant effects[227], with the study authors suggesting that the flavonol "may be used as an alternative treatment for liver diseases." Rutin has also been shown to inhibit and reduce hepatotoxicity in hypercholesterolemia-induced rats by suppressing the transforming growth factor beta (TGF-β) signaling pathway[228].

Treatment for Lung Injury: Rutin has been shown to inhibit lipopolysaccharide (LPS)-induced neutrophil infiltration in the lungs of mice, while also reducing edema (swelling) and protein leakage by suppressing the expression of vascular cell adhesion molecule and inducible nitric oxide synthase, and by reducing the activation of (NF)KB[229].

Treatment for Postmenopausal Syndrome: As the primary constituent in an extract made from the herb gardenia jasminoides (cape jasmine), rutin was found to exhibit estrogen-stimulating effects in vitro by up-regulating the FSHR-aromatase pathway, without increasing the risk of hormone-dependent breast cancer in ovarian granulosa cells[230].

[222] Ikeda NE, Novak EM, Maria DA, Velosa AS, Pereira RM. Synthesis, characterization and biological evaluation of Rutin-zinc(II) flavonoid -metal complex. Chem Biol Interact. 2015 Sep 5; 239:184-91.

[223] Wang L, Du X, Yue D, Chen X. Catechin, rutin and quercetin in Quercus mongolica Fisch leaves exert inhibitory effects on multiple cancer cells. J Food Biochem. 2022 Dec;46(12):e14486.

[224] Liu, Shangxi, Deborah Adewole, Li Yu, Victoria Sid, Blake Wang, Karmin O, and Chengbo Yang. Rutin attenuates inflammatory responses induced by lipopolysaccharide in an in vitro mouse muscle cell (C2C12) model. Poultry Science 98.7 (2019).

[225] Habtemariam, Solomon, and Abebech Belai. Natural Therapies of the Inflammatory Bowel Disease: The Case of Rutin and its Aglycone, Quercetin. Mini reviews in medicinal chemistry 18.3 (2018): 234-243.

[226] Yoo, H., Ku, S., Baek, Y. et al. Anti-inflammatory effects of rutin on HMGB1-induced inflammatory responses in vitro and in vivo. Inflamm. Res. 63, 197-206 (2014).

[227] Hafez, Mohamed M et al. Hepato-protective effect of rutin via IL-6/STAT3 pathway in CCl4-induced hepatotoxicity in rats. Biological research vol. 48,1 30. 11 Jun. 2015.

[228] Al Sharari, Shakir D et al. Rutin Attenuates Hepatotoxicity in High-Cholesterol-Diet-Fed Rats. Oxidative medicine and cellular longevity vol. 2016 (2016): 5436745.

[229] Huang, Yi-Chun, Chi-Ting Horng, Shyan-Tarng Chen, Shiuan-Shinn Lee, Ming-Ling Yang, Chien-Ying Lee, Wu-Hsien Kuo, Chung-Hsin Yeh, and Yu-Hsiang Kuan. Rutin improves endotoxin-induced acute lung injury via inhibition of iNOS and VCAM-1 expression. Environmental Toxicology 31.2 (2016): 185-191.

[230] Wang, Xueyu et al. Identification of Steroidogenic Components Derived from Gardenia jasminoides Ellis Potentially Useful for Treating Postmenopausal Syndrome. Frontiers in pharmacology vol. 9 390. 30 May. 2018.

Cardioprotective Agent: Rutin has been shown to significantly attenuate inflammatory responses in a mouse model of heart injury by relieving cardiac marker enzyme levels, mitigating fibrosis related genes, markedly increasing antioxidant enzyme activity, improving oxidative production levels, and ameliorating tumor necrosis factor alpha (TNF-a) and interleukin 6 (IL-6) activity[231].

Renal Protective Agent: In mice subjected to carbon tetrachloride-induced renal injury, rutin reduced serum biochemical markers, inflammation, and capase-3 activity' and apoptosis in kidney cells by significantly reducing reactive oxygen species, calpain and ceramide levels, decreasing TNF-a and IL-1B activities, increasing Bcl-2 protein levels, and inhibiting the release of cytochrome C[232], providing marked protection against kidney injury in mice. Rutin has also been shown to protect against gamma-irradiation and malathion (an organophosphate insecticide)-induced oxidative stress and inflammation in rat kidneys through the regulation of gene expression and protein expression[233].

Prevention of Obesity: In an article published in Pharmaceutical Research, scientists report finding that rutin protects mice from high fat diet-induced obesity, fatty liver, and insulin resistance by suppressing pro-inflammatory cytokines and the production of tumor necrosis factor alpha (TNF-a), decreasing the transcription of genes involved in chronic inflammation in white adipose (fat) tissue, and increasing the expression of genes responsible for energy expenditure in brown adipose tissue[234].

Anti-Aging Agent: Rutin has been shown to increase the longevity of fruit flies (drosophila melanogaster), causing a reduction in food intake and productiveness, increasing climbing ability, and enhancing survival upon exposure to oxidative stress[235].

Treatment for Diabetes: Rutin and metformin (an anti-diabetic medication) alone and in combination have been shown to cause significant improvements in blood glucose, cholesterol, and triglyceride levels, significantly reducing phenylephrine-induced contraction, and increasing acetylcholine-induced and sodium nitroprusside-induced relaxation in male Sprague Dawley rats[236]. In euglycemic and alloxan-diabetic rabbits, pre-treatment with rutin restored neuronal and endothelial dependent relaxations, ameliorating both endothelial dysfunction and nitrergic neuropathy[237]. As antioxidants, rutin in combination with rambutan honey (made from the flowers of a tropical tree) reduced blood glucose levels and increased insulin levels in Streptozotocin-induced rat plasma[238]. Rutin might also be useful in the treatment of diabetes-related cardiovascular disease, after it was shown to reduce the generation of oxidative stress, and to restore the structural disturbances in glycated human low-density protein

[231] Xianchu, Liu, Zheng Lan, Liu Ming, and Mo Yanzhi. Protective effects of rutin on lipopolysaccharide-induced heart injury in mice. The Journal of toxicological sciences 43.5 (2018): 329-337.

[232] Ma, Jie-Qiong, Chan-Min Liu, and Wei Yang. Protective effect of rutin against carbon tetrachloride-induced oxidative stress, inflammation and apoptosis in mouse kidney associated with the ceramide, MAPKs, p53 and calpain activities. Chemico-Biological Interactions 286 (2018): 26-33.

[233] Ismail AFM, Salem AA, Eassawy MMT. Rutin protects against gamma-irradiation and malathion-induced oxidative stress and inflammation through regulation of mir-129-3p, mir-200C-3p, and mir-210 gene expressions in rats' kidney. Environ Sci Pollut Res Int. 2023 Jun;30(28):72930-72948.

[234] Gao, Mingming, Yongjie Ma, and Dexi Liu. Rutin Suppresses Palmitic Acids-Triggered Inflammation in Macrophages and Blocks High Fat Diet-Induced Obesity and Fatty Liver in Mice. Pharmaceutical Research 30.11 (2013): 2940-2950.

[235] Chattopadhyay, Debarati, Atith Chitnis, Aishwarya Talekar, Prajakta Mulay, Manyata Makkar, Joel James, and Kavitha Thirumurugan. Hormetic efficacy of rutin to promote longevity in Drosophila melanogaster. Biogerontology 18.3 (2017): 397-411.

[236] David SR, Lai PPN, Chellian J, Chakravarthi S, Rajabalaya R. Influence of rutin and its combination with metformin on vascular functions in type 1 diabetes. Sci Rep. 2023 Aug 1;13(1):12423.

[237] de Morais Campos R, Lima LMALL, da Silva AG, Santiago RO, Paz IA, Cabral PHB, Santos CF, Fonteles MC, do Nascimento NRF. Rutin ameliorates nitrergic and endothelial dysfunction on vessels and corpora cavernosa of diabetic animals. Res Vet Sci. 2023 Aug;161:163-172.

[238] Iis Inayati Rakhmat, Euis Reni Yuslianti, Welly Ratwita, Teja Koswara, Nurul Sofiana Mutiadewi. The Effect of Rambutan Honey and Rutin on Decrease Blood Glucose and Increase Streptozotocin-Induced Rat Plasma Insulin. Proceedings of the 13th Annual Scientific Conference of Medical Faculty, Universitas Jenderal Achmad Yani (ASCMF 2022). Advances in Health Sciences Research 16 December 2022.

(LDL)[239]. Increased levels of glycated LDL are associated with a high risk of atherosclerosis in people with diabetes. Interestingly, rutin as extracted from tartary buckwheat improved glucose and lipids metabolism, alleviated colon lesions, changed the community structure of the gut microbiota, and regulated the composition of gut microbiota in diabetic mice[240]. This suggests that the gut microbiota plays an important role in diabetes management. Rutin might also prove useful in the treatment of diabetes-related muscular atrophy after researchers demonstrated that the flavonoid increased myocyte area and weight of gastrocnemius to promote muscular strength, and attenuated Atrogin-1 (a muscle-specific protein that plays a key role in atrophy) and MuRF1 (a ligase that serves an important role in muscle remodeling) expressions to improve atrophy[241].

Treatment for Huntington's Disease: In a Caenorhabditis elegans (a nematode) model of Huntington's disease, rutin exhibited protective effects, acting through mechanisms involving antioxidant and chelating properties[242], suggesting that it may be able to attenuate copper and zinc toxicity.

Delay Senescence/Aging: In rats induced by D-galactose (a metabolite of Escherichia coli bacteria), rutin enhanced antioxidant markers including superoxide dismutase-1 and glutathione peroxidase-1, significantly decreased the accumulation of various aging and senescence-related proteins and enzymes, and exhibited several other effects and actions that could contribute to reduced senescence and aging in rat brain and liver cells[243], with researchers concluding that supplementation with rutin could be a natural protective compound to delay aging and maintain health.

Treatment for Alzheimer's Disease: In a mouse model of cellular redox homeostasis in Alzheimer's disease, dietary supplementation with rutin reinstated cellular redox status and APP (amyloid-beta precursor protein, which has a role in the pathogenesis of Alzheimer's) physiological processing by regularization of APP expression and BACE1 (an enzyme involved in the formation of amyloid-beta) activity[244].

Protective Agent Against Acrylamide Exposure: Acrylamide – a probable carcinogen in humans – is a chemical that can form in some foods during high-temperature cooking processes, such as frying, roasting, and baking. In rat spinal cord motor neurons exposed to acrylamide, rutin was shown to exhibit significant antioxidant effects by up-regulating the expression of P-ERK (phosphorylated extracellular signal-regulated kinase, a protein involved in vasoconstriction and vascular smooth muscle cell growth) and Nrf2 (nuclear factor erythroid 2, which is involved in the regulation of cellular resistance to oxidants) proteins in the ERK/Nrf2 pathway[245].

[239] Wani MJ, Salman KA, Hashmi MA, Siddiqui S, Moin S. Rutin impedes human low-density lipoprotein from non-enzymatic glycation: A mechanistic insight against diabetes-related disorders. Int J Biol Macromol. 2023 May 31;238:124151.

[240] Cai C, Cheng W, Shi T, Liao Y, Zhou M, Liao Z. Rutin alleviates colon lesions and regulates gut microbiota in diabetic mice. Sci Rep. 2023 Mar 25;13(1):4897.

[241] Xianchu L, Ming L. Rutin improves diabetes-induced muscle atrophy in mice. Pak J Pharm Sci. 2023 Jan;36(1):217-221.

[242] Cordeiro LM, Soares MV, da Silva AF, Dos Santos LV, de Souza LI, da Silveira TL, Baptista FBO, de Oliveira GV, Pappis C, Dressler VL, Arantes LP, Zheng F, Soares FAA. Toxicity of copper and zinc alone and in combination in Caenorhabditis elegans model of Huntington's disease and protective effects of rutin. Neurotoxicology. 2023 Jul;97:120-132.

[243] Saafan SM, Mohamed SA, Noreldin AE, El Tedawy FA, Elewa YHA, Fadly RS, Al Jaouni SK, El-Far AH, Alsenosy AA. Rutin attenuates D-galactose-induced oxidative stress in rats' brain and liver: molecular docking and experimental approaches. Food Funct. 2023 Jun 19;14(12):5728-5751.

[244] Bermejo-Bescós P, Jiménez-Aliaga KL, Benedí J, Martín-Aragón S. A Diet Containing Rutin Ameliorates Brain Intracellular Redox Homeostasis in a Mouse Model of Alzheimer's Disease. Int J Mol Sci. 2023 Mar 2;24(5):4863.

[245] Zhang T, Zhang C, Luo Y, Liu S, Li S, Li L, Ma Y, Liu J. Protective effect of rutin on spinal motor neuron in rats exposed to acrylamide and the underlying mechanism. Neurotoxicology. 2023 Mar;95:127-135.

Treatment for Schistosomiasis: In a murine model of schistosomiasis (a tropical disease caused by trematode worms), rutin exhibited strong anti-schistosome properties in vivo, decreasing the number of eggs trapped in the tissues of the liver, and modifying the serum levels of cytokines implicated in the formation of Schistosoma granuloma, while also controlling the Th1, Th2, and Th17 (mediators of cell responses) immunological responses caused by S. mansoni infection[246].

Treatment of Gout: In quail induced to gout via genetic engineering and a high purine diet, treatment with rutin reduced inflammatory expression, reduced XOD (xanthine oxidase, an enzyme that generates ROS) activity and uric acid levels, inhibited ROS production, restored oxidative stress balance, inhibited NLRP3 inflammasome (a multiprotein complex critically involved in regulating the immune system and inflammatory signaling) activation[247], and exerted other anti-inflammatory effects.

Treatment for Alcohol-related Depression and Impairment: In ethanol-induced cognitive impairment and depression in rats, treatment with rutin significantly prevented the ethanol-mediated increase in indoleamine 2,3-dioxygenase (an enzyme involved in tryptophan metabolism) activity/expression and decrease in antioxidant enzymes, in addition to an increase in markers of inflammatory response and MDA (high levels of which are a marker for oxidative stress) production[248].

Treatment for Non-alcoholic Fatty Liver Disease: As extracted from a methanolic extract of Gardenia thunbergia leaves, rutin significantly reduced high fructose diet-induced increments in weight and hepatic damage indicators, steatosis, and hypertrophy in mice, decreased the levels of total cholesterol, LDL–C, and triglycerides in the blood, and downregulated the expressions of CYP2E1 (an enzyme involved in the metabolism of fatty acids), JNK1 (a cell activity regulator), and iNOS (inducible nitric oxide synthase, an enzyme that metabolizes reactive oxygen) in the diseased mice, with results comparable to simvastatin[249] (a drug used to treat high cholesterol and triglyceride levels).

IMPLICATIONS FOR HUMAN HEALTH & NUTRITION

Most people probably consume rutin on a daily basis as part of their regular diet. If you are looking to increase your dietary intake of this flavonol, consider eating fruits like apples, apricots, bananas, blackberries, black currants, black raspberries, blueberries, cherries, figs, grapes, grapefruit, lemons, oranges, peaches, pears, plums, pomegranates, and watermelon. In meals and snacks, vegetables to include are asparagus, bell peppers, black beans, broccoli, Brussels sprouts, capers, cayenne peppers, celery, onions, parsnips, potatoes, rhubarb, soybeans, spinach, tomatoes, watercress, and zucchinis. When preparing, cooking, or serving meals, consider working with herbs and spices like anise, basil, black pepper, coriander, dill, fennel, garlic, horseradish, marjoram, mugwort, parsley, peppermint, and tarragon.

Although most cannabis strains are relatively high in rutin (presumably purple strains have the highest concentration of this flavonol), we know for certain that this compound can be obtained in the following varieties of cannabis: Cheese, Critical Dream, Fruit Punch, Huckleberry, Northern Lights, Purple Moose, Purple Punch, Skywalker, Space Candy, Strawberry Cough, Strawberry Banana, Super Lemon Haze, and White Cookies.

[246] Hamad RS. Rutin, a Flavonoid Compound Derived from Garlic, as a Potential Immunomodulatory and Anti-Inflammatory Agent against Murine Schistosomiasis mansoni. Nutrients. 2023 Feb 28;15(5):1206.

[247] Wu H, Wang Y, Huang J, Li Y, Lin Z, Zhang B. Rutin ameliorates gout via reducing XOD activity, inhibiting ROS production and NLRP3 inflammasome activation in quail. Biomed Pharmacother. 2023 Feb;158:114175.

[248] Ebokaiwe AP, Obasi DO, Obeten U, Onyemuche T. Rutin co-treatment prevented cognitive impairment/depression-like behavior and decreased IDO activation following 35 days of ethanol administration in male Wistar rats. Alcohol. 2023 Feb;106:22-29.

[249] El-Shial EM, Kabbash A, El-Aasr M, El-Feky OA, El-Sherbeni SA. Elucidation of Natural Components of Gardenia thunbergia Thunb. Leaves: Effect of Methanol Extract and Rutin on Non-Alcoholic Fatty Liver Disease. Molecules. 2023 Jan 16;28(2):879.

Experimentation with isolated or purified rutin should be carried out with caution, as these substances can be toxic in some situations.

Rutin Review

Answer the following questions to test your knowledge of this flavonoid:

Question #1: What type of flavonoid is rutin?

      a.   Flavonol
      b.   Flavorall
      c.   Flavonol
      d.   Flavone

Question #2: The base components in the biosynthesis of rutin are:

      a.   Cannabisin and glycoside
      b.   Rutin and rutinoside
      c.   Quercetin and rutinose
      d.   Luteolin and quercetin

Question #3: How common is rutin in cannabis?

      a.   Top three
      b.   Top five
      c.   Top ten
      d.   Secondary

Question #4: What is the chemical formula for rutin?

      a. $C_{24}H_{26}O$
      b. $C_{15}H_{24}O_6$
      c. $C_{26}H_{28}O_6$
      d. $C_{27}H_{30}O_{16}$

Question #5: Name two biological roles of rutin in plants:

1 _____ 2 _____

Question #6: Name two potential medical uses of rutin:

1 _____ 2 _____

For the answer key to Rutin, please visit www.cannabischemistry.org

APIGENIN

Type: Flavone
Chemical Formula: $C_{15}H_{10}O_5$
Molecular Weight: 270.24 g/mol
Boiling Point: 555.50 °C @ 760.00 mm Hg (estimated)[250]
Flash Point: (217.10 °C) (estimated by TGSC)
Melting Point: 345 - 350 °C[251]
Solubility: Soluble in ethanol, DMSO, dimethyl formamide[252], water, ethanol, pyridine, sulfuric acid[253]
Oral TDLO: 5mg/kg (mouse)[254]
Biological Role: Insecticidal, allelopathic agent
Therapeutic Role: Anticancer, UV protection, antibacterial, anti-inflammatory
Commercial Use: Dyes, dietary supplements, antioxidants, cosmetics
Occurrence in Cannabis: Top five

Occurs in Cannabis Strains: Cheese, Critical Dream, Fruit Punch, Huckleberry, Northern Lights, Purple Moose, Purple Punch, Skywalker, Space Candy, Strawberry Cough, Strawberry Banana, Super Lemon Haze, White Cookies. Apigenin has also been found in CBD Mango Haze[255].

INTRODUCTION

Apigenin is one of the most common flavonoids in plants, including cannabis, where it occurs within the top five flavonoids by concentration in most varieties. The medical literature regarding apigenin is extensive, with the flavonoid offering numerous roles as an anticancer agent, liver and kidney protective agent, antioxidant, anti-inflammatory agent, and as an antibacterial agent. A member of the flavone subclass of flavonoids, apigenin is found in a wide variety of plants, most notably parsley and celery, and likely offers plants UV protection, as well as insecticidal and allelopathic functions, among others.

---

[250] The Good Scents Company Data Sheet for Apigenin, from: https://www.thegoodscentscompany.com/data/rw1108801.html. Accessed November 6, 2023.

[251] O'Neil, M.J. (ed.). The Merck Index - An Encyclopedia of Chemicals, Drugs, and Biologicals. Whitehouse Station, NJ: Merck and Co., Inc., 2006., p. 118

[252] Cayman Chemical Product Information Sheet for Apigenin, from: https://cdn.caymanchem.com/cdn/insert/10010275.pdf. Accessed November 6, 2023.

[253] Lide, D.R., G.W.A. Milne (eds.). Handbook of Data on Organic Compounds. Volume I. 3rd ed. CRC Press, Inc. Boca Raton ,FL. 1994., p. V2: 1566.

[254] Cayman Chemical Safety Data Sheet for Apigenin, from: https://cdn.caymanchem.com/cdn/msds/10010275m.pdf. Accessed November 6, 2023.

[255] Jin, D., Dai, K., Xie, Z. et al. Secondary Metabolites Profiled in Cannabis Inflorescences, Leaves, Stem Barks, and Roots for Medicinal Purposes. Sci Rep 10, 3309 (2020).

CHEMICAL STRUCTURE

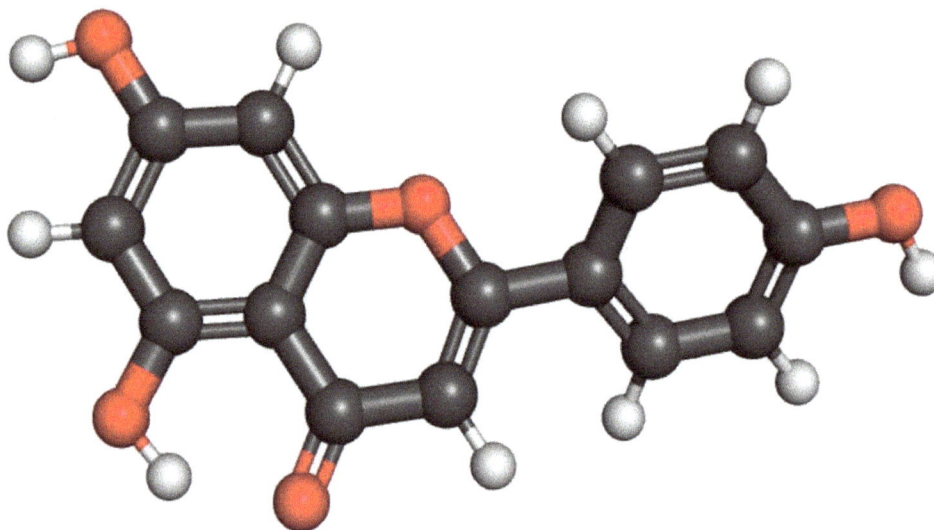

Apigenin is formed in the phenylpropanoid and flavone synthesis pathway of plants, and is derived from the source flavonoid naringenin. Technically referred to as a trihydroxyflavone, apigenin belongs to the flavone subclass of flavonoids, and is built on a carbon molecular skeleton that includes two phenyl groups with substitutions of hydroxy groups at positions 4', 5, and 7. Apigenin is comprised of fifteen carbon atoms, ten hydrogen atoms, five oxygen atoms, seven endocyclic double bonds, and one oxygenated exocyclic double bond. This molecule exists as the following glycosides:

Apigenin 7-O-apioglucoside: glycosylated at the 7th carbon atom with apioglucose (a sugar derivative). The "7-O" designation indicates that the sugar is attached to the 7th carbon atom.

Apigenin 7-glucoside: glycosylated at the 7th carbon atom with a glucose molecule.

Apigenin 8-C-glucoside: glycosylated at the 8th carbon atom with a glucose molecule.

Apigenin 6-C-glucoside: glycosylated at the 6th carbon atom with a glucose molecule.

Apigenin 7-O-neohesperidoside: glycosylated at the 7th carbon atom with neohesperidin (a sugar derivative).

Apigenin 6-C-glucoside 8-C-arabinoside: a more complex glycoside, where apigenin is glycosylated at the 6th carbon atom with a glucose molecule, and at the 8th carbon atom with arabinose (a type of sugar).

Each of these apigenin glycosides has a slightly different chemical structure due to the type of sugar and the position at which the sugar is attached to the molecule. These structural differences can affect their solubility, stability, and bioavailability, as well as their potential physiological effects in the human body. Furthermore, the sugar moiety can also influence the taste and other sensory properties of the compounds.

OCCURRENCE IN PLANTS

Primarily sourced in the natural state from chamomile, apigenin is also prominent in parsley, celery, and many other common fruits, vegetables, herbs, and spices including but not limited to:

Anise	Apple wood	Ashitaba
Basil	Black bean	Buckwheat
Cabbage	Cannabis	Carrot
Celery	Chamomile	Chicory
Coriander	Date palm	Dill
Fenugreek	Flax	Garlic
Ginkgo biloba	Lettuce	Marjoram
Olive	Oregano	Parsley
Peppermint	Pomegranate	Rosemary
Sage	Savory	Sour cherry
Spearmint	Tarragon	Tea
Thyme	Water mint	Wheat

BIOLOGICAL ACTIVITY IN PLANTS

While more research is needed to fully elucidate the biological functions of apigenin in plants, there is evidence to suggest that plants use apigenin for its insecticidal and allelopathic properties, which means it is likely that this flavone is a defensive compound.

Insecticidal Agent: As a primary constituent in a fractionation made from liverwort (marchantia linearis), apigenin contributed to significant antifeedant, larvicidal, and pupicidal effects against the tobacco cutworm (spodoptera litura)[256]. As a primary constituent in several fractionations/extractions made from ficus sarmentosa (fig tree), apigenin exhibited marked insecticidal activity against adult houseflies (musca. domestica), and the fourth instar larvae stage of tiger mosquitoes (aedes albopictus)[257].

Allelopathic Agent: As a major component in an extract made from plantago virginica (plantain), apigenin contributed to the suppression of root growth in the grasses A. matsumurae, C. dactylon, and P. annua, while also reducing the seedling height of A. matsumurae and cynodon dactylon, and inhibiting seed germination of A. matsumurae[258]. As a primary constituent in an ethyl acetate fraction of castanea sativa (sweet chestnut) leaves, and as an individual compound, apigenin contributed to and caused a reduction of seed germination and root and epicotyl growth in raphanus sativus[259] (radish).

USES IN INDUSTRY

Commercial samples of apigenin appear as a yellow crystalline solid or as yellow needle-like structures. This flavone is generally produced by various extractions of chamomile, or by enzymatic hydrolysis.

[256] Remya Krishnan, Murugan Kumara. Insecticidal Potentiality of Flavonoids from Cell Suspension Culture of Marchantia Linearis Lehm. & Lindenb Against Spodoptera Litura F. International Journal of Applied Biology and Pharmaceutical Technology 6(2). February 2015.

[257] Xue-gui WANG, Xiao-yi WEI, Xing-yan HUANG, Li-tao SHEN, Yong-qing TIAN, Han-hong XU. Insecticidal Constructure and Bioactivities of Compounds from Ficus sarmentosa var. henryi. Agricultural Sciences in China Volume 10, Issue 9, September 2011, Pages 1402-1409.

[258] Wang H, Zhou Y, Chen Y, Wang Q, Jiang L, Luo Y (2015) Allelopathic Potential of Invasive Plantago virginica on Four Lawn Species. PLoS ONE 10(4): e0125433.

[259] Basile, A, S Sorbo, S Giordano, L Ricciardi, S Ferrara, D Montesano, R Castaldo Cobianchi, M L Vuotto, and L Ferrara. Antibacterial and allelopathic activity of extract from Castanea sativa leaves. Fitoterapia 71 Suppl 1 (2003): S110-S6.

Apigenin is often used as a dye because of its yellow characteristics; this includes use as a dye in clothing, hair dyes and conditioners, and cosmetic products. Apigenin is also used in herbal and dietary supplements, and there have been some recent efforts made to incorporate this and other flavonoids into a baked snack product, thanks to their similar bioavailability to the same phytonutrients found in steamed vegetables[260].

POTENTIAL USES IN MEDICINE

Apigenin is a potent anti-cancer agent, capable of preventing and treating a wide variety of cancers, and synergizing existing cancer treatment drugs. This flavone is also a strong anti-inflammatory and antibacterial agent, capable of preventing and treating radiation poisoning, among many other potential and existing therapeutic uses.

Treatment for Uterine Cancer: Apigenin has been shown to reduce the proliferation of endometrial adenocarcinoma (uterine cancer) cells in vivo by activating nuclear receptors and blocking the genistein (an isoflavone)-stimulated increase in uterine epithelial cell height, while also stimulating the expression of Hand2 transcription factor[261] (a protein that controls the exchange of genetic information between DNA and RNA), overall acting as a progesterone receptor modulator.

Treatment for Bladder Cancer: Apigenin has been shown to suppress the proliferation of and inhibit the migration and invasion potential of T24 bladder cancer cells by modulating the Akt pathway (related to cell survival and growth) and increasing capase-3 (a protein that aids in the destruction of DNA fragments or other degradation products) activity[262], leading to apoptosis.

Treatment for Breast Cancer: Vascular endothelial growth factor (VEGF) in human breast cancer cells contributes to the development of tumor angiogenesis - the growth of blood vessels that support the tumor's survival and proliferation. Apigenin has been shown to significantly inhibit both VEGF and AKT (a protein kinase that plays a key role in cellular processes) phosphorylation, while also decreasing HIF-1a (a master cancer cell regulator) in breast cancer cells[263]. In triple-negative breast cancer (TNBC) cells, apigenin was demonstrated to significantly suppress the proliferation, colony formation, and migration of TNBC cells, while also reversing the malignant phenotype of these cells[264]. In HER2-overexpressing breast cancer, apigenin was shown to induce apoptosis and decrease the phosphorylation level of IKBa (proteins that inhibit NF-κB (nuclear factor kappa-light-chain-enhancer of activated B cells) transcription factor) in the cytosol, while also abrogating the nuclear translocation of p65 (a protein) within the nucleus[265] of the cells. Finally, apigenin suppresses senescence-associated secretory phenotype in several senescent-induced fibroblast strains, including in human breast cancer cells, where the flavone reduced the aggressive phenotype of the cells[266].

---

[260] Perez-Moral, Natalia et al. Comparative bio-accessibility, bioavailability and bioequivalence of quercetin, apigenin, glucoraphanin and carotenoids from freeze-dried vegetables incorporated into a baked snack versus minimally processed vegetables: Evidence from in vitro models and a human bioavailability study. Journal of functional foods vol. 48 (2018): 410-419.

[261] Dean, Matthew, Julia Austin, Ren Jinhong, Michael Johnson, Daniel Lantvit, and Joanna Burdette. The Flavonoid Apigenin Is a Progesterone Receptor Modulator with In Vivo Activity in the Uterus. Hormones and Cancer 9.4 (2018): 265-277.

[262] Zhu, Y., Mao, Y., Chen, H. et al. Apigenin promotes apoptosis, inhibits invasion and induces cell cycle arrest of T24 human bladder cancer cells. Cancer Cell Int 13, 54 (2013).

[263] Jin, Xue-ying, and Chang-shan Ren. VEGF expression is inhibited by apigenin in human breast cancer cells. Chinese Journal of Cancer Research 18.4 (2006): 306-311.

[264] Li, Y., Xu, J., Zhu, G. et al. Apigenin suppresses the stem cell-like properties of triple-negative breast cancer cells by inhibiting YAP/TAZ activity. Cell Death Discovery 4, 105 (2018).

[265] Seo, Hye-Sook, Han-Seok Choi, Soon-Re Kim, Youn Choi, Sang-Mi Woo, Incheol Shin, Jong-Kyu Woo, Sang- Yoon Park, Yong Shin, and Seong-Gyu Ko. Erratum to: Apigenin induces apoptosis via extrinsic pathway, inducing p53 and inhibiting STAT3 and NFKB signaling in HER2-overexpressing breast cancer cells. Molecular and Cellular Biochemistry 368.2 (2012): 215-215.

[266] Perrott, K.M., Wiley, C.D., Desprez, P. et al. Apigenin suppresses the senescence-associated secretory phenotype and paracrine effects on breast cancer cells. GeroScience 39, 161-173 (2017).

Treatment for Prostate Cancer: In experiments carried out in mice and in prostate cancer cells in vivo, apigenin suppressed the XIAP protein encoding gene, both the c-IAP1 and c-IAP2 inhibitor of apoptosis proteins, and levels of survivin, a caspase activation inhibitor protein, while also increasing the acetylation (changes a compound to include an acetyl functional group) of Ku70 (a protein related to DNA repair), and disassociating Bax[267] (a protein encoding gene associated with various cancers), among other effects, which all served to induce apoptosis in prostate cancer cells.

Treatment for Colon Cancer: In human colon carcinoma cells, apigenin increased the levels of a multifunctional cell-surface protein called CD26, which can suppress pathways involved in tumor metastasis[268]. Apigenin was also shown to synergistically increase the potency of chemotherapeutic agents normally utilized in the treatment of advanced colorectal cancer, including 5-fluorouracil, oxaliplatin, and irinotecan, increasing the potency of the latter by fourfold.

Treatment for Nasopharyngeal Cancer: Nasopharyngeal carcinoma (a rare type of head and neck cancer) is associated with reactivation of the Epstein-Barr virus (EBV), which apigenin suppressed by inhibiting the expression of EBV lytic (one of two EBV life cycles) proteins in epithelial and B cells (a type of white blood cell), while also reducing the number of EBV-reactivating cells, and was found to dramatically reduce the production of EBV virions[269] (infective state of a virus), among other effects that inhibit EBV reactivation.

Treatment for Gastric Cancer: Apigenin has been shown to inhibit the proliferation of human gastric carcinoma cells by reducing the mitochondrial membrane potential of the cells, significantly increasing caspase-3 and Bax protein expression levels, and reducing anti-apoptotic protein Bcl-2 levels, the effects of which, combined, led to cell apoptosis[270].

Treatment of Advanced HPV-Associated Cancers: After treatment with apigenin, E7-expressing (a protein found in HPV-associated non-melanoma skin cancers) TC-1 tumor cells derived from mice were more susceptible to lysis (cell death via rupture of the membrane) by cytotoxic CD8+ T cells (important for immune defense against viruses and other pathogens in the cell) that specifically target the E7 protein, and an enhanced apoptotic tumor cell death in vitro was observed[271].

Treatment for Non-small Cell Lung Cancer: Combined treatment with apigenin and naringenin caused significant cytotoxicity with cell cycle arrest at G2/M phases (essentially preventing mitosis), while also enhancing mitochondria dysfunction, elevating oxidative stress, and activating the apoptotic pathway in A549 (hypotriploid alveolar basal epithelial cells) and H1299 lung carcinoma cells[272]. Apigenin has also been shown to be a significant inhibitor of angiogenesis and tumor growth in a xenograft model of NCI-H1299 lung cancer cells[273].

[267] Shukla, Sanjeev, Pingfu Fu, and Sanjay Gupta. Apigenin induces apoptosis by targeting inhibitor of apoptosis proteins and Ku70-Bax interaction in prostate cancer. Apoptosis 19.5 (2014): 883-894.

[268] Lefort, Emilie, and Jonathan Blay. The dietary flavonoid apigenin enhances the activities of the anti-metastatic protein CD26 on human colon carcinoma cells. Clinical & Experimental Metastasis 28.4 (2011): 337-349.

[269] Wu, C., Fang, C., Cheng, Y. et al. Inhibition of Epstein-Barr virus reactivation by the flavonoid apigenin. J Biomed Sci 24, 2 (2017).

[270] Chen, J., Chen, J., Li, Z. et al. The apoptotic effect of apigenin on human gastric carcinoma cells through mitochondrial signal pathway. Tumor Biol. 35, 7719-7726 (2014).

[271] Chuang, Chi-Mu, Archana Monie, Annie Wu, and Chien-Fu Hung. Combination of apigenin treatment with therapeutic HPV DNA vaccination generates enhanced therapeutic antitumor effects. Journal of Biomedical Science 16.1 (2009): 49.

[272] Liu X, Zhao T, Shi Z, Hu C, Li Q, Sun C. Synergism Antiproliferative Effects of Apigenin and Naringenin in NSCLC Cells. Molecules. 2023 Jun 23;28(13):4947.

[273] Fu J, Zeng W, Chen M, Huang L, Li S, Li Z, Pan Q, Lv S, Yang X, Wang Y, Yi M, Zhang J, Lei X. Apigenin suppresses tumor angiogenesis and growth via inhibiting HIF-1α expression in non-small cell lung carcinoma. Chem Biol Interact. 2022 Jul 1;361:109966.

Anticancer Agent: Researchers studying cancer's energy dependence on glucose found that apigenin lowered levels of GLUT1 (a protein that transports glucose) and GLUT3 (a protein that transports glucose to dendrites and axons) mRNA, inhibiting cell migration in an anaplastic thyroid cancer cell line[274].

Liver Cancer Treatment Drug Synergist: In hepatocellular carcinoma (liver cancer) cells resistant to the cancer treatment drug doxorubicin, apigenin was shown to significantly enhance doxorubicin sensitivity in the cells, inhibiting the growth of hepatocellar carcinoma xenografts in nude mice[275].

Protection against UV Radiation: Apigenin was shown to protect cell viability at 50% against UVA, and 90% against UVB radiation[276], in large part because of the high oxygen radical absorbance capacity of the compound. The flavone has also been demonstrated to inhibit ROS production in immortalized human keratinocytes (epidermal cells), inhibiting damage in cells exposed to UVA irradiation[277].

Antibacterial Agent: Modified versions of apigenin potently inhibited gram-negative bacteria including listeria monocytogenes, pseudomonas aeruginosa, and aeromonas hydrophila by 100%[278]. Other research has shown that apigenin - as isolated from an ethanolic extract of common purslane (portulaca oleracea) leaves - significantly inhibited salmonella typhimurium and proteus mirabilis bacterial strains[279].

Anti-inflammatory Agent: Apigenin has been shown to downregulate pancreatic tumor necrosis factor alpha (TNF-α) expression and prevent pancreatic necrosis in an experimental model of acute obstructive gallstone pancreatitis[280], an inflammatory condition. In osteoarthritic rats, apigenin also reduced TNF-α, as well as interleukin 1 beta (IL-1β, a cytokine protein), and malondialdehyde[281] (MDA, a toxic molecule and marker of oxidative stress), increasing the effectiveness of allogenic synovial membrane-derived stem cell therapy. In the treatment of lupus, a disease marked by inflammation of the skin, apigenin suppressed cell responses to lupus nucleosomes by up to 98%, and inhibited the ability of lupus B cells to produce autoantibodies in the presence of nucleosomes by up to 82%, with researchers concluding that apigenin; "could be valuable for suppressing inflammation in lupus and other Th17-mediated diseases like rheumatoid arthritis, Crohn's disease, and psoriasis and in prevention of inflammation-based tumors overexpressing COX-2 (colon, breast)[282]."

[274] Heydarzadeh S, Moshtaghie AA, Daneshpour M, Hedayati M. The effect of Apigenin on glycometabolism and cell death in an anaplastic thyroid cancer cell line. Toxicol Appl Pharmacol. 2023 Sep 15;475:116626.

[275] Gao, Ai-Mei, Xiao-Yu Zhang, Juan-Ni Hu, and Zun-Ping Ke. Apigenin sensitizes hepatocellular carcinoma cells to doxorubic through regulating miR-520b/ATG7 axis. Chemico-biological interactions 280 (2018): 45-50.

[276] Noelia Sanchez-Marzo, Almudena Perez-Sanchez, Ver6nica Ruiz-Torres, Adrian Martinez-Tebar, Julian Castillo, Maria Herranz-L6pez, and Enrique Barraj6n-Catalan. Antioxidant and Photoprotective Activity of Apigenin and Its Potassium Salt Derivative in Human Keratinocytes and Absorption in Caco-2 Cell Monolayers Int J Mol Sci. 2019 May; 20(9): 2148.

[277] Hwang, Yong Pil, Kyo Nyeo Oh, Hyo Jeong Yun, and Hye Gwang Jeong. The flavonoids apigenin and luteolin suppress ultraviolet A-induced matrix metalloproteinase-1 expression via MAPKs and AP-1-dependent signaling in HaCaT cells. Journal of dermatological science 61.1 (2011): 23-31.

[278] Francis J. Osonga, Ali Akgul, Roland M. Miller, Gaddi B. Eshun, Idris Yazgan, Ayfer Akgul, and Omowunmi A. Sadik. Antimicrobial Activity of a New Class of Phosphorylated and Modified Flavonoids. ACS Omega. 2019 July 31; 4(7): 12865-12871.

[279] Hanumantappa B. Nayaka, Ramesh L. Londonkar, Madire K. Umesh, and Asha Tukappa. Antibacterial Attributes of Apigenin, Isolated from Portulaca oleracea L. Int J Bacteriol. 2014; 2014: 175851.

[280] Charalabopoulos, Alexandros et al. Apigenin Exerts Anti-inflammatory Effects in an Experimental Model of Acute Pancreatitis by Down-regulating TNF-a. In vivo (Athens, Greece) vol. 33,4 (2019): 1133-1141.

[281] Estakhri, Firoozeh, Mohammad Reza Panjehshahin, Nader Tanideh, Rasoul Gheisari, Amir Mahmoodzadeh, Negar Azarpira, and Nasser Gholijani. The effect of kaempferol and apigenin on allogenic synovial membrane- derived stem cells therapy in knee osteoarthritic male rats. The Knee (2020)1.

[282] Kang, Hee-Kap, Diane Ecklund, Michael Liu, and Syamal Datta. Apigenin, a non-mutagenic dietary flavonoid, suppresses lupus by inhibiting autoantigen presentation for expansion of autoreactive Th1 and Th17 cells. Arthritis Research & Therapy 11.2 (2009): 1-13.

Liver Protective Agent: Apigenin has been shown to protect the liver of rats from deltamethrin (a pyrethroid insecticide) toxicity[283], and exhibits a cytoprotective effect against alcohol-induced liver injury in mice by regulating hepatic CYP2E1 (a prolific membrane protein in the liver)-mediated oxidative stress and PPARa (a nuclear receptor protein)-mediated lipogenic gene expression[284].

Lung Protective Agent: Administration of apigenin markedly ameliorated paraquat (a highly toxic herbicide)-induced pulmonary edema and lung injury in mice by decreasing biochemical parameters of inflammation and oxidative stress while improving oxygenation and lung edema, in part via the inhibition of NF-KB[285] (a protein with important roles in cell survival and DNA transcription).

Treatment for Parkinson's Disease & Neurodegenerative Diseases: In a rat model of rotenone-induced Parkinson's disease, the administration of apigenin caused a significant improvement in behavioral, biochemical and mitochondrial enzyme activities by significantly attenuating the upregulation of NF-KB gene expression, inhibiting the release of pro-inflammatory cytokines TNF-a, IL-6, and pro-inflammatory enzyme iNOS-1, and preventing neuroinflammation in the substantia nigra pars compacta[286] (a portion of the midbrain), among other effects. Apigenin has also been shown to exhibit strong antioxidant and neuroprotective effects against peripheral nerve degeneration by inhibiting the degradation of myelin and peripheral axons, and the trans-dedifferentiation and proliferation of Schwann cells[287] (glial cells of the peripheral nervous system). In a transgenic Drosophila (fruit flies) model of Alzheimer's disease (AD), dietary apigenin caused a significant decrease in the oxidative stress and delay in the loss of climbing ability in the flies, while also inhibiting the activity of acetylcholinesterase and the formation of Aβ-42 (highly implicated in the pathology of AD) aggregates, potently reducing AD symptoms[288].

Treatment for Rheumatoid Arthritis: Apigenin induced marked apoptosis in rheumatoid arthritis (RA) fibroblast-like synoviocytes isolated from patients with RA; it also enhanced the cytotoxic effect of TNF-related apoptosis-inducing ligands[289].

Treatment for Osteoarthritis: In a murine model of osteoarthritis, apigenin alleviated cartilage injury (surgically induced) in mice by mediating macrophage polarization, and inhibiting chondrocyte (cells involved in the production of cartilage) inflammation and apoptosis[290].

Kidney Protective Agent: Apigenin has been shown to significantly reduce renal function markers, serum creatinine, and urea nitrogen content in Sprague-Dawley rats subjected to chemical induction of renal injury, while

[283] Amany Yosry; Mohamed Abd-Elaal; Waleed Barakat. Antiapoptotic effect of apigenin and vitamin E against deltamethrin induced toxicity in rats Zagazig Journal of Pharmaceutical Sciences Article 5, Volume 26, Issue 2, 2017, Page 67-77.

[284] Wang, Feng, Jin-Cheng Liu, Rui-Jun Zhou, Xi Zhao, Mei Liu, Hua Ye, and Mei-Lin Xie. Apigenin protects against alcohol-induced liver injury in mice by regulating hepatic CYP2E1-mediated oxidative stress and PPARa- mediated lipogenic gene expression. Chemico-biological interactions 275 (2017): 171-177.

[285] Luan, Rui-Ling, Xiang-Xi Meng, and Wei Jiang. Protective Effects of Apigenin Against Paraquat-Induced Acute Lung Injury in Mice. Inflammation 39.2 (2016): 752-758.

[286] Anusha, Chandran, Thangarajan Sumathi, and Leena Dennis Joseph. Protective role of apigenin on rotenone induced rat model of Parkinson's disease: Suppression of neuroinflammation and oxidative stress mediated apoptosis. Chemico-biological interactions 269 (2017): 67-79.

[287] Kim, Muwoong, Junyang Jung, Na Jeong, and Hyung-Joo Chung. The natural plant flavonoid apigenin is a strong antioxidant that effectively delays peripheral neurodegenerative processes. Anatomical Science International 94.4 (2019): 285-294.

[288] Siddique YH, Rahul, Ara G, Afzal M, Varshney H, Gaur K, Subhan I, Mantasha I, Shahid M. Beneficial effects of apigenin on the transgenic Drosophila model of Alzheimer's disease. Chem Biol Interact. 2022 Oct 1;366:110120.

[289] Sun, Qing-wen, Song-min Jiang, Ke Yang, Jian-ming Zheng, Li Zhang, and Wei-dong Xu. Apigenin enhances the cytotoxic effects of tumor necrosis factor-related apoptosis-inducing ligand in human rheumatoid arthritis fibroblast-like synoviocytes. Molecular Biology Reports 39.5 (2011): 5529-5535.

[290] Ji X, Du W, Che W, Wang L, Zhao L. Apigenin Inhibits the Progression of Osteoarthritis by Mediating Macrophage Polarization. Molecules. 2023 Mar 24;28(7):2915.

also modulating oxidative phosphorylation, re-establishing mitochondrial membrane potential, and reducing cytochrome C release[291], among other renal protective effects.

Prevention and Treatment of Osteoporosis: In ovariectomized mice, apigenin markedly inhibited cell proliferation and indices of osteoblast differentiation, completely inhibited the formation of multinucleated osteoclasts from mouse splenic cells, and significantly suppressed trabecular bone loss in the femurs of the mice, inhibiting osteoblastogenesis and osteoclastogenesis and preventing bone loss[292].

Treatment for Myocardial Infarction: In isoproterenol (ISO)-induced oxidative stress and myocardial infarction in rats, apigenin was shown to safeguard cardiac functions by activating and increasing the antioxidant defense system[293]. In an isolated rat heart model of ischemia/reperfusion, administration of apigenin caused an improved ischemic cardiac functional recovery, a decreased myocardial infarct size, reduced activities of creatine kinase isoenzyme and lactate dehydrogenase in the coronary flow, and a reduced number of apoptotic cardiomyocytes, among other beneficial cardioprotective effects, through the inhibition of the p38 MAPKS (stress-responsive protein kinases) signaling pathway[294].

Treatment for Vitiligo: In a melanocyte cell model of vitiligo, apigenin-treated cells exhibited enhanced viability, enhanced expression of cellular antioxidants, inhibited production of malondialdehyde, and significantly increased expression and nuclear localization of the Nrf2 transcription factor[295].

Treatment for Diabetic Retinopathy: Apigenin has been shown to inhibit the proliferation, migration, and angiogenesis of high glucose-induced human retinal microvascular endothelial cells through the upregulation of miR-140-5p (inhibitor of cell proliferation and migrations), PTEN (a gene involved in the production of a tumor-suppressing enzyme), and inhibition of the PI3K/AKT signaling pathway[296] (stimulates cells to proliferation and growth, and inhibits cell apoptosis).

Endodontic (Root Canal) Disinfection Agent: Combination treatment with apigenin and reduced graphene oxide (RGO) reduced the biomass of Enterococcus faecalis (a bacteria that often causes root canals to fail or become infected), decreasing the bio-volume of live bacteria and increasing the biofilms of dead bacteria, leading researchers to suggest that combinatorial treatment with apigenin may be a potential root canal treatment aid[297].

Delay Senescence/Aging: Because some inhibition of mitochondria can lead to significant longevity in the cells of certain microorganisms, insects, and nematodes, researchers studied the potential of combinatorial treatment using apigenin and chrysin (also a flavonoid) and found that the treatment inhibited mitochondrial respiration and induced an early ROS, thereby increasing oxidative stress capacity and cellular metabolic adaptation, leading to longevity in Caenorhabditis elegans[298] (roundworms).

[291] Zhong, Yujie, Chengni Jin, Xiaorui Wang, Xuan Li, Jiahui Han, Wei Xue, Peng Wu, Xiaoli Peng, and XiaodongXia. Protective effects of apigenin against 3-MCPD-induced renal injury in rat. Chemico-biological interactions 296 (2018): 9-17.

[292] Goto, Tadashi, Keitaro Hagiwara, Nobuaki Shirai, Kaoru Yoshida, and Hiromi Hagiwara. Apigenin inhibits osteoblastogenesis and osteoclastogenesis and prevents bone loss in ovariectomized mice. Cytotechnology 67.2 (2014): 357-365.

[293] Buwa, Chhabildas, Umesh Mahajan, Chandragouda Patil, and Sameer Goyal. Apigenin Attenuates B-Receptor- Stimulated Myocardial Injury Via Safeguarding Cardiac Functions and Escalation of Antioxidant Defence System. Cardiovascular Toxicology 16.3 (2015): 286-297.

[294] Hu, Jing, Zilin Li, Li-ting Xu, Ai-jun Sun, Xiao-yan Fu, Li Zhang, Lin-lin Jing, An-dong Lu, Yi-fei Dong, and Zheng-ping Jia. Protective Effect of Apigenin on Ischemia/Reperfusion Injury of the Isolated Rat Heart. Cardiovascular Toxicology 15.3 (2014): 241-249.

[295] Zhang, Baoxiang, Jing Wang, Guodong Zhao, Mao Lin, Yong Lang, Diancai Zhang, Dianqin Feng, and Caixia Tu. Apigenin protects human melanocytes against oxidative damage by activation of the Nrf2 pathway. Cell Stress and Chaperones 25.2 (2020): 277-285.

[296] Fu C, Peng J, Ling Y, Zhao H, Zhao Y, Zhang X, Ai M, Peng Q, Qin Y. Apigenin inhibits angiogenesis in retinal microvascular endothelial cells through regulating of the miR-140-5p/HDAC3-mediated PTEN/PI3K/AKT pathway. BMC Ophthalmol. 2023 Jul 6;23(1):302.

[297] Kim MA, Min KS. Combined effect of apigenin and reduced graphene oxide against Enterococcus faecalis biofilms. J Oral Sci. 2023 Jul 1;65(3):163-167.

[298] Cheng Y, Hou BH, Xie GL, Shao YT, Yang J, Xu C. Transient inhibition of mitochondrial function by chrysin and apigenin prolong longevity via

Treatment for Asthma: In a model of endocrine disrupting chemical-aggravated asthma in BALB/c mice induced via mono-n-butyl phthalate (MnBP), treatment with apigenin decreased activation of epithelial cells, T cells, and eosinophils, reducing all asthma features, such as airway hyperresponsiveness, airway inflammation, type 2 cytokines, and the expression of the aryl hydrocarbon receptor[299].

Treatment for Hyperuricemia: As extracted from Paeonia × suffruticosa Andrews leaves, apigenin 7-O-glucoside – a glycoside of apigenin – reduced uric acid, creatinine, and malondialdehyde serum levels, increased superoxide dismutase activity, and partially restored the spleen coefficient in hyperuricemic (a condition associated with gout involving an elevated uric acid level in the blood) mice[300].

Treatment for Epilepsy: In a kainate model of temporal lobe epilepsy, pre-treatment with apigenin exhibited a significant inhibitory effect on neural cell death, spontaneous seizure spikes, aberrant neurogenesis, mTOR (mammalian target of rapamycin (mTOR), which regulates cell proliferation, autophagy, and apoptosis) hyperactivity, and aberrant mossy fiber sprouting[301].

Treatment for Allergic Rhinitis: In BALB/c mice induced to allergic rhinitis via 48/80 (a polymer that is a potent promoter of histamine release) and lipopolysaccharide, treatment with apigenin significantly inhibited compound 48/80-induced secretion of beta-hexosaminidase (an enzyme found in lysosomes, serves crucial functions in the brain and spinal cord) and histamine, while also blocking the lipopolysaccharide-induced decrease in cell viability and increase in cell apoptosis and inflammatory cytokine secretion, among other anti-allergic rhinitis effects as reported in Environmental Toxicology[302].

Treatment for Hepatic Fibrosis: In a murine model of carbon tetrachloride-induced hepatic fibrosis, apigenin attenuated oxidative stress by restoring glutathione content and chloramphenicol acetyltransferase activity, normalized lipid peroxidation, mitigated liver inflammation by reducing the expression of proinflammatory cytokines, and inhibited vascular endothelial growth factor (a significant modulator of angiogenesis) and CD34 (a transmembrane phosphoglycoprotein that blocks cell adhesion), preventing pathological angiogenesis[303].

IMPLICATIONS FOR HUMAN HEALTH & NUTRITION

Trace or secondary amounts of apigenin may be obtained by consuming cannabis products based on the following strains: Cheese, Critical Dream, Fruit Punch, Huckleberry, Northern Lights, Purple Moose, Purple Punch, Skywalker, Space Candy, Strawberry Cough, Strawberry Banana, Super Lemon Haze, White Cookies, or chemovars with similar genetics. When preparing, cooking, or serving food, consider working with the herbs and spices anise, basil, coriander, dill, flax, garlic, marjoram, oregano, parsley, rosemary, sage, savory, tarragon, or thyme.

mitohormesis in C. elegans. Free Radic Biol Med. 2023 Jul;203:24-33.

[299] Kim SH, Quoc QL, Park HS, Shin YS. The effect of apigenin, an aryl hydrocarbon receptor antagonist, in Phthalate-Exacerbated eosinophilic asthma model. J Cell Mol Med. 2023 Jul;27(13):1900-1910.

[300] Zhang Y, Li Y, Li C, Zhao Y, Xu L, Ma S, Lin F, Xie Y, An J, Wang S. Paeonia × suffruticosa Andrews leaf extract and its main component apigenin 7-O-glucoside ameliorate hyperuricemia by inhibiting xanthine oxidase activity and regulating renal urate transporters. Phytomedicine. 2023 Sep;118:154957.

[301] Nikbakht, F., Hashemi, P., Vazifekhah, S. et al. Investigating the mechanism of antiepileptogenic effect of apigenin in kainate temporal lobe epilepsy: possible role of mTOR. Exp Brain Res 241, 753–763 (2023).

[302] Li H, Zhang H, Zhao H. Apigenin attenuates inflammatory response in allergic rhinitis mice by inhibiting the TLR4/MyD88/NF-κB signaling pathway. Environ Toxicol. 2023 Feb;38(2):253-265.

[303] Melaibari M, Alkreathy HM, Esmat A, Rajeh NA, Shaik RA, Alghamdi AA, Ahmad A. Anti-Fibrotic Efficacy of Apigenin in a Mice Model of Carbon Tetrachloride-Induced Hepatic Fibrosis by Modulation of Oxidative Stress, Inflammation, and Fibrogenesis: A Preclinical Study. Biomedicines. 2023 May 2;11(5):1342.

Vegetables and other foods that are high in apigenin include black beans, buckwheat, cabbage, carrots, celery, lettuce, olives, and wheat, while fruits like pomegranate and sour cherries are another potential source of this flavone. Many teas also contain apigenin, especially teas based on peppermint or spearmint.

Apigenin Review

Answer the following questions to test your knowledge of this flavonoid:

Question #1: What type of flavonoid is apigenin?

       a.   Flavanol
       b.   Flavorall
       c.   Flavonol
       d.   Flavone

Question #2: The base component in the biosynthesis of apigenin is:

       a.   Cannabisin
       b.   Rutinoside
       c.   Naringenin
       d.   Quercetin

Question #3: How common is apigenin in cannabis?

       a.   Top three
       b.   Top five
       c.   Top ten
       d.   Secondary

Question #4: What is the chemical formula for apigenin?

       a. $C_{24}H_{26}O$
       b. $C_{15}H_{10}O_5$
       c. $C_{26}H_{28}O_6$
       d. $C_{27}H_{30}O_{16}$

Question #5: Name two biological roles of apigenin in plants:

1 _____ 2 _____

Question #6: Name two potential medical uses of apigenin:

1 _____ 2 _____

For the answer key to Apigenin, please visit www.cannabischemistry.org

QUERCETIN

Type: Flavonol
Chemical Formula: $C_{15}H_{10}O_7$
Molecular Weight: 302.23 g/mol
Boiling Point: 642.00 to 643.00 °C @ 760.00 mm Hg (estimated by TGSC)
Flash Point: 248.10 °C (estimated by TGSC)
Melting Point: 316.50 °C @ 760.00 mm Hg[304]
Solubility: Practically insoluble in water, soluble in aqueous alkaline solutions, very soluble in ether, methanol, soluble in ethanol, acetone, pyridine, acetic acid
Oral LD50: 159 mg/kg (mouse)[305]
Biological Role: Allelopathic agent
Therapeutic Role: Chemotherapeutic agent synergist, anticancer agent, neurological agent
Commercial Use: Dietary supplements, food additive
Occurrence in Cannabis: Top ten

Occurs in Cannabis Strains: Cheese, Fruit Punch, Northern Lights, Purple Moose, Purple Punch, Skywalker, Space Candy, Strawberry Cough, Strawberry Banana, Super Lemon Haze, White Cookies. Most likely occurs in the majority of cannabis and hemp varieties.

INTRODUCTION

Quercetin is one of the most common flavonoids in plants, and in fact it likely occurs more prolifically than any other flavonoid compound. In cannabis, quercetin generally occurs within the top ten flavonoids by concentration, typically at number six. A well-known antioxidant with a bitter flavor, this flavonol is a potent anticancer agent, capable of preventing and treating a wide variety of carcinomas.

Some evidence suggests that this flavonol contributed to the evolution of land-based plants, including findings that the ABA signaling pathway (regulates flavonoid biosynthesis) - modulated by quercetin - led water-based plants to adapt to land[306], with supporting evidence showing that water-based plants used flavonols as UV radiation protection to further this adaptation[307].

---

[304] The Good Scents Company Data Sheet for Quercetin, from: https://www.thegoodscentscompany.com/data/rw1343701.html#:~:text=Quercetin%20shows%20anti%2Dinflammatory%20action,by%20basophils%20and%20mast%20cells. Accessed November 9, 2023.

[305] Sigma Aldrich Safety Data Sheet for Quercetin dihydrate, from: https://www.sigmaaldrich.com/US/en/sds/sigma/q0125. Accessed November 10, 2023.

[306] Brunetti, Cecilia, Federico Sebastiani, and Massimiliano Tattini. Review: ABA, flavonols, and the evolvability of land plants. Plant science - an international journal of experimental plant biology 280 (2019): 448-454.

[307] Jiao, Chen, Iben S0rensen, Xuepeng Sun, Honghe Sun, Hila Behar, Saleh Alseekh, Glenn Philippe, Kattia Palacio Lopez, Li Sun, Reagan Reed, Susan Jeon, Reiko Kiyonami, Sheng Zhang, Alisdair R Fernie, Harry Brumer, David S Domozych, Zhangjun Fei, and Jocelyn K C Rose. The Penium margaritaceum Genome: Hallmarks of the Origins of Land Plants. Cell (2020).

CHEMICAL STRUCTURE

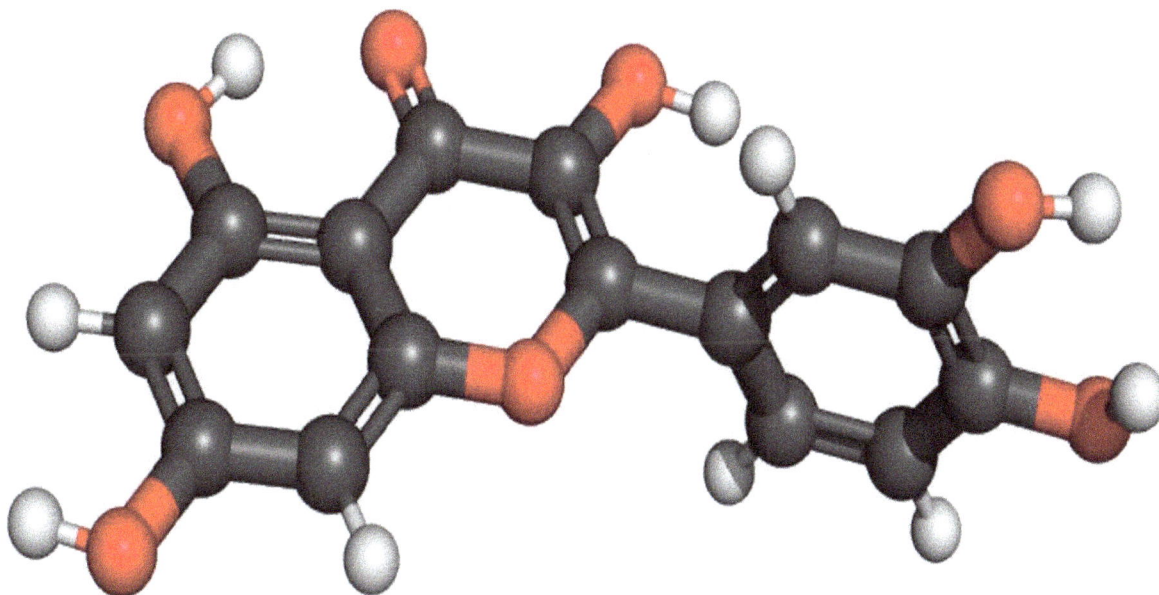

Derived from naringenin, and closely related to kaempferol, the biosynthesis of quercetin is rooted in the phenylpropanoid metabolic pathway of plants. Comprised of fifteen carbon atoms, ten hydrogen atoms, and seven oxygen atoms notated as $C_{15}H_{10}O_7$, quercetin is arranged in the classic 3-ring formation of flavonoids, while its attachment of ring B at position 2 of ring C further classifies this compound as a flavonol. Interesting features of quercetin's molecular skeleton include eight double bonds: seven endocyclic, and one double-bonded exocyclic oxygen atom. This molecule features 5 hydroxy groups, further classifying the compound as a pentahydroxyflavone.

The quercetin molecule can be modified to produce different variations of this flavonol, each with unique features, including the following:

Quercetin-3,4'-dimethyl ether: two methyl groups (CH3) attached at the 3 and 4' positions.

Quercetin caprylate: an attached caprylate group (a fatty acid with eight carbon atoms).

Quercetin 3-methyl ether: a methyl group attached to the 3-position.

Quercetin-3-O-glucose-6"-acetate: a glucose molecule attached to the 3-position and an acetate group attached to the 6" position.

Quercetin-3-O-glucopyranoside: a glucose molecule attached at the 3-position.

Quercetin dihydrate: contains two water molecules as part of its crystal structure.

Quercetin 3-neohesperidoside: neohesperidoside (a disaccharide, a class of sugar) attached to the 3-position.

Quercetin 3-alpha-D-galactoside: a galactose molecule attached at the 3-position.

Quercetin 3-xyloside: a xylose molecule attached at the 3-position.

Quercetin 3-gentiobioside: gentiobioside (a disaccharide composed of two glucose molecules) attached to the 3-position.

Isoquercetin: a glucose molecule attached at the 3-position; this is one of the most common quercetin derivatives found in nature.

Interestingly, in 2023 a new glycoside of quercetin was discovered in four species of the genus Prunella, where researchers determined the novel quercetin variation was quercetin 3-O-(4″-O-β-Dxylopyranosyl-6″-O-α-L-rhamnopyranosyl)-β-D-glycopyranoside[308], which features several sugar moiety groups.

OCCURRENCE IN PLANTS

Quercetin is found in many common foods, including common grains, fruits, vegetables, herbs, spices, and nuts.

Almond	Anise	Apple
Apricot	Arrowroot	Ashitaba
Beet	Bilberry	Blueberry
Black currant	Black elder	Black pepper
Buckwheat	Cabbage	Cacao
Cannabis	Caper	Caraway
Chestnut	Chicory	Chive
Clove	Coriander	Corn
Cotton	Cranberry	Dandelion
Date palm	Dill	Endive
Fennel	Fenugreek	Fig
Garlic	Ginger	Ginkgo biloba
Grapefruit	Grapes	Guava
Hazelnut	Horseradish	Kale
Kiwi	Lemon	Lemongrass
Lettuce	Loquat	Lotus
Mango	Marango	Mung bean
Nutmeg	Oat	Okra
Olive	Onion	Orange
Oregano	Parsley	Parsnip
Peach	Peanut	Pear
Persimmon	Plum	Pomegranate
Potato	Primrose	Raspberry
Rice	Rose	Saffron
Sorrel	Soybean	Spinach
Sunflower	Tarragon	Tea
Tomato	Turmeric	Walnut
Wheat		

[308] Olennikov, D.N., Shamilov, A.A. & Kashchenko, N.I. New Glycoside of Quercetin from the Genus Prunella. Chem Nat Compd 59, 647–650 (2023).

BIOLOGICAL ACTIVITY IN PLANTS

As discussed previously, quercetin and other flavonoids are likely used by plants as UV radiation protection agents. Outside of this biological role, there is scant information available as of late 2023 regarding other functions this compound might serve in plants.

Allelopathic Agent: The growth and survival rates of oedaleus asiaticus (grasshoppers) fed a substrate coated in quercetin were reduced by up to 65% in a study carried out in Mongolia, China. Researchers showed that quercetin significantly increased reactive oxygen species (ROS) production, negatively regulated the TLP (insulin-like signaling pathway), and reduced both gene expression and protein phosphorylation levels of a cascade-related stress response in the TLP[309].

USES IN INDUSTRY

Commercial preparations of quercetin appear as yellow crystalline needles or powder. This flavonol is used as a food additive in pastas and grains, prepared beverages, candies, processed fruits, and juice. Quercetin is also found in dietary supplements, and cosmetic and skin conditioning products.

Quercetin is used extensively as a research chemical, with interesting recent work focusing on using this flavonol as a type of molecular scaffolding from which to develop new and more potent variations of the compound[310]. Researchers have also found that quercetin supplementation significantly reduced the mortality of crayfish caused by white spot syndrome virus infection, increased the expression of immune-related genes, affected the activity of six immune-related enzymes and increased the total number of hemocytes, and significantly reduced the rate of hemocyte apoptosis in commercial crayfish[311]. Other work has demonstrated the protective effect of quercetin on avermectin-induced splenic toxicity in carp[312], and the ameliorative effects of quercetin against ochratoxin-induced toxicity in broiler chickens[313].

POTENTIAL USES IN MEDICINE

Quercetin offers the potential for numerous new types of medical treatment, especially in the treatment of cancer, and neurological conditions. However, quercetin does not easily cross the blood brain barrier, which means that its absorption rate in humans is limited. One group of researchers has sought to correct this using superparamagnetic iron oxide nanoparticles as a delivery system, which increased the bioavailability of quercetin in the brains of rats by tenfold[314]. Other work has shown that Quercetin aglycone derived from onion skin extract powder was 4.8 times more bioavailable via oral administration in humans than pure quercetin dihydrate[315]. Therefore, it seems likely that

[309] Cui, Boyang, Xunbing Huang, Shuang Li, Kun Hao, Hussain BabarChang, Xiongbing Tu, Baoping Pang, and Zehua Zhang. Quercetin Affects the Growth and Development of the Grasshopper Oedaleus asiaticus (Orthoptera: Acrididae). Journal of Economic Entomology 112.3 (2019).

[310] Chatziathanasiadou, M.V., Geromichalou, E.G., Sayyad, N. et al. Amplifying and broadening the cytotoxic profile of quercetin in cancer cell lines through bioconjugation. Amino Acids 50, 279-291 (2018).

[311] Gong J, Pan X, Zhou X, Zhu F. Dietary quercetin protects Cherax quadricarinatus against white spot syndrome virus infection. J Invertebr Pathol. 2023 Jun;198:107931.

[312] Pan E, Chen H, Wu X, He N, Gan J, Feng H, Sun Y, Dong J. Protective effect of quercetin on avermectin induced splenic toxicity in carp: Resistance to inflammatory response and oxidative damage. Pestic Biochem Physiol. 2023 Jun;193:105445.

[313] Abdelrahman RE, Khalaf AAA, Elhady MA, Ibrahim MA, Hassanen EI, Noshy PA. Quercetin ameliorates ochratoxin A-Induced immunotoxicity in broiler chickens by modulation of PI3K/AKT pathway. Chem Biol Interact. 2022 Jan 5;351:109720.

[314] Enteshari Najafabadi, R., Kazemipour, N., Esmaeili, A. et al. Using superparamagnetic iron oxide nanoparticles to enhance bioavailability of quercetin in the intact rat brain. BMC Pharmacol Toxicol 19, 59 (2018).

[315] Burak, C., BrUll, V., Langguth, P. et al. Higher plasma quercetin levels following oral administration of an onion skin extract compared with pure quercetin dihydrate in humans. Eur J Nutr 56, 343-353 (2017).

the poor bioavailability of quercetin can be overcome soon. Additionally, despite the many potential therapeutic uses of quercetin listed below, excessive intake of this flavonol could promote the growth of cancerous tumors, especially in the upper intestine[316]. Finally, quercetin has been shown to antagonize the sedative effects of linalool[317] – an important consideration for cannabis product formulators.

Treatment for Brain Cancer & Gliomas: In human glioma cells, quercetin was shown to reduce phosphorylation of ERK (extracellular signal-regulated kinase) and Akt (protein kinase B), and reduce the expression of survivin, which induced cell death via these and other capase-dependent mechanisms[318]. Quercetin also significantly suppressed the growth and migration of human GBM T98G cells (glioblastoma multiforme), inducing apoptosis, and arresting cells in the S-phase cell cycle[319] (a point where DNA replication and repair activity occurs in cells).

Treatment for Breast Cancer: Quercetin has been shown to induce apoptosis in triple-negative breast cancer cells, and significantly inhibit tumor growth when encapsulated and injected in ethylene glycol-polylactide nanoparticles[320]. As isolated from the leaves of Acalypha indica, quercetin also inhibited MCF-7 and MDA-MB-231 breast cancer cells, increasing activity in pro-apoptotic activity[321]. Other research has found that naringenin (a "mother" flavonoid compound), quercetin, and the flavonoid fisetin act synergistically, reducing cell growth, suppressing cell migration, and inducing apoptosis, also in MCF-7 and MDA-MB-231 breast cancer cell lines[322].

Treatment for Neuroblastoma & Adrenal Gland Cancer: In the N2a mouse neuroblastoma cell line, quercetin markedly decreased anti-apoptotic gene expression, increased tumor suppressor gene p53, and induced apoptosis in more than 50% of cells after treatment with the flavonol at 40 uM[323].

Treatment for Gastric Cancer: Quercetin has been shown to increase ROS production, decrease levels of mitochondrial membrane potential, increase the apoptotic cell number in human gastric cancer cells, decrease anti-apoptotic proteins while increasing pro-apoptotic proteins, and increase expressions of tumor necrosis factor receptor superfamily member 10d[324].

Chemotherapeutic Agent Synergist & Enhancer: Cisplatin is a common chemotherapeutic drug. When combined with quercetin, a synergistic effect was observed, with the combination of compounds being significantly more effective in suppressing growth and inducing apoptosis in HepG2 human hepatocellular carcinoma cells[325] than the

[316] Matsukawa, Yoshizumi, Hoyoku Nishino, Mitsunori Yoshida, Hiroyuki Sugihara, Kanade Katsura, Tetsurou Takamatsu, Junichi Okuzumi, Katsuhiko Matsumoto, Fumiko Sato-Nishimori, and Toshiyuki Sakai. Quercetin enhances tumorigenicity induced by N-ethyl-N'-nitro-N-nitrosoguanidine in the duodenum of mice. Environmental Health and Preventive Medicine 6.4 (2008): 235-239.

[317] Bappi MH, Prottay AAS, Kamli H, Sonia FA, Mia MN, Akbor MS, Hossen MM, Awadallah S, Mubarak MS, Islam MT. Quercetin Antagonizes the Sedative Effects of Linalool, Possibly through the GABAergic Interaction Pathway. Molecules. 2023 Jul 24;28(14):5616.

[318] Kim, Eui, Chang Choi, Ji Park, Soo Kang, and Yong Kim. Underlying Mechanism of Quercetin-induced Cell Death in Human Glioma Cells. Neurochemical Research 33.6 (2008): 971-979.

[319] Wang W, Yuan X, Mu J, Zou Y, Xu L, Chen J, Zhu X, Li B, Zeng Z, Wu X, Yin Z, Wang Q. Quercetin induces MGMT+ glioblastoma cells apoptosis via dual inhibition of Wnt3a/β-Catenin and Akt/NF-κB signaling pathways. Phytomedicine. 2023 Sep;118:154933.

[320] Sharma, G., Park, J., Sharma, A. et al. Methoxy Poly(ethylene glycol)-Poly(lactide) Nanoparticles Encapsulating Quercetin Act as an Effective Anticancer Agent by Inducing Apoptosis in Breast Cancer. Pharm Res 32, 723-735 (2015).

[321] Chekuri S, Vyshnava SS, Somisetti SL, Cheniya SBK, Gandu C, Anupalli RR. Isolation and anticancer activity of quercetin from Acalypha indica L. against breast cancer cell lines MCF-7 and MDA-MB-231. 3 Biotech. 2023 Aug;13(8):289.

[322] Jalalpour Choupanan M, Shahbazi S, Reiisi S. Naringenin in combination with quercetin/fisetin shows synergistic anti-proliferative and migration reduction effects in breast cancer cell lines. Mol Biol Rep. 2023 Sep;50(9):7489-7500.

[323] Sugantha Priya, E., K. Selvakumar, S. Bavithra, P. Elumalai, R. Arunkumar, P. Raja Singh, A. Brindha Mercy, and J. Arunakaran. Anti-cancer activity of quercetin in neuroblastoma: an in vitro approach. Neurological Sciences 35.2 (2013): 163-170.

[324] Shang, Hung-Sheng, Hsu-Feng Lu, Ching-Hsiao Lee, Han-Sun Chiang, Yung-Lin Chu, Ann Chen, Yuh-Feng Lin, and Jing-Gung Chung. Quercetin induced cell apoptosis and altered gene expression in AGS human gastric cancer cells. Environmental Toxicology 33.11 (2018): 1168-1181.

[325] Zhao, Ji-ling, Jing Zhao, and Hong-jun Jiao. Synergistic Growth-Suppressive Effects of Quercetin and Cisplatin on HepG2 Human Hepatocellular

single agent alone. In H1299 lung cancer cells, quercetin synergized the effects of trichostatin A (a histone deacetylase inhibitor), increasing apoptosis in the cell line by as much as 88%[326]. In a murine model of breast cancer, quercetin enhanced the effects of the common chemotherapeutic agent doxorubicin, suppressing tumor growth and prolonging survival in mice[327]. When combined with cisplatin or vincristine, quercetin increased the sensitivity and partly revised the resistance of A549 human lung cancer cells[328]. When combined with rituximab, quercetin increased inhibitory activity against the STAT3 pathway (modulates gene transcription) and downregulated the expression of survival genes, inhibited cell growth, induced apoptosis, and enhanced the sensitivity of human diffuse large B-cell lymphoma to rituximab[329]. Additionally, quercetin has been shown to sensitize human myeloid leukemia KG-1 cells against TNF-related apoptosis-inducing ligand (TRAIL) apoptosis by increasing the messenger RNA expression levels of the death receptor genes, reducing the expression of antiapoptotic proteins, and decreasing the expression of the NF-KB subunit[330]. Quercetin has also been found to synergize and significantly improve the effectiveness of the cancer treatment drug 5-fluorouracil[331], mitigate cyclophosphamide (used in the treatment of several types of cancer)-induced organ toxicity[332], and overcome the acquired resistance of Sorafenib (a targeted anticancer drug for liver and kidney malignancies) in Sorafenib-resistant hepatocellular carcinoma cells[333]

Treatment for Ovarian Cancer: Quercetin has been shown to decrease antiapoptotic molecules while decreasing proapoptotic molecules, thereby inhibiting the growth of PA-1 human metastatic ovarian cancer cells[334].

Treatment for Cervical Cancer: At low concentrations, quercetin has been demonstrated to inhibit the proliferation of growth and induce apoptosis in the HeLa (Henrietta Lacks cervical cancer) cell line in a dose-dependent manner[335].

Prevention of Skin Cancer: The combination of titanium dioxide and quercetin was shown to significantly reduce the number and volume of tumors in an animal model of UVB-induced skin photocarcinogenesis[336], offering potential as a chemopreventive agent for skin cancer.

Carcinoma Cells. Applied Biochemistry and Biotechnology 172.2 (2013): 784-791.

[326] Chuang, Cheng-Hung, Shu-Ting Chan, Chao-Hsiang Chen, and Shu-Lan Yeh. Quercetin enhances the antitumor activity of trichostatin A through up-regulation of p300 protein expression in p53 null cancer cells. Chemico-biological interactions 306 (2019): 54-61.

[327] Du, Gangjun, Haihong Lin, Mei Wang, Shuo Zhang, Xianchuang Wu, Linlin Lu, Liyan Ji, and Lijuan Yu. Quercetin greatly improved therapeutic index of doxorubicin against 4T1 breast cancer by its opposing effects on HIF-1a in tumor and normal cells. Cancer Chemotherapy and Pharmacology 65.2 (2010): 277-287.

[328] Zhan, Xuejun, Runxiang Zhang, Yanping Xu, Shuhua Yang, Daze Xie, and Liwei Tan. Empirical studies about quercetin increasing chemosensitivity on human lung adenocarcinoma cell line A549. The Chinese-German Journal of Clinical Oncology 11.7 (2012): 380-383.

[329] Li, Xin, Xinhua Wang, Mingzhi Zhang, Aimin Li, Zhenchang Sun, and Qi Yu. Quercetin Potentiates the Antitumor Activity of Rituximab in Diffuse Large B-Cell Lymphoma by Inhibiting STAT3 Pathway. Cell Biochemistry and Biophysics 70.2 (2014): 1357-1362.

[330] Naimi, Adel, Atefeh Entezari, Majid Farshdousti Hagh, Ali Hassanzadeh, Raedeh Saraei, and Saeed Solali. Quercetin sensitizes human myeloid leukemia KG-1 cells against TRAIL-induced apoptosis. Journal of Cellular Physiology 234.8 (2019): 13233-13241.

[331] Tang Z, Wang L, Chen Y, Zheng X, Wang R, Liu B, Zhang S, Wang H. Quercetin reverses 5-fluorouracil resistance in colon cancer cells by modulating the NRF2/HO-1 pathway. Eur J Histochem. 2023 Aug 7;67(3):3719.

[332] Onaolapo AY, Ojo FO, Onaolapo OJ. Biflavonoid quercetin protects against cyclophosphamide-induced organ toxicities via modulation of inflammatory cytokines, brain neurotransmitters, and astrocyte immunoreactivity. Food Chem Toxicol. 2023 Aug;178:113879.

[333] Zhang Z, Wu H, Zhang Y, Shen C, Zhou F. Dietary antioxidant quercetin overcomes the acquired resistance of Sorafenib in Sorafenib-resistant hepatocellular carcinoma cells through epidermal growth factor receptor signaling inactivation. Naunyn Schmiedebergs Arch Pharmacol. 2023 Jul 25.

[334] Teekaraman, Dhanaraj, Sugantha Priya Elayapillai, Mangala Priya Viswanathan, and Arunakaran Jagadeesan. Quercetin inhibits human metastatic ovarian cancer cell growth and modulates components of the intrinsic apoptotic pathway in PA-1 cell line. Chemico-biological interactions 300 (2019): 91-100.

[335] Wang, Yijun, Wei Zhang, Qiongying Lv, Juan Zhang, and Dingjun Zhu. The critical role of quercetin in autophagy and apoptosis in HeLa cells. Tumor Biology 37.1 (2015): 925-929.

[336] Bagde, Arvind, Ketan Patel, Arindam Mondal, Shallu Kutlehria, Nusrat Chowdhury, Aragaw Gebeyehu, Nilkumar Patel, Nagi Kumar, and Mandip Singh. Combination of UVB Absorbing Titanium Dioxide and Quercetin Nanogel for Skin Cancer Chemoprevention. AAPS PharmSciTech 20.6 (2019): 240-240.

Treatment for Retinoblastoma: Quercetin has been shown to cause arrest of the G1 (cell growth) phase in cell cycle, a collapse in mitochondrial membrane potential, and apoptosis in Y79 retinoblastoma cells[337] (eye cancer).

Treatment for Colon Cancer: Quercetin in combination with curcumin exhibited significant synergistic effects in HT-29 and HCT-116 colorectal cancer cells, largely due to the combined antioxidant effects of the compounds when administered in nanocapsules[338].

Treatment for Neurological Disorders: In irradiated cultured dorsal root ganglion neurons, quercetin was shown to decrease the expression of endoplasmic reticulum stress marker genes, downregulate tumor necrosis factor alpha and JNK, and increase the Tuj1 protein[339], among other neuroprotective effects. In Sprague Dawley rats subjected to chlorpyrifos (an organophosphate insecticide)-induced apoptotic events, treatment with quercetin protected mitochondrial integrity via the inhibition of apoptosis in neuron cells and by limiting oxidative stress[340]; important work considering the prolific worldwide use of chlorpyrifos in the control of insects. Quercetin has also been shown to decrease cell apoptosis in the focal cerebral ischemia rat brain[341] with induced middle cerebral artery occlusion. In rotenone (an isoflavone often used as a pesticide)-induced neuroinflammation and alterations in mice, treatment with quercetin reversed neuroinflammation, decreased immobility time in the forced swim test, and reversed reduced muscular strength, improving memory and cognitive function in the animals[342]. Another similar study investigated rotenone-induced Parkinsonism in male Wistar rats, where treatment with dihydroquercetin significantly suppressed rotenone-induced upregulation of several proteins and related transcription factor expression, improving Parkinsonian symptoms including bradykinesia, catalepsy, postural instability, impaired locomotor behavior, and tremor, while also ameliorating neurochemical dysfunctions by the modulation of genes involved in activation of the canonical pathway of NF-κB-mediated inflammation[343]. An additional study that investigated rotenone-induced Parkinsonism in Wistar rats found that treatment with the flavonoids catechin and quercetin attenuated striatal redox stress and neurochemical dysfunction, optimized disturbed dopamine metabolism, and improved depletion of neuron density caused by rotenone toxicity[344].

Antiviral Agent: As a primary constituent in the ethyl acetate fraction of houttuynia cordata (chameleon plant), quercetin significantly contributed to the inhibition of infectivity of coronavirus, dengue virus, and hepatitis in mice, exhibiting excellent antiviral efficacy with no cytotoxicity both in vitro and in vivo[345].

---

[337] Liu, H., Zhou, M. Antitumor effect of Quercetin on Y79 retinoblastoma cells via activation of JNK and p38 MAPK pathways. BMC Complement Altern Med 17, 531 (2017).

[338] Jain S, Lenaghan S, Dia V, Zhong Q. Co-delivery of curcumin and quercetin in shellac nanocapsules for the synergistic antioxidant properties and cytotoxicity against colon cancer cells. Food Chem. 2023 Dec 1;428:136744.

[339] Chatterjee, Jit, Jaldeep Langhnoja, Prakash P Pillai, and Mohammed S Mustak. Neuroprotective effect of quercetin against radiation-induced endoplasmic reticulum stress in neurons. Journal of Biochemical and Molecular Toxicology 33.2 (2019).

[340] Fereidouni, Soheil, Ravi Ranjan Kumar, Vijayta D. Chadha, and Devinder Kumar Dhawan. Quercetin plays protective role in oxidative induced apoptotic events during chronic chlorpyrifos exposure to rats. Journal of Biochemical and Molecular Toxicology 33.8 (2019).

[341] Yao, Rui-Qin, Da-Shi Qi, Hong-Li Yu, Jing Liu, Li-Hua Yang, and Xiu-Xiang Wu. Quercetin Attenuates Cell Apoptosis in Focal Cerebral Ischemia Rat Brain Via Activation of BDNF-TrkB-PI3K/Akt Signaling Pathway.

[342] Jain J, Hasan W, Biswas P, Yadav RS, Jat D. Neuroprotective effect of quercetin against rotenone-induced neuroinflammation and alterations in mice behavior. J Biochem Mol Toxicol. 2022 Oct;36(10):e23165.

[343] Akinmoladun AC, Famusiwa CD, Josiah SS, Lawal AO, Olaleye MT, Akindahunsi AA. Dihydroquercetin improves rotenone-induced Parkinsonism by regulating NF-κB-mediated inflammation pathway in rats. J Biochem Mol Toxicol. 2022 May;36(5):e23022.

[344] Josiah SS, Famusiwa CD, Crown OO, Lawal AO, Olaleye MT, Akindahunsi AA, Akinmoladun AC. Neuroprotective effects of catechin and quercetin in experimental Parkinsonism through modulation of dopamine metabolism and expression of IL-1β, TNF-α, NF-κB, IκKB, and p53 genes in male Wistar rats. Neurotoxicology. 2022 May;90:158-171.

[345] Chiow, K.H., M.C. Phoon, Thomas Putti, Benny K.H. Tan, and Vincent T. Chow. Evaluation of antiviral activities of Houttuynia cordata Thunb. extract, quercetin, quercetrin and cinanserin on murine coronavirus and dengue virus infection. Asian Pacific Journal of Tropical Medicine 9.1 (2015): 1-7.

Treatment of Obesity-related Cardiovascular Disease: Quercetin has been shown to strongly suppress elevated ERK1/2 phosphorylation and NFkB activation and decrease TNF-alpha secretion, suggesting that quercetin modulates leptin-induced inflammation[346] and may be used to treat obesity-related cardiovascular disease.

Treatment for Macular Degeneration: In Rhesus monkeys, quercetin has been shown to significantly inhibit choroidal and retinal angiogenesis[347], the primary cause of age-related macular degeneration (refers to the gradual loss of eyesight).

Treatment for Depression: Quercetin has been shown to exhibit antioxidant and anti-inflammatory effects in a murine model of depression (unpredictable chronic mild stress), while also augmenting 5-HT (serotonin receptor) levels[348]. In lipopolysaccharide-induced depression in rats, treatment with quercetin attenuated the decreased mobility and sucrose preference of the animals, decreasing the induced elevated levels of inflammasomes, proinflammatory cytokines, and microglia positive cells in the hippocampus and prefrontal cortex[349].

Treatment for Allergic Conjunctivitis: In animal models of conjunctivitis and in cultured human mast cells, quercetin has been shown to inhibit allergic conjunctivitis by lyn kinase inhibition and signaling, which dampens the inflammatory response[350] caused by this condition.

Treatment for Rheumatoid Arthritis: Quercetin has been shown to attenuate inflammation and joint destruction in rheumatoid arthritis, inhibiting unstimulated and interleukin-1 (a cytokine protein)-induced proliferation of rheumatoid synovial fibroblasts and matrix metalloproteinases 1 and 3 (cellular degradation enzymes), cyclooxygenase (prostanoid enzymes) and related prostaglandin E2 (inflammatory mediator) production[351].

Prevention & Treatment for Acute Liver Injury: Researchers using liposomal nanoparticles as a drug delivery mechanism of quercetin found that the method resulted in high bioavailability of the flavonol, which was particularly effective at attenuating induced acute liver injury in rats[352].

Prevention & Treatment for Lung Injury: Quercetin has been shown to inhibit the release of inflammatory cytokines, block neutrophil recruitment, reduce albumin leakage, and increase the cyclic amp (an intracellular messenger) and epac (modulator of several cellular processes) levels in lung tissue in mice subjected to lipopolysaccharide (LPS)-induced acute lung injury[353].

[346] Indra, M.R., Karyono, S., Ratnawati, R. et al. Quercetin suppresses inflammation by reducing ERK1/2 phosphorylation and NF kappa B activation in Leptin-induced Human Umbilical Vein Endothelial Cells (HUVECs). BMC Res Notes 6, 275 (2013).

[347] Chen, Yi, Xiao-xin Li, Nian-zeng Xing, and Xiao-guang Cao. Quercetin inhibits choroidal and retinal angiogenesis in vitro. Graefe's Archive for Clinical and Experimental Ophthalmology 246.3 (2007): 373-378.

[348] Khan, Khadeeja, Abul Kalam Najmi, and Mohd Akhtar. A Natural Phenolic Compound Quercetin Showed the Usefulness by Targeting Inflammatory, Oxidative Stress Markers and Augment 5-HT Levels in One of the Animal Models of Depression in Mice. Drug research 69.7 (2019): 392-400.

[349] Adeoluwa OA, Olayinka JN, Adeoluwa GO, Akinluyi ET, Adeniyi FR, Fafure A, Nebo K, Edem EE, Eduviere AT, Abubakar B. Quercetin abrogates lipopolysaccharide-induced depressive-like symptoms by inhibiting neuroinflammation via microglial NLRP3/NFκB/iNOS signaling pathway. Behav Brain Res. 2023 Jul 26;450:114503.

[350] Ding, Yuanyuan, Chaomei Li, Yongjing Zhang, Pengyu Ma, Tingting Zhao, Delu Che, Jiao Cao, Jue Wang, Rui Liu, Tao Zhang, and Langchong He. Quercetin as a Lyn kinase inhibitor inhibits IgE-mediated allergic conjunctivitis. Food and chemical toxicology: an international journal published for the British Industrial Biological Research Association 135 (2020): 110924-110924.

[351] Sung, Myung-Soon, Eun-Gyeong Lee, Hyun-Soon Jeon, Han-Jung Chae, Seoung Park, Yong Lee, and Wan-Hee Yoo. Quercetin Inhibits IL-1-Induced Proliferation and Production of MMPs, COX-2, and PGE2 by Rheumatoid Synovial Fibroblast. Inflammation 35.4 (2012): 1585-1594.

[352] Liu, X., Zhang, Y., Liu, L. et al. Protective and therapeutic effects of nanoliposomal quercetin on acute liver injury in rats. BMC Pharmacol Toxicol 21, 11 (2020).

[353] Wang, X., Song, S., Li, Y. et al. Protective Effect of Quercetin in LPS-Induced Murine Acute Lung Injury Mediated by cAMP-Epac Pathway. Inflammation 41, 1093-1103 (2018).

Treatment for Herpes (HSV-1): Quercetin was shown to significantly lower herpes simplex virus-1 infectivity and inhibit the expression of HSV proteins and genes, while also inhibiting inflammatory transcriptional factors in raw 264.7 cells[354].

Treatment for Tardive Dyskinesia: Chronic use of neuroleptic drugs can cause rapid, repetitive, and uncontrollable movements (such as chewing, blinking, wincing, grimacing, etcetera) a condition referred to as tardive dyskinesia. In rats subjected to haloperidol-induced orofacial dyskinesia, treatment with quercetin significantly reduced associated vacuous chewing and tongue protrusions, reduced lipid peroxidation, and restored GSH, SOD, and catalase levels[355].

Wound Protective Agent: Researchers have developed a nanofiber scaffolding loaded with ciprofloxacin hydrochloride and quercetin, which suppressed infection and oxidative damage and displayed accelerated healing with complete re-epithelialization and improved collagen deposition in a full thickness wound model in rats[356].

Treatment for Ulcerative Colitis: Researchers investigating the traditional Chinese herbal formula, Xiang-lian Pill, to determine its use in treating ulcerative colitis found that oral administration of the pill in mice ameliorated dextran sulfate sodium-induced colitis in mice, and that quercetin was the major active compound[357].

Treatment for Toxoplasmosis: Quercetin has been shown to significantly inhibit pyrimethamine/sulfadiazine-resistant Toxoplasma gondii tachyzoites (parasite that causes toxoplasmosis), and to synergize strongly with azithromycin[358], an antiparasitic and antibiotic agent.

Treatment for Systemic Sclerosis: In a murine model of bleomycin-induced systemic sclerosis, combined treatment with 3'5-dimaleamylbenzoic acid (also called mesitylenic acid, a benzoic acid) and quercetin regulated cell proliferation, inflammation, and oxidative stress, attenuating induced histological alterations and pulmonary fibrosis as evidenced by the recovery of alveolar spaces and decreased collagen deposits[359].

Antibacterial Agent: At a concentration of 125.0 μg/mL, quercetin had both inhibitory and bactericidal effects against Streptococcus pneumoniae, and these effects were increased when quercetin was combined with ampicillin[360]. Quercetin also inhibited biofilm formation, adhesion, and invasion of Candida albicans in vitro, and ameliorated C. albicans-induced inflammation and protected the integrity of vaginal mucosa in vivo[361].

---

[354] Lee, S., Lee, H.H., Shin, Y.S. et al. The anti-HSV-1 effect of quercetin is dependent on the suppression of TLR-3 in Raw 264.7 cells. Arch. Pharm. Res. 40, 623-630 (2017).

[355] Naidu, Pattipati S., Amanpreet Singh, and Shrinivas K. Kulkarni. Reversal of haloperidol-induced orofacial dyskinesia by quercetin, a bioflavonoid. Psychopharmacology 167.4 (2003): 418-423.

[356] Ajmal, Gufran, Gunjan Vasant Bonde, Sathish Thokala, Pooja Mittal, Gayasuddin Khan, Juhi Singh, Vivek Kumar Pandey, and Brahmeshwar Mishra. Ciprofloxacin HCl and quercetin functionalized electrospun nanofiber membrane: fabrication and its evaluation in full thickness wound healing. Artificial cells, nanomedicine, and biotechnology 47.1 (2019): 228-240.

[357] Zhou HF, Yang C, Li JY, He YY, Huang Y, Qin RJ, Zhou QL, Sun F, Hu DS, Yang J. Quercetin serves as the major component of Xiang-lian Pill to ameliorate ulcerative colitis via tipping the balance of STAT1/PPARγ and dictating the alternative activation of macrophage. J Ethnopharmacol. 2023 Sep 15;313:116557.

[358] Abugri DA, Wijerathne SVT, Sharma HN, Ayariga JA, Napier A, Robertson BK. Quercetin inhibits Toxoplasma gondii tachyzoite proliferation and acts synergically with azithromycin. Parasit Vectors. 2023 Aug 3;16(1):261.

[359] Reyes-Jiménez E, Ramírez-Hernández AA, Santos-Álvarez JC, Velázquez-Enríquez JM, González-García K, Carrasco-Torres G, Villa-Treviño S, Baltiérrez-Hoyos R, Vásquez-Garzón VR. Coadministration of 3'5-dimaleamylbenzoic acid and quercetin decrease pulmonary fibrosis in a systemic sclerosis model. Int Immunopharmacol. 2023 Sep;122:110664.

[360]Willian de Alencar Pereira E, Fontes VC, da Fonseca Amorim EA, de Miranda RCM, Carvalho RC, de Sousa EM, Cutrim SCPF, Alves Lima CZGP, de Souza Monteiro A, Neto LGL. Antimicrobial effect of quercetin against Streptococcus pneumoniae. Microb Pathog. 2023 Jul;180:106119.

[361] Tan Y, Lin Q, Yao J, Zhang G, Peng X, Tian J. In vitro outcomes of quercetin on Candida albicans planktonic and biofilm cells and in vivo effects on vulvovaginal candidiasis. Evidences of its mechanisms of action. Phytomedicine. 2023 Jun;114:154800.

Treatment for Osteoporosis: Isoquercetin was shown to increase cell viability and proliferation while also promoting osteogenic differentiation in bone marrow mesenchymal stem cells in an osteoporosis model in mice[362].

Protection Against Cypermethrin-Induced Lung Toxicity: In cypermethrin (a toxic synthetic insecticide)-induced lung toxicity in Sprague Dawley rats, treatment with quercetin provided significant protection against oxidative stress, inflammation, apoptosis, endoplasmic reticulum stress, and autophagy in lung tissue[363].

IMPLICATIONS FOR HUMAN HEALTH & NUTRITION

Most people obtain at least some quercetin in their regular diet. Increasing daily intake of this flavonol can be accomplished by eating fruits that are high in this compound, which means apples, apricots, bananas, blueberries, black currants, cherries, cranberries, date palms, and figs. Vegetables that may be a good source of quercetin include asparagus, avocados, broccoli, Brussels sprouts, cabbage, capers, carrots, celery, chives, corn, dandelions, and fenugreek. When preparing, cooking, or serving foods, consider working with spices and herbs like anise, basil, bay laurel, black pepper, caraway, cayenne, cloves, coriander, dill, fennel, garlic, and ginger. You can also snack on almonds, black walnuts, and chestnuts, all of which are likely to contain quercetin.

Cannabis strains that have been shown to contain quercetin include but are not limited to Cheese, Fruit Punch, Northern Lights, Purple Moose, Purple Punch, Skywalker, Space Candy, Strawberry Cough, Strawberry Banana, Super Lemon Haze, and White Cookies.

As always, extreme care should be used when working with purified quercetin.

---

[362] Wu M, Qin M, Wang X. Therapeutic effects of isoquercetin on ovariectomy-induced osteoporosis in mice. Nat Prod Bioprospect. 2023 Jun 8;13(1):20.

[363] Ileriturk M, Kandemir O, Kandemir FM. Evaluation of protective effects of quercetin against cypermethrin-induced lung toxicity in rats via oxidative stress, inflammation, apoptosis, autophagy, and endoplasmic reticulum stress pathway. Environ Toxicol. 2022 Nov;37(11):2639-2650.

Quercetin Review

Answer the following questions to test your knowledge of this flavonoid:

Question #1: What type of flavonoid is quercetin?

      a.   Flavanol
      b.   Flavorall
      c.   Flavonol
      d.   Flavone

Question #2: The base component in the biosynthesis of quercetin is:

      a.   Cannabisin
      b.   Rutinoside
      c.   Naringenin
      d.   Quercetin

Question #3: How common is quercetin in cannabis?

      a.   Top three
      b.   Top five
      c.   Top ten
      d.   Secondary

Question #4: What is the chemical formula for quercetin?

      a. $C_{24}H_{26}O$
      b. $C_{15}H_{10}O_5$
      c. $C_{15}H_{10}O_7$
      d. $C_{27}H_{30}O_{16}$

Question #5: What is the primary known biological role of quercetin in plants?

_____

Question #6: Name two potential medical uses of quercetin:

1 _____ 2 _____

For the answer key to Quercetin, please visit www.cannabischemistry.org

CANNFLAVIN C

Type: Prenylflavonoid
Chemical Formula: $C_{26}H_{28}O_6$
Molecular Weight: 436.50 g/mol
Boiling Point: Not reported
Flash Point: Not reported
Melting Point: Not reported
Solubility: Soluble in DMSO[364]
Biological Role: Unknown
Therapeutic Role: Research ongoing
Commercial Use: None
Occurrence in Cannabis: Top ten

Occurs in Cannabis Strains: Cheese, Critical Dream, Fruit Punch, Huckleberry, Northern Lights, Purple Moose, Purple Punch, Skywalker, Space Candy, Strawberry Banana, Strawberry Cough, Super Lemon Haze, White Cookies

INTRODUCTION

Cannflavin C is a prenylated flavonoid that is similar to cannflavin A, with the only difference being the location of a geranyl group in the molecular skeleton.

Discovered in 2008 by Mohamed Radwan, et al, little information about this compound was available as of the publication of this text. While some research is ongoing regarding the potential therapeutic potential of this flavonoid, the literature regarding cannflavin C is virtually non-existent.

However, it is interesting to note that of thirteen cannabis strains we sent to the laboratory for flavonoid testing, all of them contained moderate amounts of cannflavin C, with the compound occurring as the number seven constituent on average across the varieties.

---

[364] ChemScene Data Sheet for Cannflavin C, from: https://www.chemscene.com/1086463-05-9.html. Accessed November 14, 2023.

CHEMICAL STRUCTURE

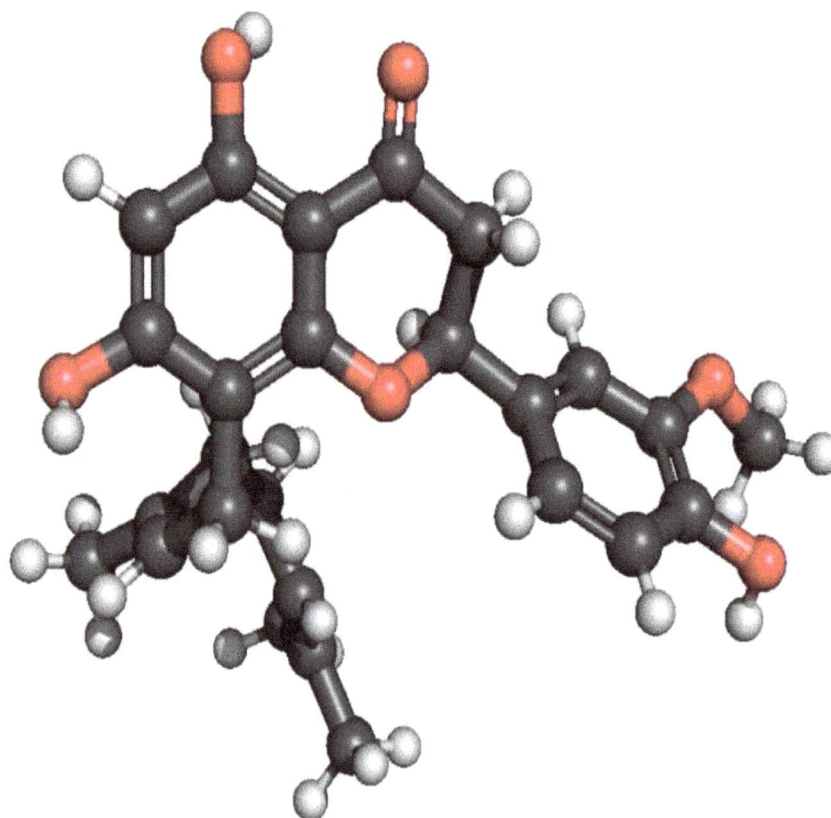

Cannflavin C is biosynthesized in the phenylpropanoid metabolic pathway from geranyl diphosphate. As a prenylated flavonoid, cannflavin C belongs to the flavone subclass, with hydroxyl groups at C-4' and C-7, a methoxyl group at C-3', and a geranyl group at C-8[365], whereas cannflavin A features the geranyl group at C-6[366]. Cannflavin C contains twenty-six carbon atoms, twenty-eight hydrogen atoms, and six oxygen atoms notated as $C_{26}H_{28}O_6$. The molecular skeleton of this flavone contains nine double-bonds, six of which are endocyclic, and one double-bonded oxygen atom.

OCCURRENCE IN PLANTS

As of early 2025, the only plant cannflavin C has been found to occur in is various chemovars of cannabis.

USES IN INDUSTRY

Cannflavin C appears as a yellow amorphous powder and is only used as a research chemical.

---

[365] Radwan, Mohamed M et al. Non-cannabinoid constituents from a high potency Cannabis sativa variety. Phytochemistry vol. 69,14 (2008): 2627-33.

[366] Maged S. Abdel-Kader, Mohamed M. Radwan, Ahmed M. Metwaly, Ibrahim H. Eissa, Arno Hazekamp, and Mahmoud A. Sohly. Chemistry and Biological Activities of Cannflavins of the Cannabis Plant. Cannabis and Cannabinoid Research, ahead of print. Accessed November 14, 2023.

BIOLOGICAL ACTIVITY IN PLANTS

At the time of this publication, there were no studies or research papers available regarding the biological roles of cannflavin C in cannabis. However, we can assume some of these roles based on the general activities of flavonoids in plants, which include UV protection, antioxidant activity, and allelopathic activity, among other potential functions. More research is still needed for this molecule.

POTENTIAL USES IN MEDICINE

There are no known therapeutic uses for cannflavin C, however, this compound is likely to offer medical benefits like those discovered in cannflavin A and B, particularly as an anti-inflammatory agent, neuroprotective agent, antileishmanial agent, and as a treatment for some types of cancer, particularly pancreatic carcinoma.

IMPLICATIONS FOR HUMAN HEALTH & NUTRITION

Because cannflavin C is only known to occur in cannabis, consumption of cannabis strains and products remains the only way to dietarily obtain this phytonutrient. Strains that we have tested and know to contain cannflavin C include Cheese, Critical Dream, Fruit Punch, Huckleberry, Northern Lights, Purple Moose, Purple Punch, Skywalker, Space Candy, Strawberry Cough, Strawberry Banana, Super Lemon Haze, and White Cookies, however, it is probable that most cannabis varieties contain this flavonoid in the top ten flavonoid constituents by concentration.

Cannflavin C Review

Answer the following questions to test your knowledge of this flavonoid:

Question #1: What type of flavonoid is cannflavin C?

    a.   Flavanol
    b.   Flavorall
    c.   Flavone
    d.   Flavonol

Question #2: Which functional groups are contained in cannflavin C?

    a.   Prenyl
    b.   Geranyl
    c.   Methoxyl
    d.   Hydroxyl
    e.   All of the above

Question #3: How common is cannflavin C in cannabis?

    a.   Top three
    b.   Top five
    c.   Top ten
    d.   Secondary

Question #4: What is the chemical formula for cannflavin C?

    a. $C_{24}H_{26}O$
    b. $C_{15}H_{10}O_5$
    c. $C_{26}H_{28}O_6$
    d. $C_{27}H_{30}O_{16}$

Question #5: What year was cannflavin C discovered?

_____

For the answer key to Cannflavin C, please visit www.cannabischemistry.org

LUTEOLIN

Type: Flavone
Chemical Formula: $C_{15}H_{10}O_6$
Molecular Weight: 286.239 g/mol
Boiling Point: 616.1 °C @ 760 mmHg[367]
Flash Point: 239.50 °C
Melting Point: 328 – 330 °C[368]
Solubility: Insoluble in water, soluble in alcohol, ethanol, DMSO, dimethyl formamide
Oral LD50: Very safe, >5000 mg/kg (rat)[369]
Biological Role: Allelopathic agent, host-symbiont signaling agent
Therapeutic Role: Anticancer agent, cancer drug synergist, neurological agent, anti-coronavirus agent
Commercial Use: Food additive
Occurrence in Cannabis: Top ten

Occurs in Cannabis Strains: Cheese, Fruit Punch, Huckleberry, Northern Lights, Purple Moose, Purple Punch, Skywalker, Space Candy, Strawberry Cough, Strawberry Banana, Super Lemon Haze, White Cookies. Luteolin has also been found in the hemp variety called Futura 75[370].

INTRODUCTION

Luteolin is a phytonutrient that has been researched heavily for its anti-cancer properties, among other medical and therapeutic uses described below, including the fascinating potential for this compound to be used in the treatment of meth addiction. Luteolin is also known as the primary constituent in a natural yellow dye called weld, which has been obtained from the immature flowering parts of reseda luteola (dyer's rocket) for thousands of years. Known to occur in many plant species, luteolin is a secondary flavonoid constituent in cannabis, occurring often as the eighth flavonoid by concentration.

---

[367] Chemical Book Data Sheet for Luteolin, from: https://www.chemicalbook.com/ChemicalProductProperty_EN_CB7282616.htm. Accessed November 14, 2023.

[368] FoodDB Compound Sheet for Luteolin, from: https://foodb.ca/compounds/FDB013255. Accessed November 14, 2023.

[369] Alzaabi MM, Hamdy R, Ashmawy NS, Hamoda AM, Alkhayat F, Khademi NN, Al Joud SMA, El-Keblawy AA, Soliman SSM. Flavonoids are promising safe therapy against COVID-19. Phytochem Rev. 2022;21(1):291-312.

[370] Cásedas G, Moliner C, Maggi F, Mazzara E, López V. Evaluation of two different Cannabis sativa L. extracts as antioxidant and neuroprotective agents. Front Pharmacol. 2022 Sep 13;13:1009868.

CHEMICAL STRUCTURE

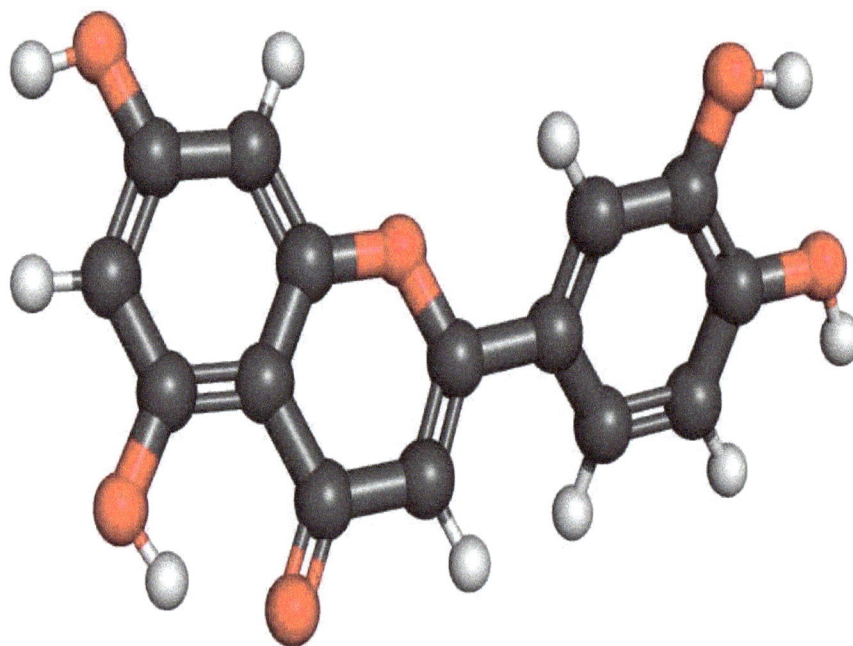

Like many flavonoids in cannabis, luteolin is derived first from naringenin, but biosynthesis of this compound can then take two paths to result in the development of luteolin: via the pathway to apigenin, or eridictyol[371]. The specific biosynthesis pathway of luteolin in cannabis has not been elucidated as of the publishing of this text.

Luteolin is classified as a flavone thanks to its three-ring backbone containing fifteen carbon atoms, ten hydrogen atoms, and six oxygen atoms notated as $C_{15}H_{10}O_6$. Luteolin contains eight double bonds: seven endocyclic and one exocyclic double-bonded oxygen atom, with hydroxy groups at positions 3', 4', 5, and 7.

Luteolin can be modified to create many different variations, including:

Luteolin-7,3'-di-O-glucoside: glucose molecules attached at both the 7th and 3rd carbon positions.

Luteolin-7-glucoside / Luteolin 7-O-glucoside / Luteolin 7-O-beta-D-glucoside(1-): a glucose molecule attached at the 7th carbon position.

Luteolin-4'-O-glucoside: a glucose molecule attached at the 4th carbon position.

Luteolin-7-O-gentiobioside: a gentiobiose unit (a disaccharide composed of two glucose molecules) attached at the 7th carbon position.

Luteolin 7-O-(6"-malonylglucoside): a malonylglucoside moiety attached at the 7th carbon position.

Luteolin-7-O-rutinoside: a rutinose (a disaccharide composed of the sugar rutinose and glucose) group attached at the 7th carbon position.

---

[371] Foerster H, The Arabidopsis Information Resource, from MetaCyc Pathway: Luteolin Biosynthesis https://biocyc.org/META/NEW-IMAGE?type=PATHWAY&object=PWY-5060. Accessed September 8, 2020.

Luteolin 7-O-beta-D-Glucuronide: a glucuronic acid group attached at the 7th carbon position.

Luteolin 4'-sulfate: a sulfate group attached at the 4th carbon position.

Luteolin 7-(6"-acetylglucoside): an acetylglucoside moiety attached at the 7th carbon position.

Luteolin-7-O-alpha-L-rhamnoside: an alpha-L-rhamnose unit attached at the 7th carbon position.

5,7,3',4'-tetramethyl luteolin: methyl groups attached to the 5th, 7th, 3rd, and 4th carbon positions.

OCCURRENCE IN PLANTS

Luteolin is found in a wide variety of plants, including many edible and readily available plant-based food sources including but not limited to the following:

Aiphanes acul	Alfalfa	Anise
Apple heartwood	Ashitaba	Basil
Bell pepper	Black bean	Broccoli
Cacao	Cannabis	Carrot
Celery	Chamomile	Cornmint
Cumin	Dandelion	Fenugreek
Flax	Ginkgo biloba	Grape
Green pepper	Lemon	Lemongrass
Lentil	Lettuce	Lotus
Olive	Orange	Oregano
Parsley	Peanut	Peppermint
Perilla	Pomegranate	Potato
Rosemary	Sage	Salsify
Salvia	Spearmint	Tarragon
Thyme	Yarrow	

BIOLOGICAL ACTIVITY IN PLANTS

Little research has been carried out to examine the specific roles that luteolin plays in plants, but some work suggests that the primary uses of this flavonoid include host-symbiont signaling, and allelopathic activities.

Host-Symbiont Signaling: A symbiotic relationship between the gram-negative bacterium rhizobium meliloti and alfalfa has been observed which induces the accelerated establishment of nitrogen-fixing root nodules in the plant, with researchers noting that "This regulatory role for a flavone contrasts with the function of some flavonoids as defense compounds[372]."

Allelopathic Agent: Luteolin has been shown to significantly inhibit the growth of the bloom of the algae microcystis aeruginosa by inducing enhanced oxidative cell damage and inhibiting photosynthesis[373]. As isolated from the leaves of chrysanthemum morifolium, luteolin was shown to significantly reduce the number of fronds and the chlorophyll content of lemna gibba[374], an aquatic plant known commonly as various types of duckweed.

[372] Peters, N K, J W Frost, and S R Long. A plant flavone, luteolin, induces expression of Rhizobium meliloti nodulation genes. Science (New York, N.Y.) 233.4767 (1986): 977-980.

[373] Linrong Cao H and Jieming Li Plant-Originated Kaempferol and Luteolin as Allelopathic Algaecides Inhibit Aquatic Microcystis Growth Through Affecting Cell Damage, Photosynthetic and Antioxidant Responses Bioremediate Biodegrade 9:431, Vol 9(2) Feb 28, 2018.

[374] Beninger, Clifford W., Hall, Christopher J. Allelopathic activity of luteolin 7-O-B-glucuronide isolated from Chrysanthemum morifolium L. Biochemical Systematics and Ecology Volume 33, Issue 2, February 2005, Pages 103-111.

USES IN INDUSTRY

The use of dyer's rocket to produce yellow dye has become outdated, so most modern preparations of luteolin are used in medicine or food. Appearing as a yellow crystalline powder, luteolin is used in baked goods, beverages, breakfast cereals, chewing gum, confectionary, frozen dairy, hard candies, imitation dairy and coffee products, seasonings, snack foods, sauces, sugar substitutes, and other prepared foods.

Researchers have found that luteolin exhibits protective effects via antioxidative properties in rabbit semen[375], which may prove useful in protecting the integrity of sperm used in breeding projects, which is often subject to cryopreservation. Other work has examined the potential application of luteolin as an antibacterial hand sanitizer ingredient after it was shown to induce cell dysfunction, change membrane permeability, promote the leakage of cellular contents, and disrupt cell membrane integrity in Escherichia coli and Staphylococcus aureus bacteria[376].

POTENTIAL USES IN MEDICINE

Although luteolin is known for its applications in the treatment of cancer, this flavone has also been shown to offer promise in the treatment of multiple sclerosis as a cardio, neurological, and renal protective agent, and, interestingly, as a potential agent to treat meth addiction. Recently, it was observed that luteolin binds with DNA[377], which could have important implications regarding the medical and therapeutic application of this compound.

Treatment for Laryngeal Cancer: Luteolin has been shown to inhibit cell proliferation and induce apoptosis in human laryngeal carcinoma cells via activation of the Fas signaling pathway, and caspase-3 and caspase-8 activation[378].

Treatment for Brain & Spinal Cord Cancer: Luteolin has been shown to induce a severe accumulation of reactive oxygen species (ROS), causing a lethal endoplasmic reticulum stress response and mitochondrial dysfunction in glioblastoma cells[379] (an aggressive type of brain cancer). Luteolin has also been shown to inhibit the migration and invasion of glioblastoma cells by interfering with the activation and expression of several proteins, causing their degradation[380].

Treatment for Lung Cancer: In non-small cell lung cancer, luteolin inhibited both the invasion and migration of cells, and the epithelial-mesenchymal transition of cells[381] (a feature of malignancy) via several mechanisms. In hexavalent chromium induced lung cancer, luteolin was observed to reduce the number of tumors, protect BEAS-2B cells (bronchial epithelial cell line) from malignant transformation, inhibit the production of pro-inflammatory

[375] Akarsu SA, Acısu TC, Güngör IH, Çakır Cihangiroğlu A, Koca RH, Türk G, Sönmez M, Gür S, Fırat F, Esmer Duruel HE. The effect of luteolin on spermatological parameters, apoptosis, oxidative stress rate in freezing rabbit semen. Pol J Vet Sci. 2023 Mar;26(1):91-98.

[376] Xi M, Hou Y, Wang R, Ji M, Cai Y, Ao J, Shen H, Li M, Wang J, Luo A. Potential Application of Luteolin as an Active Antibacterial Composition in the Development of Hand Sanitizer Products. Molecules. 2022 Oct 28;27(21):7342.

[377] Zou, Na, Xueliang Wang, and Guifang Li. Spectroscopic and electrochemical studies on the interaction between luteolin and DNA. Journal of Solid State Electrochemistry 20.6 (2016): 1775-1782.

[378] Zhang, Hui, Xiuguo Li, Yuanyuan Zhang, and Xinyong Luan. Luteolin induces apoptosis by activating Fas signaling pathway at the receptor level in laryngeal squamous cell line Hep-2 cells. European Archives of Oto- Rhino-Laryngology 271.6 (2014): 1653-1659.

[379] Wang, Qiang, Handong Wang, Yue Jia, Hao Pan, and Hui Ding. Luteolin induces apoptosis by ROS/ER stress and mitochondrial dysfunction in gliomablastoma. Cancer Chemotherapy and Pharmacology 79.5 (2017): 1031-1041.

[380] Cheng, Wen-Yu, Ming-Tsang Chiao, Yea-Jiuen Liang, Yi-Chin Yang, Chiung-Chyi Shen, and Chiou-Ying Yang. Luteolin inhibits migration of human glioblastoma U-87 MG and T98G cells through downregulation of Cdc42 expression and PI3K/AKT activity. Molecular Biology Reports 40.9 (2013): 5315-5326.

[381] Yu, Qian, Minda Zhang, Qidi Ying, Xin Xie, Shuwen Yue, Bending Tong, Qing Wei, Zhaoshi Bai, and Lingman Ma. Decrease of AIM2 mediated by luteolin contributes to non-small cell lung cancer treatment. Cell Death & Disease 10.3 (2019).

cytokines, and inhibit multiple genes linked to survival, inflammation, and angiogenesis[382] (the development of blood vessels serving a tumor). In A549 lung adenocarcinoma cells, luteolin inhibited proliferation, induced apoptosis, and dramatically inhibited cell motility and migration[383]. Other work has shown that luteolin inhibited the growth of A549 lung cancer cells by inducing G1 phase cycle arrest and apoptosis, and by suppressing stress fiber assembly and cell migration[384].

Treatment for Colon Cancer: Luteolin has been shown to inhibit proliferation, induce cell cycle arrest, and induce apoptosis in human colon carcinoma cells[385]. Luteolin has also been shown to decrease IGF-II (insulin-like growth factor 2) production and down regulate insulin-like growth factor-I receptor signaling in HT29 colon cancer cells[386]. In metastatic human colon cancer, luteolin reduced the viability and proliferation of SW620 cells, increased the expression of antioxidant enzymes, and reversed the epithelial-mesenchymal transition process[387] (a characteristic of metastasis).

Treatment for Gastric Cancer: In gastric cancer cells, luteolin was demonstrated to significantly increase levels of pro-apoptotic proteins while simultaneously decreasing anti-apoptotic proteins, thereby inducing apoptosis[388]. Luteolin has also been shown to suppress gastric cancer tumor growth in vivo, while also inhibiting cell proliferation, invasion, and migration[389]. In cMet-overexpressing (a cell-regulating protein) patient-derived tumor xenograft models of gastric cancer, luteolin was shown to inhibit proliferation and invasiveness, while also inducing apoptosis in cells[390].

Treatment for Melanoma: Specially developed luteolin-encapsulated nanoparticles were shown to significantly inhibit tumor cell proliferation, migration, and invasion in melanoma cells[391].

Cancer Treatment Synergist: When combined with paclitaxel (one of the most common breast cancer treatment drugs), luteolin increased apoptosis and significantly reduced tumor size and weight in an orthotropic tumor model in mice[392]. In combination with celecoxib - a common anti-inflammatory drug - luteolin significantly increased the

[382] Pratheeshkumar, Poyil et al. Luteolin inhibits Cr(VI)-induced malignant cell transformation of human lung epithelial cells by targeting ROS mediated multiple cell signaling pathways. Toxicology and applied pharmacology vol. 281,2 (2014): 230-41.

[383] Meng, Guanmin, Kequn Chai, Xinda Li, Yongqiang Zhu, and Weihua Huang. Luteolin exerts pro-apoptotic effect and anti-migration effects on A549 lung adenocarcinoma cells through the activation of MEK/ERK signaling pathway. Chemico-biological interactions 257 (2017): 26-34.

[384] Zhao, Yunxue, Guotao Yang, Dongmei Ren, Xiumei Zhang, Qiuwei Yin, and Xuefei Sun. Luteolin suppresses growth and migration of human lung cancer cells. Molecular Biology Reports 38.2 (2010): 1115-1119.

[385] Sulaiman, Ghassan. In vitro study of molecular structure and cytotoxicity effect of luteolin in the human colon carcinoma cells. European Food Research and Technology 241.1 (2015): 83-90.

[386] Lim, Do, Han Cho, Jongdai Kim, Chu Nho, Ki Lee, and Jung Park. Luteolin decreases IGF-II production and downregulates insulin-like growth factor-I receptor signaling in HT-29 human colon cancer cells. BMC Gastroenterology 12.1 (2012): 1-10.

[387] Potocnjak, Iva, Lidija Simic, Ivana Gobin, Iva Vukelic, and Robert Domitrovic. Antitumor activity of luteolin in human colon cancer SW620 cells is mediated by the ERK/FOXO3a signaling pathway. Toxicology in vitro: an international journal published in association with BIBRA 66 (2020): 104852-104852.

[388] Wu, Bin, Qiang Zhang, Weiming Shen, and Jun Zhu. Anti-proliferative and chemosensitizing effects of luteolin on human gastric cancer AGS cell line. Molecular and Cellular Biochemistry 313.2 (2008): 125-132.

[389] Zang, Ming-de, Lei Hu, Zhi-yuan Fan, He-xiao Wang, Zheng-lun Zhu, Shu Cao, Xiong-yan Wu, Jian-fang Li, Li-ping Su, Chen Li, Zheng-gang Zhu, Min Yan, and Bing-ya Liu. Luteolin suppresses gastric cancer progression by reversing epithelial-mesenchymal transition via suppression of the Notch signaling pathway. Journal of Translational Medicine 15.1 (2017): 1-11.

[390] Lu, Jun, Guangliang Li, Kuifeng He, Weiqin Jiang, Cong Xu, Zhongqi Li, Haohao Wang, Weibin Wang, Haiyong Wang, Xiaodong Teng, and Lisong Teng. Luteolin exerts a marked antitumor effect in cMet-overexpressing patient-derived tumor xenograft models of gastric cancer. Journal of Translational Medicine 13.1 (2015): 1-11.

[391] Fu QT, Zhong XQ, Chen MY, Gu JY, Zhao J, Yu DH, Tan F. Luteolin-Loaded Nanoparticles for the Treatment of Melanoma. Int J Nanomedicine. 2023 Apr 20;18:2053-2068.

[392] Yang, Mon-Yuan, Chau-Jong Wang, Nai-Fang Chen, Wen-Hsin Ho, Fung-Jou Lu, and Tsui-Hwa Tseng. Luteolin enhances paclitaxel-induced apoptosis in human breast cancer MDA-MB-231 cells by blocking STAT3. Chemico-biological interactions 213 (2014): 60-68.

inhibition of cell growth in four different breast cancer cell lines[393]. Luteolin has also been shown to sensitize ovarian cancer cells to cisplatin - a common cancer treatment drug - resulting in the inhibition of migration and invasion of these cells[394], among other effects. Additionally, when combined with 5-fluorouracil (an antitumor agent), luteolin reduced tumor volume and weight and restored oxidant indices in solid ehrlich carcinoma cells[395] to previous levels.

Apoptotic Agent: Luteolin has been shown to enhance apoptosis in human skin cells and was even more effective when administered via PEGylated liposomes[396] - a delivery method that involves the attachment of polyethylene glycol to a therapeutic agent (in this case, luteolin).

Prolong Skin Graft Survival: Luteolin was confirmed as a potent immunosuppressant when it was shown to significantly increase the survival time of allografts (a transplant involving tissue from a donor of the same species) in mice via numerous mechanisms, including the amelioration of cellular infiltration and the downregulation of proinflammatory cytokine gene expression[397].

Vascular Protective Agent: Luteolin has been shown to act directly on vascular endothelial cells, inducing nitric oxide production and relaxing arteries in rats[398].

Renoprotective Agent: In ischemia reperfusion injury (tissue damage that occurs after a lack of oxygen) in rats, luteolin was shown to significantly improve creatinine and blood urea nitrogen levels, and improved enzymatic activity of superoxide dismutase, glutathione peroxidase and catalase[399], conferring significant renoprotective effects.

Diuretic Agent: Luteolin has been shown to exhibit significant diuretic (lowers blood pressure) and natriuretic (excretion of sodium by the kidneys) effects in rats[400].

Adjunctive Therapeutic Agent for Multiple Sclerosis: In human multiple sclerosis patients, administration of luteolin reduced the proliferation of peripheral blood mononuclear cells (specialized immune cells), modulated the levels of interleukin 1-beta and tumor necrosis factor-alpha, and contributed to the modulation of cell proliferation[401], among other effects.

[393] Jeon, Ye, Young Ahn, Won Chung, Hyun Choi, and Young Suh. Synergistic effect between celecoxib and luteolin is dependent on estrogen receptor in human breast cancer cells. Tumor Biology 36.8 (2015): 6349-6359.

[394] Wang, Haixia, Youjun Luo, Tiankui Qiao, Zhaoxia Wu, and Zhonghua Huang. Luteolin sensitizes the antitumor effect of cisplatin in drug-resistant ovarian cancer via induction of apoptosis and inhibition of cell migration and invasion. Journal of Ovarian Research 11.1 (2018): 1-12.

[395] Soliman, Nema A, Rania N Abd-Ellatif, Amira A ELSaadany, Shahinaz M Shalaby, and Asmaa E Bedeer. Luteolin and 5-flurouracil act synergistically to induce cellular weapons in experimentally induced Solid Ehrlich Carcinoma: Realistic role of P53; a guardian fights in a cellular battle. Chemico-biological interactions 310 (2019): 108740-108740.

[396] Sinha, Abhishek, and Suresh P. K. Enhanced Induction of Apoptosis in HaCaT Cells by Luteolin Encapsulated in PEGylated Liposomes-Role of Caspase-3/Caspase-14. Applied Biochemistry and Biotechnology 188.1 (2018): 147-164.

[397] Ye, Shulin, Huazhen Liu, Yuchao Chen, Feifei Qiu, Chun-Ling Liang, Qunfang Zhang, Haiding Huang, Sumei Wang, Zhong-De Zhang, Weihui Lu, and Zhenhua Dai. A Novel Immunosuppressant, Luteolin, Modulates Alloimmunity and Suppresses Murine Allograft Rejection. Journal of immunology (Baltimore, Md.: 1950) 203.12 (2020): 3436-3446.

[398] Si, Hongwei, Richard Wyeth, and Dongmin Liu. The flavonoid luteolin induces nitric oxide production and arterial relaxation. European Journal of Nutrition 53.1 (2013): 269-275.

[399] Kalbolandi, Sanaz, Armita Gorji, Hossein Babaahmadi-Rezaei, and Esrafil Mansouri. Luteolin confers renoprotection against ischemia-reperfusion injury via involving Nrf2 pathway and regulating miR320. Molecular Biology Reports 46.4 (2019): 4039-4047.

[400] Boeing, Thaise, Luisa Mota da Silva, Mariha Mariott, Sergio Faloni de Andrade, and Priscila de Souza. Diuretic and natriuretic effect of luteolin in normotensive and hypertensive rats: Role of muscarinic acetylcholine receptors. Pharmacological reports: PR 69.6 (2018): 1121-1124.

[401] Sternberg, Zohara, Kailash Chadha, Alicia Lieberman, Allison Drake, David Hojnacki, Bianca Weinstock-Guttman, and Frederick Munschauer. Immunomodulatory responses of peripheral blood mononuclear cells from multiple sclerosis patients upon in vitro incubation with the flavonoid luteolin:

Cardioprotective Agent Against High Fat Diet: In Wistar rats fed a high-fat diet, administration of luteolin caused a significant improvement in cardiac function and tissue integrity, a reduction in collagen deposition, fibrosis percentage, lipid peroxidation, and inflammatory cells, decreased lipid peroxidation, and elevated endogenous antioxidant biomarkers[402].

Treatment for Myocardial Diseases: Luteolin protects against lipopolysaccharide (LPS)-induced production of TNF-a expression in neonatal rat cardiac myocytes via the inhibition of the NF-KB signaling pathway[403], offering potential as a therapeutic agent in the treatment of inflammation-related heart diseases.

Weight Loss Agent: In a study that sought to determine the effects of six flavonoids on fat accumulation in Caenorhabditis elegans (a soil-dwelling nematode), luteolin exhibited the strongest inhibitory activity, working primarily by promoting the central serotonin pathway[404].

Treatment for Neurological and Psychological Diseases: Luteolin has been shown to inhibit GABA-mediated currents (inhibits neural firing) and slowed the activation kinetics of several recombinant receptors, while also significantly reducing the amplitude and slowing the rise time of miniature inhibitory postsynaptic currents in hippocampal pyramidal neurons[405] (multipolar neurons found in the brain) in mice.

Treatment for Arthritis: In induced arthritis in rats, luteolin reduced the severity of the disease by suppressing tumor necrosis factor-alpha, interleukin 1-beta, interleukin 6, and interleukin 17, while also ameliorating the invasion of inflammatory cells and synovial hyperplasia (an increase in cells), and suppressing several proteins related to joint destruction[406].

Treatment for Rheumatoid arthritis: As the primary active constituent in Siegesbeckia orientalis (a Chinese medicinal herb), luteolin exhibited anti-rheumatoid arthritis activities and potently inhibited TLR4 signaling[407] (toll-like receptor 4, a transmembrane protein involved in the production of ROS and cytokine production) both in vitro and in vivo.

Treatment for Osteonecrosis: In dexamethasone-induced osteonecrosis of the femoral head, treatment with luteolin reduced necroptosis in bone microvascular endothelial cells (BMECs) via the RIPK1 (receptor-interacting protein kinase 1)/RIPK3 (receptor-interacting protein kinase 3)/MLKL (mixed lineage kinase domain-like protein) pathway, also referred to as the necroptotic pathway, resulting in increased proliferation, migration, and angiogenesis ability, and decreased necroptosis of BMECs[408].

additive effects of IFN-B. Journal of Neuroinflammation 6 (2009): 28.

[402] Abu-Elsaad, Nashwa, and Amr El-Karef. The Falconoid Luteolin Mitigates the Myocardial Inflammatory Response Induced by High-Carbohydrate/High-Fat Diet in Wistar Rats. Inflammation 41.1 (2017): 221-231.

[403] Lv, Lihua, Linhua Lv, Yubi Zhang, and Qiuhuan Kong. Luteolin Prevents LPS-Induced TNF-a Expression in Cardiac Myocytes Through Inhibiting NF-KB Signaling Pathway. Inflammation 34.6 (2010): 620-629.

[404] Lin, Yan, Nan Yang, Bin Bao, Lu Wang, Juan Chen, and Jian Liu. Luteolin reduces fat storage in Caenorhabditis elegans by promoting the central serotonin pathway. Food & Function 11.1 (2020): 730-740.

[405] Shen, Mei-Lin, Chen-Hung Wang, Rita Yu-Tzu Chen, Ning Zhou, Shung-Te Kao, and Dong Chuan Wu. Luteolin inhibits GABAA receptors in HEK cells and brain slices. Scientific Reports 6.1 (2016): 1-11.

[406] Shi, Fengchao, Dun Zhou, Zhongqiu Ji, Zhaofeng Xu, and Huilin Yang. Anti-arthritic activity of luteolin in Freund's complete adjuvant-induced arthritis in rats by suppressing P2X4 pathway. Chemico-biological interactions 226 (2015): 82-87.

[407] Xiao B, Li J, Qiao Z, Yang S, Kwan HY, Jiang T, Zhang M, Xia Q, Liu Z, Su T. Therapeutic effects of Siegesbeckia orientalis L. and its active compound luteolin in rheumatoid arthritis: network pharmacology, molecular docking and experimental validation. J Ethnopharmacol. 2023 Dec 5;317:116852.

[408] Xu X, Fan X, Wu X, Xia R, Liang J, Gao F, Shu J, Yang M, Sun W. Luteolin ameliorates necroptosis in Glucocorticoid-induced osteonecrosis of the femoral head via RIPK1/RIPK3/MLKL pathway based on network pharmacology analysis. Biochem Biophys Res Commun. 2023 Jun 18;661:108-118.

Protection against Influenza: Luteolin has been found to suppress the replication of influenza A virus by interfering with viral replication and suppressing a protein related to virus invasion[409].

Treatment for Coronavirus: As isolated from Juncus acutus (spiny rush) stems, luteolin exhibited strong antiviral activity against the alphacoronavirus HCoV-229E[410]. Luteolin has also been shown to destroy the stable interaction of S-protein (transmembrane fusion protein) RBD (receptor binding domain)-ACE2 (human angiotensin-converting enzyme-2 receptor) and inhibit SARS-CoV-2 spike protein-induced platelet spread[411], inhibiting the binding of the spike protein to ACE2. Finally, co-treatment with palmitoylethanolamide (an endogenous fatty acid amide that acts as a lipid modulator) and luteolin combined with olfactory training in patients with post-COVID smell loss led to a 92% improvement rate in the treated group[412].

Neuroprotective Agent: In rats subjected to induced cobalt chloride intoxication, administration of luteolin increased exploratory activities, decreased anxiety, significantly increased hanging latency, prevented cobalt chloride-induced increases in hydrogen peroxide, malondialdehyde, and nitric oxide in the brain, restored the activities of acetylcholinesterase, glutathione S-transferase, and superoxide dismutase, and reduced levels of interleukin 1-beta and tumor necrosis factor-alpha[413], among other effects.

Treatment for Meth Addiction: Luteolin has been shown to alleviate methamphetamine-induced neurotoxicity in rats by attenuating apoptosis and autophagy via the suppression of the PI3K (phosphatidylinositol 3-kinase) / Akt (protein kinase B) pathway[414], which regulates cell survival and growth.

Prevention of Subretinal Fibrosis: In laser-induced choroidal neovascularization (a major cause of vision loss) and subretinal fibrosis (fibrosis underneath the retina) in C57BL/6J mice, treatment with luteolin inhibited epithelial-mesenchymal transition of retinal pigment epithelial cells by deactivating Smad2/3 (proteins that mediate signals related to cell proliferation, differentiation, and death) and YAP (yes-associated protein, which plays a role in cell growth, regeneration, and other cellular functions) signaling[415].

Treatment for Cirrhosis of the Liver: Researchers investigating Deduhonghua-7 powder (a traditional Chinese medicine) found that one of the primary active ingredients – luteolin – attenuated liver fibrosis and decreased the liver fibrosis index level in vivo in rat hepatic stellate cells[416].

[409] Yan, Haiyan, Linlin Ma, Huiqiang Wang, Shuo Wu, Hua Huang, Zhengyi Gu, Jiandong Jiang, and Yuhuan Li. Luteolin decreases the yield of influenza A virus in vitro by interfering with the coat protein I complex expression. Journal of Natural Medicines 73.3 (2019): 487-496.

[410] Hakem A, Desmarets L, Sahli R, Malek RB, Camuzet C, François N, Lefèvre G, Samaillie J, Moureu S, Sahpaz S, Belouzard S, Ksouri R, Séron K, Rivière C. Luteolin Isolated from Juncus acutus L., a Potential Remedy for Human Coronavirus 229E. Molecules. 2023 May 23;28(11):4263.

[411] Zhu J, Yan H, Shi M, Zhang M, Lu J, Wang J, Chen L, Wang Y, Li L, Miao L, Zhang H. Luteolin inhibits spike protein of severe acute respiratory syndrome coronavirus-2 (SARS-CoV-2) binding to angiotensin-converting enzyme 2. Phytother Res. 2023 Aug;37(8):3508-3521.

[412] Di Stadio A, D'Ascanio L, Vaira LA, Cantone E, De Luca P, Cingolani C, Motta G, De Riu G, Vitelli F, Spriano G, De Vincentiis M, Camaioni A, La Mantia I, Ferreli F, Brenner MJ. Ultramicronized Palmitoylethanolamide and Luteolin Supplement Combined with Olfactory Training to Treat Post-COVID-19 Olfactory Impairment: A Multi-Center Double-Blinded Randomized Placebo- Controlled Clinical Trial. Curr Neuropharmacol. 2022;20(10):2001-2012.

[413] Akinrinde, A S, and O E Adebiyi. Neuroprotection by luteolin and gallic acid against cobalt chloride-induced behavioural, morphological and neurochemical alterations in Wistar rats. Neurotoxicology 74 (2020): 252-263.

[414] Tan, Xiao-Hui, Kai-Kai Zhang, Jing-Tao Xu, Dong Qu, Li-Jian Chen, Jia-Hao Li, Qi Wang, Hui-Jun Wang, and Xiao-Li Xie. Luteolin alleviates methamphetamine-induced neurotoxicity by suppressing PI3K/Akt pathway- modulated apoptosis and autophagy in rats. Food and chemical toxicology : an international journal published for the British Industrial Biological Research Association 137 (2020): 111179-111179.

[415] Zhang C, Zhang Y, Hu X, Zhao Z, Chen Z, Wang X, Zhang Z, Jin H, Zhang J. Luteolin inhibits subretinal fibrosis and epithelial-mesenchymal transition in laser-induced mouse model via suppression of Smad2/3 and YAP signaling. Phytomedicine. 2023 Jul 25;116:154865.

[416] Batudeligen, Han Z, Chen H, Narisu, Xu Y, Anda, Han G. Luteolin Alleviates Liver Fibrosis in Rat Hepatic Stellate Cell HSC-T6: A Proteomic Analysis. Drug Des Devel Ther. 2023 Jun 17;17:1819-1829.

Treatment for Lung Injury: As a primary constituent in the essential oil of Perilla frutescens (beefsteak plant), luteolin improved the lung pathological structure and reduced the wet/dry weight ratio, bronchoalveolar protein, and inflammatory cytokines, and upregulated the expression level of the epithelial sodium channel (ENaC) in both the primary alveolar epithelial type 2 cells and three-dimensional alveolar epithelial organoid models in a lipopolysaccharide-induced acute lung injury model in mice[417].

Treatment for Corneal Alkali Burns: In rats with induced corneal alkali burn (in humans, this can be caused by exposure to drain cleaner, ammonia, cleaning solutions, etc.), intraperitoneal injection of luteolin ameliorated alkali burn-elicited corneal opacity, corneal epithelial defects, collagen degradation, neovascularization, and the infiltration of inflammatory cells, while also downregulating the mRNA expressions of several proteins, vascular endothelial growth factor, and MMPs in corneal tissue[418].

Treatment for Diabetic Dyslipidemia: In Wistar rats induced to type-2 diabetes mellitus via a high fat diet or streptozotocin, administration of luteolin significantly ameliorated dyslipidemia levels and improved atherogenic index of plasma, increased levels of malondialdehyde and diminished levels of superoxide dismutase, catalase, and glutathione, intensified PPARα expression (controls the expression of numerous genes involved in lipid metabolic pathways) while decreasing expression of acyl-coenzyme A (coenzymes that metabolize fatty acids), cholesterol acyltransferase-2 (proteins involved in cholesterol ester formation), and sterol regulatory element binding protein-2 proteins (regulates lipid homeostasis), and alleviated hepatic impairment in diabetic rats to near-normal[419].

Treatment for Neuropathic Pain: In a murine model of experimental neuropathy, treatment with luteolin reduced sensory deficits related to mechanical and thermal hypersensitivity, reduced oxidative stress, and inhibited cellular responses including reactive astrocytes[420].

Treatment for COPD: Researchers who developed a cigarette smoke-induced model of COPD (chronic obstructive pulmonary disease) in mice and in A549 (lung cancer) cells found that treatment with luteolin inhibited the inflammation factors level and oxidative stress, while also mediating parts of the NF-κB signaling pathway, alleviating the effects of cigarette smoke both in vivo and in vitro[421].

Treatment for Ischemia/Reperfusion: In ischemia/reperfusion (I/R)-induced ferroptosis (an iron-dependent type of cell death) in rat cardiomyocytes, co-treatment with baicalein (also a flavonoid) and luteolin decreased ROS and malondialdehyde generation and the protein levels of ferroptosis markers, restored glutathione peroxidase 4 protein levels, reduced the I/R-induced myocardium infarction, and decreased the levels of Acsl4 (an enzyme associated with ischemic stroke) and Ptgs2 (a stress response mediator) mRNA[422], thereby inhibiting ferroptosis.

[417] Chen L, Yu T, Zhai Y, Nie H, Li X, Ding Y. Luteolin Enhances Transepithelial Sodium Transport in the Lung Alveolar Model: Integrating Network Pharmacology and Mechanism Study. Int J Mol Sci. 2023 Jun 14;24(12):10122.

[418] Wang H, Guo Z, Liu P, Yang X, Li Y, Lin Y, Zhao X, Liu Y. Luteolin ameliorates cornea stromal collagen degradation and inflammatory damage in rats with corneal alkali burn. Exp Eye Res. 2023 Jun;231:109466.

[419] Shehnaz SI, Roy A, Vijayaraghavan R, Sivanesan S. Luteolin Mitigates Diabetic Dyslipidemia in Rats by Modulating ACAT-2, PPARα, SREBP-2 Proteins, and Oxidative Stress. Appl Biochem Biotechnol. 2023 Aug;195(8):4893-4914.

[420] Negah SS, Hajinejad M, Nemati S, Roudbary SMJM, Forouzanfar F. Stem cell therapy combined with luteolin alleviates experimental neuropathy. Metab Brain Dis. 2023 Aug;38(6):1895-1903.

[421] Li M, Wang H, Lu Y, Cai J. Luteolin suppresses inflammation and oxidative stress in chronic obstructive pulmonary disease through inhibition of the NOX4-mediated NF-κB signaling pathway. Immun Inflamm Dis. 2023 Apr;11(4):e820.

[422] Wang IC, Lin JH, Lee WS, Liu CH, Lin TY, Yang KT. Baicalein and luteolin inhibit ischemia/reperfusion-induced ferroptosis in rat cardiomyocytes. Int J Cardiol. 2023 Mar 15;375:74-86.

Treatment for Alzheimer's Disease: In induced cognitive impairment in mice, treatment with luteolin combined with exercise was shown to significantly improve the performance of Alzheimer's disease model mice in the novel object recognition test, reversing the increase of amyloid-beta[423], a hallmark of the disease.

Obese Sarcopenia Protective Agent: Luteolin has been shown to exhibit suppressive effects on obesity, inflammation, and protein degradation in high fat diet-fed obese mice, inhibiting lipid infiltration into the muscle and decreasing p38 (a protein kinase that plays a role in regulating cellular processes) activity and the mRNA expression of inflammatory factors in the muscle[424].

Treatment for Hypertension: In hypertensive rats, administration of luteolin improved elevated blood pressure and heart rate, reduced the circulating levels of NE and EPI, alleviated oxidative stress, reduced the levels of inflammatory cytokines, decreased the activity of NF-κB, decreased the production of GAD67 (glutamate decarboxylase, an enzyme), and lowered the levels of p-PI3K and p-Akt[425] (enzymes involved in cellular functions such as cell growth, proliferation, differentiation, motility, survival and intracellular trafficking), among other hypertension-ameliorating effects.

Treatment for Liver Injury: In quail induced to liver injury by administration of mercuric chloride ($HgCl_2$), treatment with luteolin significantly ameliorated oxidative stress, the release of inflammatory factors, and liver damage caused by $HgCl_2$, and reduced the accumulation of $Hg2+$ in quail liver, while also increasing levels of protein kinase-C-α, nuclear factor-erythroid-2-related factor 2 and its downstream proteins, and inhibiting NF-κB production[426].

Treatment for Ulcerative Colitis: Luteolin has been shown to prevent body weight loss, decrease disease activity index and intestinal damages, increase NCR+ILC3 (natural cytotoxicity receptor group 3 innate lymphoid cells) levels, promote the production of IL-22 while decreasing the levels of IL-17a and INF-γ in the intestines, and promoting intestinal barrier function recovery by promoting the expression of ZO-1 (a tight junction protein that regulates endothelial cell migration and angiogenic potential) and occludin[427] (an important protein in tight junction function), among other effects in mice induced to ulcerative colitis.

Reverse Damage Caused by Blue Light Irradiation: In fruit flies (Drosophila) irradiated by blue light, treatment with luteolin alleviated the damage suffered by the flies under blue light irradiation, significantly prolonged survival time, prolonged the survival time of male flies in the heat stress assay, increased the activity of female flies in the spontaneous activity assay, and increased the egg production of females at the highest concentration[428].

Treatment for Lupus Nephritis: Luteolin has been shown to ameliorate pathological abnormalities and improve renal function in mice by reducing renal oxidative stress and urinary protein levels, offering potential therapeutic

[423] Tao X, Zhang R, Wang L, Li X, Gong W. Luteolin and Exercise Combination Therapy Ameliorates Amyloid-β1-42 Oligomers-Induced Cognitive Impairment in AD Mice by Mediating Neuroinflammation and Autophagy. J Alzheimer's Dis. 2023;92(1):195-208.

[424] Kim JW, Shin SK, Kwon EY. Luteolin Protects Against Obese Sarcopenia in Mice with High-Fat Diet-Induced Obesity by Ameliorating Inflammation and Protein Degradation in Muscles. Mol Nutr Food Res. 2023 Mar;67(6):e2200729.

[425] Gao HL, Yu XJ, Feng YQ, Yang Y, Hu HB, Zhao YY, Zhang JH, Liu KL, Zhang Y, Fu LY, Li Y, Qi J, Qiao JA, Kang YM. Luteolin Attenuates Hypertension via Inhibiting NF-κB-Mediated Inflammation and PI3K/Akt Signaling Pathway in the Hypothalamic Paraventricular Nucleus. Nutrients. 2023 Jan 18;15(3):502.

[426] Liu Y, Guo X, Yu L, Huang Y, Guo C, Li S, Yang X, Zhang Z. Luteolin alleviates inorganic mercury-induced liver injury in quails by resisting oxidative stress and promoting mercury ion excretion. Mol Biol Rep. 2023 Jan;50(1):399-408.

[427] Xie X, Zhao M, Huang S, Li P, Chen P, Luo X, Wang Q, Pan Z, Li X, Chen J, Chen B, Zhou L. Luteolin alleviates ulcerative colitis by restoring the balance of NCR-ILC3/NCR+ILC3 to repairing impaired intestinal barrier. Int Immunopharmacol. 2022 Nov;112:109251.

[428] Zhong, L., Tang, H., Xu, Y. et al. Luteolin alleviated damage caused by blue light to Drosophila. Photochem Photobiol Sci 21, 2085–2094 (2022).

options for the prevention and treatment of lupus nephritis by suppressing HIF-1α (hypoxia-inducible factor-1, a regulator of oxygen homeostasis) expression in macrophages[429].

IMPLICATIONS FOR HUMAN HEALTH & NUTRITION

Because luteolin is found in many common natural foods, it is not difficult to include this phytonutrient as a dietary compound. However, it should be noted that some research has indicated that cooking vegetables or fruits might deactivate some of the therapeutic potential of luteolin, particularly in relation to human glioblastoma cells[430]; therefore, we recommend eating or juicing of these plants raw.

When cooking, preparing, or serving food, consider working with the spices and herbs anise, basil, cacao, chamomile, cumin, flax, lemongrass, parsley, peppermint, rosemary, sage, spearmint, tarragon, or thyme. Vegetables that contain luteolin include alfalfa, bell pepper, black beans, broccoli, carrots, celery, dandelion, fenugreek, lentils, lettuce, olives, and potatoes. Snacking on peanuts, or grapes, lemons, navel oranges, and pomegranate fruits may be another way to obtain this flavone.

Cannabis strains known to contain luteolin within the top ten flavonoids by concentration include those listed at the beginning of this section: Cheese, Fruit Punch, Huckleberry, Northern Lights, Purple Moose, Purple Punch, Skywalker, Space Candy, Strawberry Cough, Strawberry Banana, Super Lemon Haze, and White Cookies.

Experimentation or use of purified or isolated compounds like luteolin should be conducted with extreme care, as these compounds can be toxic in some situations.

---

[429] Ding T, Yi T, Li Y, Zhang W, Wang X, Liu J, Fan Y, Ji J, Xu L. Luteolin attenuates lupus nephritis by regulating macrophage oxidative stress via HIF-1α pathway. Eur J Pharmacol. 2023 Aug 15;953:175823.

[430] El Gueder, Dorra, Mouna Maatouk, Zahar Kalboussi, Zaineb Daouefi, Hind Chaaban, Irina Ioannou, Kamel Ghedira, Leila Ghedira, and Jose Luis. Heat processing effect of luteolin on anti-metastasis activity of human glioblastoma cells U87. Environmental Science and Pollution Research 25.36 (2018): 36545-36554.

Luteolin Review

Answer the following questions to test your knowledge of this flavonoid:

Question #1: What type of flavonoid is luteolin?

        a.   Flavanol
        b.   Flavorall
        c.   Flavonol
        d.   Flavone

Question #2: The base component in the biosynthesis of luteolin is:

        a.   Cannabisin
        b.   Rutinoside
        c.   Naringenin
        d.   Quercetin

Question #3: How common is luteolin in cannabis?

        a.   Top three
        b.   Top five
        c.   Top ten
        d.   Secondary

Question #4: What is the chemical formula for luteolin?

        a. $C_{24}H_{26}O$
        b. $C_{15}H_{10}O_6$
        c. $C_{15}H_{10}O_7$
        d. $C_{27}H_{30}O_{16}$

Question #5: What are the two primary known biological roles of luteolin in plants?

1 _____ 2 _____

Question #6: Name two potential medical uses of luteolin:

1 _____ 2 _____

For the answer key to Luteolin, please visit www.cannabischemistry.org

VITEXIN

Type: Flavone
Chemical Formula: $C_{21}H_{20}O_{10}$
Molecular Weight: 432.38 g/mol
Boiling Point: 767.70 °C @ 760.00 mm Hg (estimated)[431]
Flash Point: 273.10 °C (estimated by ACD Labs)[432]
Melting Point: 203 to 204 °C
Solubility: Soluble in water, dimethyl sulfoxide, dimethyl formamide
Oral TDLO: 1ml/kg (mouse)[433]
Biological Role: Allelopathic agent
Therapeutic Role: Anticancer agent, neuroprotective agent, anti-diabetes agent
Commercial Use: Research chemical
Occurrence in Cannabis: Top ten

Occurs in Cannabis Strains: Cheese, Critical Dream, Fruit Punch, Huckleberry, Strawberry Cough, Strawberry Banana, Super Lemon Haze

INTRODUCTION

Vitexin is a well-known flavone to cancer researchers, who have collectively suggested its use in the treatment of many types of tumors and carcinomas. Although it is also known as a potential treatment for neurological conditions like Alzheimer's disease, Vitexin is relatively unknown in the cannabis industry, despite the compound generally occurring in many cannabis strains within the top ten flavonoids by concentration. Some research found vitexin to be the most abundant flavonoid in three chemovars sampled[434], while other work has found the flavone to be more prevalent in CBD-predominant varieties than THC-predominant varieties[435].

[431]The Good Scents Company Data Sheet for Vitexin, from: https://www.thegoodscentscompany.com/data/rw1667761.html. Accessed November 22, 2023.

[432] ChemSpider Data Sheet for Vitexin, from: https://www.chemspider.com/Chemical-Structure.4444098.html. Accessed November 22, 2023.

[433] Cayman Chemical Safety Data Sheet for Vitexin, from: https://cdn.caymanchem.com/cdn/msds/15116m.pdf. Accessed November 25, 2023.

[434] Jin D, Dai K, Xie Z, Chen J. Secondary Metabolites Profiled in Cannabis Inflorescences, Leaves, Stem Barks, and Roots for Medicinal Purposes. Sci Rep. 2020 Feb 24;10(1):3309.

[435] Jin D, Henry P, Shan J, Chen J. Identification of Chemotypic Markers in Three Chemotype Categories of Cannabis Using Secondary Metabolites Profiled in Inflorescences, Leaves, Stem Bark, and Roots. Front Plant Sci. 2021 Jul 1;12:699530.

CHEMICAL STRUCTURE

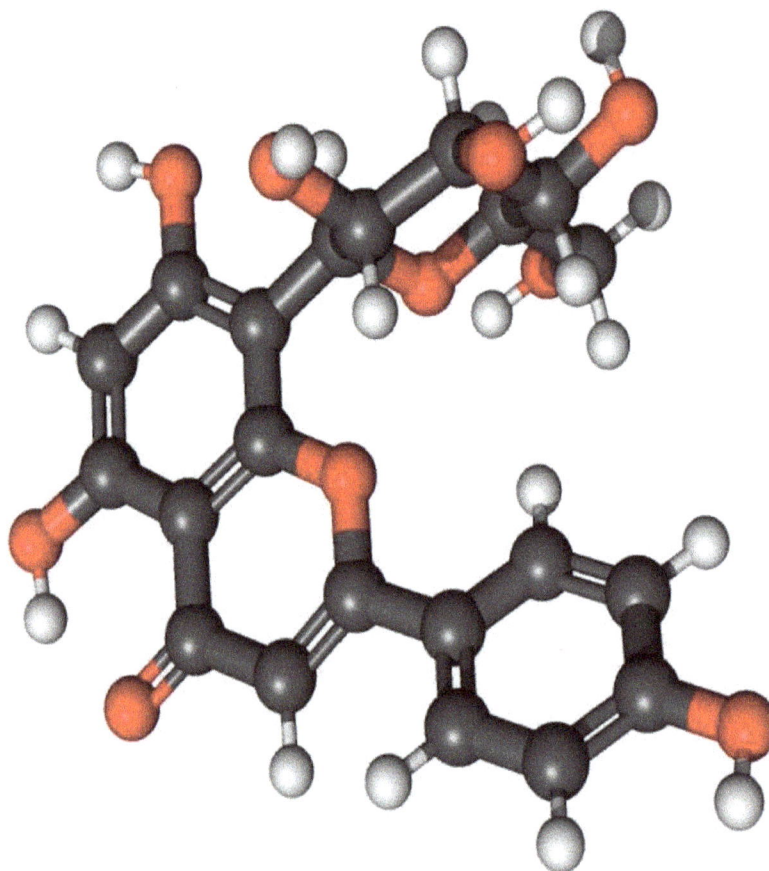

Based on the classic fifteen carbon flavonoid molecular skeleton, vitexin is a glycosylated compound derived from apigenin, consisting in total of twenty-one carbon atoms, twenty hydrogen atoms, and ten oxygen atoms scientifically notated as $C_{21}H_{20}O_{10}$. The molecular arrangement of this flavone includes three rings, with ring B attached at position 2 of ring C, and with seven endocyclic double-bonds throughout and one exocyclic oxygenated double-bond in ring C. Vitexin can occur as one of several variations, including:

Vitexin (vitexin 8-D-glucosyl-4',5,7-trihydroxyflavone): hydroxyl (-OH) groups at positions 4', 5, and 7, and a glucose molecule attached at position 8.

Vitexin-2-O-rhamnoside: a rhamnose molecule is attached to the flavone backbone at position 2.

Isovitexin (covered as an individual chapter in this book): two glucose molecules are attached - one each at positions 8 and 6.

In the ficus plant (and likely in other plants), biosynthesis of vitexin occurs along the typical pathway to apigenin, then concludes with the "transfer of sugar moiety by C-glycosyltransferase, followed by dehydration to produce flavone-8-C-glucosides[436]."

---

[436] Abdullah, F.I., Chua, L.S. & Rahmat, Z. Prediction of C-glycosylated apigenin (vitexin) biosynthesis in Ficus deltoidea based on plant proteins identified by LC-MS/MS. Front. Biol. 12, 448-458 (2017).

OCCURRENCE IN PLANTS

Although vitexin is not as widely found in plants as some of the flavonoids previously discussed in this book, it is known to occur in significant quantities in a multitude of common edible plants, vegetables, and roots, including:

Bamboo	Black gram	Buckwheat
Cacao	Cannabis	Chasteberry
Corn silk	Cucumber	Fenugreek
Flax	Golden queen	Hawthorn berry
Marjoram	Mung bean	Oregano
Passion flower	Pea	Pearl millet
Sorrel	Soybean	Tamarind
Tea	Turnip	

BIOLOGICAL ACTIVITY IN PLANTS

Although it is probable that vitexin offers some degree of UV radiation protection, several studies indicate that the primary biological role of this flavone might be as an allelopathic agent:

Allelopathic Agent: As isolated from an ethanolic extract of mung bean, vitexin was shown to significantly inhibit the activities of mushroom tyrosinase[437] (an enzyme produced by mushrooms and many other organisms). As a primary constituent in an extract made from oat, vitexin contributed to phytotoxic activity against several weed species, and one species of lettuce[438]. As isolated from several mosses, vitexin was shown to decrease the percentage of spore germination, protonemal development, and root growth, while also causing morphological changes in screw moss (tortula muralis) and radish (raphanus sativus)[439] plants.

USES IN INDUSTRY

Commercial preparations of vitexin are sold primarily for research purposes only. Purified vitexin appears as a crystalline solid or light-yellow powder and can be extracted from several distinct sources. Recently, scientists in China developed a method for the extraction of vitexin from Golden Queen (trollius chinensis) flowers using an ultrasonic circulating extraction technique[440] that is more efficient than other extraction methods.

POTENTIAL USES IN MEDICINE

Vitexin shows promise in the treatment of various cancers including lung, uterine, esophageal, and kidney carcinomas, as well as leukemia and gliomas. Because of its ability to inhibit or reduce amyloid beta toxicity - a hallmark of Alzheimer's disease - vitexin might one day have a functional role in the treatment of this and other neurological conditions. This flavone is also potentially useful in the treatment of a surprisingly varied list of diseases and conditions as detailed below.

---

[437] Yao, Yang, Xuzhen Cheng, Lixia Wang, Suhua Wang, and Guixing Ren. Mushroom tyrosinase inhibitors from mung bean (Vigna radiatae L.) extracts. International journal of food sciences and nutrition 63.3 (2012): 358-361.

[438] M De Leo, Claudia de Bertoldi, Alessandra Braca, Laura Ercoli Allelopathic potential of Avena sativa L. (oat) var. Argentina: bioassay-guided isolation of allelochemicals Planta Medica 74(09) July 2008.

[439] Basile, Adriana, Sergio Sorbo, Jose Antonio Lopez-Saez, and Rosa Castaldo Cobianchi. Effects of seven pure flavonoids from mosses on germination and growth of Tortula muralis HEDW (Bryophyta) and Raphanus sativus L. (Magnoliophyta). Phytochemistry 62.7 (2003): 1145-51.

[440] Chen, Fengli, Qiang Zhang, Junling Liu, Huiyan Gu, and Lei Yang. An efficient approach for the extraction of orientin and vitexin from Trollius chinensis flowers using ultrasonic circulating technique. Ultrasonics sonochemistry 37 (2018): 267-278.

Interestingly, it has been found that vitexin can inhibit or induce the activities of cytochrome P450 enzymes in rats[441], which has potentially broad-reaching implications for the concomitant administration of herbal supplements and pharmaceutical drugs in humans.

Treatment for Non-small Cell Lung Cancer: Vitexin has been shown to increase apoptosis, decrease the Bcl-2/Bax (protein that regulates apoptosis) ratio, and induce the release of cytochrome c (a heme protein that regulates apoptosis), while significantly reducing the levels of p-PI3K, p-Akt and p-mTOR (proliferation and metastasis signaling pathways), resulting in the reduction of viability of A549 non-small cell lung cancer cells[442].

Treatment for Leukemia: As a primary constituent in a methanolic extract made from the branches and leaves of luehea candicans (a tree native to Brazil), vitexin contributed to the inhibition of growth of leukemia and kidney cancer cells[443]. Other research has shown that vitexin can inhibit cell proliferation, induce apoptosis and morphological changes, and reduce the percentage of viable U937 human leukemia cells[444].

Treatment for Uterine Cancer: As isolated from Chinese chaste tree (vitex negundo) seeds, vitexin has been shown to inhibit cell proliferation, induce apoptosis, and inhibit mTOR signaling in a mouse model of human choriocarcinoma[445] (uterine cancer).

Treatment for Esophageal Cancer: As isolated from Golden Queen (trollius chinensis) flowers, vitexin was shown to markedly inhibit the proliferation, induce apoptosis, and regulate the p53 and bcl-2 gene expression levels in EC-109 esophageal cancer cells[446].

Treatment for Brain Cancer: When combined with hyperbaric oxygen therapy and radiotherapy, vitexin has been shown to synergistically lower the volume, weight, and weight coefficient of paw-transplanted glioma in nude mice[447]. Vitexin was also found to inhibit cell proliferation, colony formation, and invasion, and promote apoptosis by suppressing the JAK (Janus kinases)/STAT3 (signal transducer and activator of transcription proteins) signaling pathway in U251 (glioblastoma) cells[448].

Treatment for Melanoma: As extracted from Chinese chaste tree seeds, vitexin has been shown to significantly induce the accumulation of DNA-damaging ROS, arrest the cell cycle in G2/M phase (regulated chromosomes and DNA), induce apoptosis, and block melanoma cell growth in vitro and in vivo[449].

[441] Wang, Xin-shuai, Xiao-chen Hu, Gui-ling Chen, Xiang Yuan, Rui-na Yang, Shuo Liang, Jing Ren, Jia-chun Sun, Guo-qiang Kong, She-gan Gao, and Xiao-shan Feng. Effects of Vitexin on the Pharmacokinetics and mRNA Expression of CYP Isozymes in Rats. Phytotherapy Research 29.3 (2015): 366-372.

[442] Liu, Xiaoli, Qingfeng Jiang, Huaimin Liu, and Suxia Luo. Vitexin induces apoptosis through mitochondrial pathway and PI3K/Akt/mTOR signaling in human non-small cell lung cancer A549 cells. Biological Research 52 (2019).

[443] da Silva DA, Alves VG, Franco DM, Ribeiro LC, de Souza MC, Kato L, de Carvalho JE, Kohn LK, de Oliveira CM, da Silva CC. Antiproliferative activity of Luehea candicans Mart. et Zucc. (Tiliaceae). Nat Prod Res.

[444] Lee, Chao-Ying, Yung-Shin Chien, Tai-Hui Chiu, Wen-Wen Huang, Chi-Cheng Lu, Jo-Hua Chiang, and Jai- Sing Yang. Apoptosis triggered by vitexin in U937 human leukemia cells via a mitochondrial signaling pathway. Oncology Reports 28.5 (2012): 1883-1888.

[445] Zhihui Tan, Yi Zhang, Jun Deng, Guangyao Zeng, Yu Zhang Purified vitexin compound 1 suppresses tumor growth and induces cell apoptosis in a mouse model of human choriocarcinoma Int J Gynecol Cancer. 2012 Mar; 22(3):360-6. (2012): 364-9.

[446] An, Fang, Shuhua Wang, Qingqing Tian, and Dengxiang Zhu. Effects of orientin and vitexin from Trollius chinensis on the growth and apoptosis of esophageal cancer EC-109 cells. Oncology letters 10.4 (2020): 2627-2633.

[447] Xie, T., J.-R. Wang, C.-G. Dai, X.-A. Fu, J. Dong, and Q. Huang. Vitexin, an inhibitor of hypoxia-inducible factor-1a, enhances the radiotherapy sensitization of hyperbaric oxygen on glioma. Clinical and Translational Oncology 22.7 (2020): 1086-1093.

[448] Huang J, Zhou Y, Zhong X, Su F, Xu L. Effects of Vitexin, a Natural Flavonoid Glycoside, on the Proliferation, Invasion, and Apoptosis of Human U251 Glioblastoma Cells. Oxid Med Cell Longev. 2022 Mar 2;2022:3129155.

[449] Liu N, Wang KS, Qi M, Zhou YJ, Zeng GY, Tao J, Zhou JD, Zhang JL, Chen X, Peng C. Vitexin compound 1, a novel extraction from a Chinese herb, suppresses melanoma cell growth through DNA damage by increasing ROS levels. J Exp Clin Cancer Res. 2018 Nov 6;37(1):269.

Treatment for Endometrial Cancer: Vitexin has been shown to inhibit the proliferation, angiogenesis, and stemness capacity (refers to the ability of a cell to perpetuate its lineage and interact with its environment) of HEC-1B (endometrial cancer occurring in the lining of the uterus) cells via inactivation of the PI3K/AKT (major cell regulators) pathway[450].

Protective Agent Against Doxorubicin Cytotoxicity: Doxorubicin is a common cancer treatment compound that often causes significant cardiotoxicity in patients. Vitexin has been shown to inhibit lipid peroxidation, increase antioxidant enzyme activities, and elevate FOXO3a protein expression levels[451], significantly protecting rats against doxorubicin-induced myocardial damage.

Treatment for Preeclampsia: Recent work has shown that vitexin can decrease high systolic blood pressure and urinary protein, alleviate oxidative stress, and improve low birth weight and low pup/placenta ratio in L-NG-Nitro arginine methyl ester-induced pregnant Sprague-Dawley rats[452].

Accelerated Wound Healing Agent: When combined with chitosan-based gel, vitexin contributed to significant cell proliferation and skin regeneration in vitro and in vivo in male Sprague-Dawley rats[453], accelerating the healing process in constructed and induced wounds.

Treatment for Alcohol-Induced Liver Injury: In ethanol-induced liver damage in mice, vitexin was shown to significantly suppress the elevation of aminotransferase and blood lipid levels, restore Sirt1/Bcl-2 (apoptotic signaling proteins) expression levels, and inhibit the elevation of caspase-3 and cleaved caspase-3 protease enzymes and p53 protein expression levels[454].

Treatment for Anesthetic-Induced Impairment: Vitexin has been used to treat rats with isoflurane-induced neurotoxicity, reversing significant increases in escape latency periods and apoptosis of hippocampus neuron cells, while also decreasing the expression of inflammatory cytokines and ROS production, and increasing the expression of miR-409[455] (RNA gene).

Treatment for Liver Injury: In induced ulcerative colitis in mice, vitexin was shown to decrease liver levels of alanine aminotransferase and total cholesterol, reduce pro-inflammatory cytokines, and inhibit the activation of the TLR4/NF-KB inflammatory-signaling pathway[456].

Protective Agent Against Ischemia/Reperfusion Injury and Stroke: Vitexin has been shown to significantly reduce lactate dehydrogenase (an energy-production enzyme) and caspase-3 (regulates apoptosis) levels, alleviate inflammation, inhibit matrix metalloproteinases, preserve eNOS phosphorylation, and possibly maintain the integrity of the blood-brain-barrier, among other effects in a human brain microvascular endothelial cells

[450] Liang C, Jiang Y, Sun L. Vitexin suppresses the proliferation, angiogenesis and stemness of endometrial cancer through the PI3K/AKT pathway. Pharm Biol. 2023 Dec;61(1):581-589.

[451] Sun, Zhan, Bin Yan, Wen Yan Yu, Xueping Yao, Xiaojuan Ma, Geli Sheng, and Qi Ma. Vitexin attenuates acute doxorubicin cardiotoxicity in rats via the suppression of oxidative stress, inflammation and apoptosis and the activation of FOXO3a. Experimental and therapeutic medicine 12.3 (2020): 1879-1884.

[452] Zheng, Lili, Jing Huang, Yuan Su, Fang Wang, Hongfang Kong, and Hong Xin. Vitexin ameliorates preeclampsia phenotypes by inhibiting TFPI-2 and HIF-1a/VEGF in a l-NAME induced rat model. Drug Development Research 80.8 (2019): 1120-1127.

[453] Bektas N, Senel B, Yenilmez E, Ozatik O, Arslan R. Evaluation of wound healing effect of chitosan-based gel formulation containing vitexin. Saudi Pharm J. 2020 Jan;28(1):87-94.

[454] Yuan, Huiqi, Shuni Duan, Ting Guan, Xin Yuan, Jizong Lin, Shaozhen Hou, Xiaoping Lai, Song Huang, Xianhua Du, and Shuxian Chen. Vitexin protects against ethanol-induced liver injury through Sirt1/p53 signaling pathway. European journal of pharmacology 873 (2020): 173007-173007.

[455] Qi Y, Chen L, Shan S, Nie Y, Wang Y. Vitexin improves neuron apoptosis and memory impairment induced by isoflurane via regulation of miR-409 expression. Adv Clin Exp Med. 2020 Jan;29(1):135-145.

[456] Duan S, Du X, Chen S, Liang J, Huang S, Hou S, Gao J, Ding P. Effect of vitexin on alleviating liver inflammation in a dextran sulfate sodium (DSS)-induced colitis model. Biomed Pharmacother. 2020 Jan; 121:109683.

ischemia/reperfusion injury model[457]. In an oxygen-glucose deprivation and reoxygenation neuron cell and middle cerebral artery occlusion/reperfusion rat model, treatment with vitexin attenuated oxidative injury and ferroptosis, thereby improving cerebral ischemia-reperfusion injury[458].

UV Protective Agent: As the primary constituent in an ethyl acetate soluble fraction of an acer palmatum (Japanese maple) extract, vitexin was shown to inhibit superoxide radicals by nearly 70%, and showed potent free radical scavenging activity in UVB-irradiated human dermal fibroblasts[459] (connective tissue skin cells).

Analgesic Agent: Vitexin has been shown to reduce and inhibit pain-like behavior in mice by preventing the decrease of reduced glutathione levels, and, interestingly, inhibiting the production of hyperalgesic cytokines while simultaneously up regulating the levels of an anti-hyperalgesic cytokine[460], among other effects. Vitexin has also been shown to suppress pain-like behavior in mice subjected to hot plate, thermal infiltration, glacial acetic acid twisting, and formalin tests, likely by modulating TRPV4 (transient receptor potential cation channel subfamily V member 4) activity[461].

Antiadipogenesis Agent – Fat Blocker: As a primary constituent in an ethanolic extract of common duckweed (spirodela polyrhiza), vitexin contributed to potent antiadipogenesis (adipogenesis refers to the development and accumulation of fat) activity in 3T3-L1 mouse cells[462].

Treatment for Alzheimer's Disease: As isolated from the leaves of the Brazilian plant commonly called cinco folhas (serjania erecta radlk), vitexin protected PC12 rat adrenal medulla cells against amyloid-B25-35 peptide-induced toxicity[463]. This work is important because amyloid beta (AB) development is a primary neuropathological feature of Alzheimer's disease. Similar research has shown that vitexin can inhibit free radical production, suppress ROS mediated lipid peroxidation, inhibit Bax protein expression, and regulate the expression of several key neuroprotective genes in neuro-2a cells[464] (mouse neural crest cells), thereby inhibiting AB 25-35 induced toxicity.

Treatment for Diseases Related to Brain Aging: As isolated from fenugreek seeds, vitexin was shown to exhibit antiperoxidative activity, and to modulate hypercholesterolemia, hypertriglyceridemia, and hyperglycemia[465], imparting significant neuroprotective effects in a model of aluminum chloride-mediated toxicity.

[457] Cui, Yu-Huan, Xiao-Qing Zhang, Nai-Dong Wang, Mao-Dong Zheng, and Juan Yan. Vitexin protects against ischemia/reperfusion-induced brain endothelial permeability. European journal of pharmacology 853 (2019): 210-219.

[458] Guo L, Shi L. Vitexin Improves Cerebral ischemia-reperfusion Injury by Attenuating Oxidative Injury and Ferroptosis via Keap1/Nrf2/HO-1signaling. Neurochem Res. 2023 Mar;48(3):980-995.

[459] Kim, J.H., Lee, B.C., Kim, J.H. et al. The isolation and antioxidative effects of vitexin from Acer palmatum . Arch Pharm Res 28, 195 (2005).

[460] Borghi, Sergio M, Thacyana T Carvalho, Larissa Staurengo-Ferrari, Miriam S N Hohmann, Phileno Pinge-Filho, Rubia Casagrande, and Waldiceu A Verri. Vitexin inhibits inflammatory pain in mice by targeting TRPV1, oxidative stress, and cytokines. Journal of natural products 76.6 (2013): 1141-9.

[461] Qin Z, Xiang L, Zheng S, Zhao Y, Qin Y, Zhang L, Zhou L. Vitexin inhibits pain and itch behavior via modulating TRPV4 activity in mice. Biomed Pharmacother. 2023 Sep;165:115101.

[462] Kim, JinPyo, IkSoo Lee, JeongJu Seo, MunYhung Jung, YoungHee Kim, NamHui Yim, and KiHwan Bae. Vitexin, orientin and other flavonoids from Spirodela polyrhiza inhibit adipogenesis in 3T3-L1 cells. Phytotherapy Research 24.10 (2010): 1543-1548.

[463] Guimaraes, Camila Carla, Denise Dias Oliveira, Mayara Valdevite, Ana Lucia Fachin Saltoratto, Sarazete Izidia Vaz Pereira, Suzelei de Castro Fran;:a, Ana Maria Soares Pereira, and Paulo Sergio Pereira. The glycosylated flavonoids vitexin, isovitexin, and quercetrin isolated from Serjania erecta Radlk (Sapindaceae) leaves protect PC12 cells against amyloid-B25-35 peptide-induced toxicity. Food and chemical toxicology: an international journal published for the British Industrial Biological Research Association 86 (2016): 88-94.

[464] Dicson Sheeja Malar, Venkatesan Suryanarayanan, Mani Iyer Prasanth, Sanjeev Kumar Singh, Krishnaswamy Balamurugan, Kasi Pandima Devi. Vitexin inhibits AB 25-35 induced toxicity in Neuro-2a cells by augmenting Nrf- 2/HO-1 dependent antioxidant pathway and regulating lipid homeostasis by the activation of LXR-a Toxicol In Vitro. 2018 Aug;50:160-171.

[465] Belard-Nouira, Yosra, Hayfa Bakhta, Mohamed Bouaziz, Imen Flehi-Slim, Zohra Haouas, and Hassen Ben Cheikh. Study of lipid profile and parieto-temporal lipid peroxidation in AlCl3 mediated neurotoxicity. modulatory effect of fenugreek seeds. Lipids in Health and Disease 11.1 (2012): 1-8.

Treatment for Diarrhea Caused by Rotavirus: After screening 150 plant extracts, vitexin as isolated from rooibos (aspalathus linearis) was found to exhibit strong antiviral activity against rotavirus, with researchers suggesting that nutritional supplementation with this and other edible plants may be efficacious in the treatment of rotavirus-related diarrhea[466].

Treatment for Epilepsy & Seizures: In pentylenetetrazole-induced seizures in rats, vitexin was shown to minimize clonic seizures, while also significantly prolonging the onset time of seizures[467].

Treatment for Diabetes: As the primary constituent in a bioactive guided fractionation of mistletoe fig (ficus deltoidei), vitexin was shown to reduce the postprandial blood glucose level (plasma glucose concentrations after eating) in sucrose loaded normoglycemic mice[468] (mice with normal glucose levels being fed a high-sucrose diet). In streptozotocin (an antineoplastic agent that is toxic to the insulin-producing mammalian cells)-induced diabetic mice, vitexin increased reproductive organ weight and improved testicular pathological structure damage, significantly modulated several key hormones, and significantly improved sexual behavior and fertility levels[469]. Vitexin may also be useful in the treatment of diabetic nephropathy (DN), after it was shown to alleviate renal fibrosis, damage, and ferroptosis in rats with induced DN, most likely by activating glutathione peroxidase 4[470] (protects cells against membrane lipid peroxidation).

Antibacterial Agent: Vitexin has been shown to reduce Staphylococcus aureus surface hydrophobicity and membrane permeability, obstructing its ability to produce biofilm via modulation of cytokines including IL-10 and IL-12p40[471]. Vitexin was also shown to inhibit the production of ROS by promoting PPARγ (peroxisome proliferator-activated receptor gamma, a protein that plays a major regulatory role in energy homeostasis and metabolic function) activity, increasing the activity of antioxidant enzymes, and reducing inflammatory cytokines and apoptosis, possibly by alleviating endoplasmic reticulum stress and inactivation of MAPKs and the NF-κB signaling pathway in MAC-T (primary bovine mammary alveolar) cells and mouse mammary tissues infected with Staphylococcus aureus[472].

Treatment for Rheumatoid Arthritis: In a collagen-induced arthritis model in male Sprague Dawley rats, treatment with vitexin significantly reduced inflammatory enzyme markers interleukin (IL)-1β, IL-6, IL-17, IL-4, IL-10, tumor necrosis factor-α, interferon-γ, and iNOS levels, improved collagen-induced arthritic histological changes, reduced JAK/STAT expressions, while reverting aberration in apoptosis, inflammatory mediators, C-reactive protein, and rheumatoid factor levels to normal[473].

[466] Knipping, Karen, Johan Garssen, and Belinda van't Land. An evaluation of the inhibitory effects against rotavirus infection of edible plant extracts. Virology Journal 9.1 (2012): 1-8.

[467] Abbasi, Esmail, Marjan Nassiri-Asl, Mahsa Shafeei, and Mehdi Sheikhi. Neuroprotective Effects of Vitexin, a Flavonoid, on Pentylenetetrazole-Induced Seizure in Rats. Chemical Biology & Drug Design 80.2 (2012).

[468] Choo, C Y, N Y Sulong, F Man, and T W Wong. Vitexin and isovitexin from the Leaves of Ficus deltoidea with in-vivo a-glucosidase inhibition. Journal of ethnopharmacology 142.3 (2012): 776-781.

[469] Li, Zhi-Mei, Ning Liu, Ya-Ping Jiang, Jia-Mei Yang, Jie Zheng, Miao Sun, Yu-Xiang Li, Tao Sun, Jing Wu, and Jian-Qiang Yu. Vitexin alleviates streptozotocin-induced sexual dysfunction and fertility impairments in male mice via modulating the hypothalamus-pituitary-gonadal axis. Chemico-biological interactions 297 (2019): 119-129.

[470] Zhang S, Zhang S, Wang H, Chen Y. Vitexin ameliorated diabetic nephropathy via suppressing GPX4-mediated ferroptosis. Eur J Pharmacol. 2023 Jul 15;951:175787.

[471] Das MC, Samaddar S, Jawed JJ, Ghosh C, Acharjee S, Sandhu P, Das A, Daware AV, De UC, Majumdar S, Das Gupta SK, Akhter Y, Bhattacharjee S. Vitexin alters Staphylococcus aureus surface hydrophobicity to obstruct biofilm formation. Microbiol Res. 2022 Oct;263:127126.

[472] Chen Y, Yang J, Huang Z, Yin B, Umar T, Yang C, Zhang X, Jing H, Guo S, Guo M, Deng G, Qiu C. Vitexin Mitigates Staphylococcus aureus-Induced Mastitis via Regulation of ROS/ER Stress/NF-κB/MAPK Pathway. Oxid Med Cell Longev. 2022 Jun 27;2022:7977433.

[473] Zhang D, Ning T, Wang H. Vitexin alleviates inflammation and enhances apoptosis through the regulation of the JAK/STAT/SOCS signaling pathway in the arthritis rat model. J Biochem Mol Toxicol. 2022 Dec;36(12):e23201.

Treatment for Non-alcoholic Fatty Liver Disease: In high-fat-diet-induced NAFLD mice, daily administration of vitexin for four weeks markedly improved hepatic architecture, attenuated lipid accumulation, regulated lipid abnormalities, reduced endoplasmic reticulum stress, restored mitochondrial biological proteins, increased autophagy, and increased peroxisome proliferator-activated receptor-γ (PPAR-γ) protein, which was inhibited by the high fat diet[474].

Treatment for Ulcerative Colitis: In a dextran sodium sulfate-induced model of acute colitis in mice, treatment with vitexin altered the structure of gut microbiota by decreasing harmful bacteria and increasing beneficial bacteria, significantly improving colitis symptoms, maintaining the intestinal barrier, and down-regulating the expression of inflammatory factors, while also improving the diversity of gut microbiota[475]. In a similar mouse model of ulcerative colitis, treatment with vitexin markedly inhibited the production of pro-inflammation cytokines, markedly down-regulated the phosphorylation levels of p65 (protein involved in NF-κB heterodimer formation, nuclear translocation and activation), IκB (an enzyme complex that is involved in propagating the cellular response to inflammation), and STAT1 (signal transducer and activator of transcription 1, a protein coding gene), dose-dependently increased the expressions of muc-2 (a gene that protects the intestinal tract from self-digestion and microorganisms), ZO-1 (zonula occludens, a tight junction protein) and occludin proteins (important protein in tight junction stability and barrier function) in colonic tissues, and dramatically modulated the disturbed intestinal flora[476].

IMPLICATIONS FOR HUMAN HEALTH & NUTRITION

According to current research, vitexin is not found in many common foods. Consider eating foods that contain buckwheat, flax, or pearl millet, and use or procure foods with marjoram, oregano, and sorrel to obtain dietary vitexin. Common foods to include in regular meals that contain vitexin include bamboo, black gram, cucumber, fenugreek, mung bean, peas, soybean, tamarind, and turnip. Many teas - particularly those made with passion flower - also contain vitexin. Research has shown that mung bean soup exhibits significant antioxidant activity, with attribution of this activity going to vitexin and isovitexin[477].

Deliberate dietary consumption or supplementation with this flavone is unlikely but not impossible when consuming certain preparations of cannabis, especially the strains noted at the beginning of this chapter: Cheese, Critical Dream, Fruit Punch, Huckleberry, Strawberry Cough, Strawberry Banana, and Super Lemon Haze. In general, eating raw cannabis will be the most likely way, if any, to obtain vitexin from these strains.

[474] Jiang Y, Gong Q, Gong Y, Zhuo C, Huang J, Tang Q. Vitexin Attenuates Non-alcoholic Fatty Liver Disease Lipid Accumulation in High Fat-Diet Fed Mice by Activating Autophagy and Reducing Endoplasmic Reticulum Stress in Liver. Biol Pharm Bull. 2022 Mar 1;45(3):260-267.

[475] Li S, Luo L, Wang S, Sun Q, Zhang Y, Huang K, Guan X. Regulation of gut microbiota and alleviation of DSS-induced colitis by vitexin. Eur J Nutr. 2023 Dec;62(8):3433-3445.

[476] Zhang J, Liang F, Chen Z, Chen Y, Yuan J, Xiong Q, Hou S, Huang S, Liu C, Liang J. Vitexin Protects against Dextran Sodium Sulfate-Induced Colitis in Mice and Its Potential Mechanisms. J Agric Food Chem. 2022 Sep 28;70(38):12041-12054.

[477] Li, He, Dongdong Cao, Jianyong Yi, Jiankang Cao, and Weibo Jiang. Identification of the flavonoids in mungbean (Phaseolus radiatus L.) soup and their antioxidant activities. Food chemistry 135.4 (2013): 2942-6.

Vitexin Review

Answer the following questions to test your knowledge of this flavonoid:

Question #1: What type of flavonoid is vitexin?

    a. Flavane
    b. Flavone
    c. Flavonel
    d. Flavoned

Question #2: The base component in the biosynthesis of vitexin is:

    a. Cannabisin
    b. Rutinoside
    c. Apigenin
    d. Quercetin

Question #3: How common is vitexin in cannabis?

    a. Top three
    b. Top five
    c. Top ten
    d. Secondary

Question #4: What is the chemical formula for vitexin?

    a. $C_{21}H_{20}O_{10}$
    b. $C_{24}H_{10}O_{15}$
    c. $C_{15}H_{10}O_{78}$
    d. $C_{27}H_{30}O_{16}$

Question #5: What is the primary known biological role of vitexin in plants?

_____

Question #6: Name two potential medical uses of vitexin:

1 _____ 2 _____

For the answer key to Vitexin, please visit www.cannabischemistry.org

KAEMPFEROL

Type: Flavonol
Chemical Formula: $C_{15}H_{10}O_6$
Molecular Weight: 286.239 g/mol
Boiling Point: 582.00 to 583.00 °C @ 860.00 mm Hg (estimated)[478]
Flash Point: 226 °C
Melting Point: 276-278 °C[479]
Solubility: Soluble in alcohol, ethanol, ethers, acetone, acetic acid, alkalies, and DMSO, slightly soluble in water, insoluble in benzene.
Oral LD50: 192.84 mg/kg (mouse)[480]
Biological Role: Insecticidal agent, fungicidal agent, UVB radiation protection
Therapeutic Role: Anti-inflammatory, anticancer, cardioprotective agent, chemotherapeutic protective agent
Commercial Use: Research chemical
Occurrence in Cannabis: Top ten

Occurs in Cannabis Strains: Northern Lights, Purple Moose, White Cookies

INTRODUCTION

Kaempferol is another well-studied flavonoid, particularly in cancer research, where it has been shown not only to aid in the treatment of various carcinomas, but also to attenuate toxicity caused by widely administered chemotherapeutic agents. Kaempferol is found in an enormous variety of common edible plants, fruits, vegetables, herbs, and spices, where the compound can impart a bitter flavor.

Derived from the base compound for many flavonoids - naringenin - kaempferol likely occurs as approximately the number ten flavonoid by concentration in many cannabis chemovars, according to laboratory testing conducted by this author. Researchers first quantified kaempferol in a variety of cannabis strains in 1979[481], and recent work found kaempferol to be more prevalent in high-THC varieties than CBD strains[482].

It is interesting to note that, of the hundreds of studies reviewed for this chapter, there were many instances where kaempferol occurred in plants simultaneously with quercetin, particularly when both were the primary constituents. Further research to elucidate this connection is needed.

---

[478] The Good Scents Company Data Sheet for Kaempferol, from: https://www.thegoodscentscompany.com/data/rw1588991.html. Accessed November 23, 2023.

[479] PubChem Compound Summary for Kaempferol, from: https://pubchem.ncbi.nlm.nih.gov/compound/Kaempferol#section=Structures. Accessed November 23, 2023.

[480] Cayman Chemical Safety Data Sheet for Kaempferol, from: https://cdn.caymanchem.com/cdn/msds/11852m.pdf. Accessed December 4, 2023.

[481] M. N. CLARK, B. A. BOHM, Flavonoid variation in Cannabis L., Botanical Journal of the Linnean Society, Volume 79, Issue 3, October 1979, Pages 249–257.

[482] Jin D, Henry P, Shan J and Chen J (2021) Identification of Chemotypic Markers in Three Chemotype Categories of Cannabis Using Secondary Metabolites Profiled in Inflorescences, Leaves, Stem Bark, and Roots. Front. Plant Sci. 12:699530.

CHEMICAL STRUCTURE

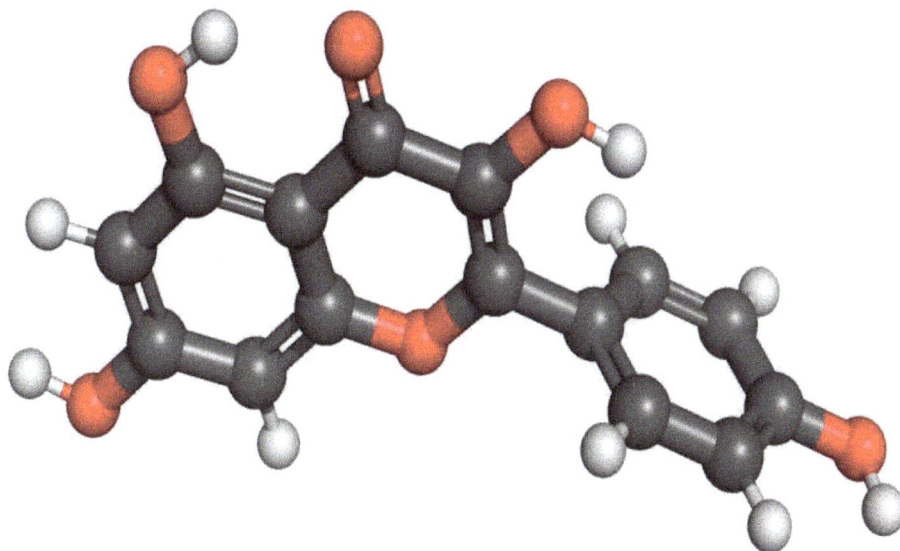

The biosynthesis of kaempferol occurs in the phenylpropanoid pathway, which produces naringenin, and then through hydroxylation and the addition of a double bond, becomes kaempferol, with hydroxy groups at positions 3, 5, 7, and 4', classifying this flavonoid as a flavonol. Structurally similar to the isoflavonoid genistein, kaempferol contains fifteen carbon atoms, ten hydrogen atoms, and six oxygen atoms scientifically notated as $C_{15}H_{10}O_6$. The carbon skeleton of this flavonol contains eight double bonds, seven of which are endocyclic, and one is an exocyclic oxygen atom in the C ring.

There are many variations of kaempferol, although currently it is not clear which variation(s) occur in cannabis – more testing is needed using specific standards for different variations of this molecule. Variants of kaempferol have distinct chemical structures due to different attachments or modifications at various positions on the kaempferol molecule. Each variation involves distinct sugar molecules, methyl groups, or other substituents, resulting in diverse chemical structures and potentially different properties or activities. In fact, there can be a virtually unlimited number of variations of this and similar compounds, including:

Kaempferol Rhamnoside: contains a rhamnose sugar moiety.

Kaempferol 7-O-Glucoside: a glucose molecule is attached at position 7.

Kaempferol-3-O-Glucuronide: contains a glucuronic acid moiety at position 3.

Kaempferol Trimethyl Ether: contains three methyl (CH_3) groups attached to the core structure of the molecule.

Kaempferol 7-Neohesperidoside: neohesperidose, a sugar molecule, is attached at position 7.

Kaempferol 3-Neohesperidoside: similar to 7-neohesperidoside, but attached at position 3.

Kaempferol 3-Sophoroside 7-Rhamnoside: contains sophorose (a disaccharide composed of two glucose units) attached at position 3, and rhamnose attached at position 7.

Kaempferol-3-O-Rutinoside: contains rutinose, a disaccharide composed of glucose and rhamnose, attached at position 3.

Kaempferol-7-Rhamnoside: has a rhamnose sugar attached at position 7.

Kaempferol 3-O-Beta-D-Glucoside: contains a beta-D-glucose molecule at position 3.

Kaempferol Oxoanion: likely refers to the anionic form of kaempferol resulting from deprotonation of one of its hydroxyl groups.

Kaempferol-3-O-Glucosyl(1-2)Rhamnoside: has a glucosylrhamnoside group attached at position 3.

Kaempferol 4'-Methyl Ether 3-(4Rha-Rhamnosylrutinoside): contains methyl ether at the position 4' and a rhamnosylrutinoside group attached at position 3.

Kaempferol 3-O-Glucosyl-Rhamnosyl-Glucoside: contains glucose and rhamnose moieties attached at differing positions.

Kaempferol 4'-Glucoside: contains a glucose molecule attached at position 4'.

Kaempferol 5-Methyl Ether: contains a methyl ether group attached at position 5.

Kaempferol-3-O-(Apiofuranosyl-(1'''-2''))-Galactopyranoside: contains an apiofuranosyl-galactopyranoside group attached at position 3.

Kaempferol 3-O-(2'',6''-Di-O-Rhamnopyranosyl): contains two rhamnose units attached at position 3.

Kaempferol 3-Isorhamninoside-7-Rhamnoside: contains isorhamninoside and rhamnose attached at positions 3 and 7, respectively.

Kaempferol 3-(2''-Acetylrhamnoside): contains an acetylrhamnoside group attached at position 3.

Kaempferol 3,7,4'-O-Triglucoside: contains three glucose molecules attached at positions 3, 7, and 4'.

Recently, researchers investigating phytochemicals as a potential treatment for SARS-CoV-2 discovered a rare novel variation of kaempferol in Calligonum tetrapterum (phog) named kaempferol 3-O-(6''-O-acetyl)-glucoside, which was found to have similar antiviral properties as remdesivir[483] (a broad-spectrum antiviral medication).

OCCURRENCE IN PLANTS

Kaempferol is found as a secondary metabolite in many common plants and foods:

Almond	Aloe vera	Alpinia nigra
Apple	Apricot	Ashitaba
Asparagus	Banana	Basil
Bay laurel	Black pepper	Borage
Broccoli	Brussels sprouts	Buckthorn
Butterbur	Cabbage	Cacao

[483] Suleimen YM, Jose RA, Mamytbekova GK, Suleimen RN, Ishmuratova MY, Dehaen W, Alsfouk BA, Elkaeed EB, Eissa IH, Metwaly AM. Isolation and In Silico Inhibitory Potential against SARS-CoV-2 RNA Polymerase of the Rare Kaempferol 3-O-(6''-O-acetyl)-Glucoside from Calligonum tetrapterum. Plants. 2022; 11(15):2072.

Cannabis	Caper	Chestnut
Chicory	Chive	Clove
Collar	Cotton	Cucumber
Dill	Endive	Fava bean
Ginkgo biloba	Ginger	Grape
Green beans	Guava	Hazelnut
Horseradish	Kale	Leek
Lettuce	Loquat	Lotus
Lovage	Mango	Marango
Mung bean	Nutmeg	Olive
Onion	Oregano	Parsley
Parsnip	Pea	Peach
Persimmon	Plum	Pomegranate
Potato	Primrose	Raspberry
Rosemary	Saffron	Soybean
Spinach	Squash	St. John's wort
Strawberry	Tarragon	Tea
Thyme	Tomato	Walnut

BIOLOGICAL ACTIVITY IN PLANTS

Kaempferol's biological roles in plants include acting as an insecticidal agent, as well as typical flavonoid-based UV radiation protective activity, among several other functions:

Insecticidal Agent: As a primary constituent in an alcoholic foliar extract made from apple of Sodom (calotropis procera), kaempferol contributed to excellent insecticidal activity against callosobruchus chinensis L[484] (a type of bean weevil). As a primary constituent in an aqueous leaf extract of castor bean plant (ricinus communis), kaempferol again showed strong insecticidal activity against the bean weevil, including ovicidal and oviposition deterrent activities, reducing the number of viable emerging adults to almost zero[485].

Allelopathic Agent: Kaempferol has been shown to inhibit the growth of the algal bloom microcystis aeruginosa up to 95% by causing enhanced cell oxidative damage and inhibited photosynthesis[486].

Fungicidal Agent: Kaempferol has been shown to decrease the metabolic activity and biofilm biomass of the candida parapsilosis complex[487] (a type of yeast that can cause sepsis and infection).

UVB Radiation Protection: Researchers have demonstrated that several lines of Mitchell petunia - one wild-type and two with transgenic lines - exhibit increased levels of kaempferol in response to UVB radiation[488].

---

[484] Mendki, Prashant & Salunke, Bipinchandra & Kotkar, Hemlata & Maheshwari, Vijay & Mahulikar, Pramod & Kothari, Raman. (2005). Antimicrobial and Insecticidal Activities of Flavonoid Glycosides from Calotropis procera L. for Post-harvest Preservation of Pulses. 1. 193-200.

[485] Upasani, Shripad M, Hemlata M Kotkar, Prashant S Mendki, and Vijay L Maheshwari. Partial characterization and insecticidal properties of Ricinus communis L foliage flavonoids. Pest Management Science 59.12 (2003): 1349-1354.

[486] Linrong Cao, Jieming Li Plant-Originated Kaempferol and Luteolin as Allelopathic Algaecides Inhibit Aquatic Microcystis Growth Through Affecting Cell Damage, Photosynthetic and Antioxidant Responses January 2018 Journal of Bioremediation & Biodegradation 09(02).

[487] Marcos Fabio Gadelha Rocha, Jamille Alencar Sales, Maria Gleiciane da Rocha, Livia Maria Galdino, Lara de Aguiar, Waldemiro de Aquino Pereira-Neto, Rossana de Aguiar Cordeiro, Debora de Souza Collares Maia Castelo- Branco, Jose Julio Costa Sidrim & Raimunda Samia Nogueira Brilhante (2019) Antifungal effects of the flavonoids kaempferol and quercetin: a possible alternative for the control of fungal biofilms, Biofouling, 35:3, 320-328.

[488] Ryan, Ken G., Kenneth R. Markham, Stephen J. Bloor, J. Marie Bradley, Kevin A. Mitchell, and Brian R. Jordan. UVB Radiation Induced Increase in Quercetin: Kaempferol Ratio in Wild-Type and Transgenic Lines of Petunia. Photochemistry & Photobiology 68.3 (1998): 323-330.

Antiviral Agent: In thale cress (Arabidopsis thaliana), kaempferol has been proposed to mediate or regulate the auxin-dependent defense response, which limited systemic movement of particles of an attacking cucumber mosaic virus[489].

USES IN INDUSTRY

Kaempferol is used in a variety of commercial and industrial applications, including as an ingredient in cosmetics, dyes, emollients, and conditioners, where its primary purpose is often to lend antioxidant properties to product formulations. Commercial samples of kaempferol appear as yellow crystalline solids or yellow needles; they are primarily derived from the rhizome of Kaempferia galanga, an herb used in traditional Chinese medicine. Significant work has been undertaken to produce kaempferol using microorganisms. Some researchers have produced this flavonol by reconstructing the flavonoid biosynthetic pathway in Escherichia coli[490], while other scientists have produced the compound via the biosynthesis of naringenin[491] in the same microorganism. Kaempferol has also been produced in a microbial cell factory consisting of the budding yeast saccharomyces cerevisiae[492]. Recent research has also shown that kaempferol could be useful in the eradication of Microcystis-dominated cyanobacterial blooms, after it was shown to exhibit toxic effects against the bacteria when combined with luteolin[493].

POTENTIAL USES IN MEDICINE

There are many potential therapeutic and medicinal uses of kaempferol, including use in treating cancer, cardiovascular disease, diabetes, arthritis, allergic disorders, depression, and obesity, among many others:

Treatment for Colon Cancer: When administered together with sulindac (an anti-inflammatory drug), kaempferol has been shown to decrease thiobarbituric acid reactive substances level, tissue nitric oxide, serum, and tissue B-catenin, while also down regulating the proliferation of cell nuclear antigen (a protein that plays a role in nucleic acid metabolism) and cyclooxygenase-2 (an enzyme involved in tumor invasiveness and angiogenesis) in 1,2-dimethyl hydrazine-induced preneoplastic lesions in rats[494] - a model of colon cancer. As a primary constituent in an extract made from the blueberry plant, kaempferol reduced total oxidant levels and oxidative stress index levels, while also causing a two-fold reduction in several apoptotic proteins in HCT-116 human colon cancer cells[495].

Treatment for Esophageal Cancer: Kaempferol has been shown to induce G0/G01 phase arrest (a state where cells do not divide or grow) and dramatically change the expression of a cell cycle regulation protein, while also

[489] Likic, Sasa, Ivana Sola, Jutta Ludwig-Milller, and Gordana Rusak. Involvement of kaempferol in the defence response of virus infected Arabidopsis thaliana. European Journal of Plant Pathology 138.2 (2013): 257-271.

[490] Yang, So-Mi, So Han, Bong-Gyu Kim, and Joong-Hoon Ahn. Production of kaempferol 3-O-rhamnoside from glucose using engineered Escherichia coli. Journal of Industrial Microbiology Biotechnology 41.8 (2014): 1311-1318.

[491] Pei, J., Chen, A., Dong, P. et al. Modulating heterologous pathways and optimizing fermentation conditions for biosynthesis of kaempferol and astragalin from naringenin in Escherichia coli. J Ind Microbiol Biotechnol 46, 171-186 (2019).

[492] Duan, Lijin, Wentao Ding, Xiaonan Liu, Xiaozhi Cheng, Jing Cai, Erbing Hua, and Huifeng Jiang. Biosynthesis and engineering of kaempferol in Saccharomyces cerevisiae. Microbial Cell Factories 16 (2017).

[493] Li J, Cao L, Guo Z. Joint effects and mechanisms of luteolin and kaempferol on toxigenic Microcystis growth-Comprehensive analysis on two isomers interaction in binary mixture. J Environ Manage. 2022 Jun 15;312:114904.

[494] Hassanein NMA, Hassan ESG, Hegab AM, Elahl HMS. Chemopreventive effect of sulindac in combination with epigallocatechin gallate or kaempferol against 1,2-dimethyl hydrazine-induced preneoplastic lesions in rats: A Comparative Study. J Biochem Mol Toxicol. 2018 Oct;32(10):e22198. Epub 2018 Jul 12. Erratum in: J Biochem Mol Toxicol. 2019 Feb; 33(2):e22284.

[495] Sezer, Ebru Demirel, Latife Merve Oktay, Elif Karadada, Hikmet Memmedov, Nur Selvi Gunel, and Eser Sozmen. Assessing Anticancer Potential of Blueberry Flavonoids, Quercetin, Kaempferol, and Gentisic Acid, Through Oxidative Stress and Apoptosis Parameters on HCT-116 Cells. Journal of Medicinal Food 22.11 (2019): 1118-1126.

significantly inhibiting tumor glycolysis, and markedly suppressing epidermal growth factor receptor (a protein involved in cell division and survival) downstream signaling pathways in esophagus squamous cell carcinoma[496].

Treatment for Radiation-Induced Skin Cancer: Kaempferol has been shown to inhibit the activity of ribosomal S6 kinase (involved in signal transduction) and mitogen and stress activated protein kinase, which potently inhibited the carcinogenesis of solar UV radiation-induced skin cancer in mice[497].

Treatment for Gastric Cancer: In gastric cancer cells, researchers have demonstrated that kaempferol promotes autophagic cell death and mediates epigenetic change via the inhibition of G9a[498] (a methyltransferase regulator of gene expression).

Treatment for Estrogen Imbalances: Kaempferol has been shown to induce a strong antiproliferative effect in MCF-7 breast cancer cells, blocking the focus formation induced by estradiol, potentially inhibiting the malignant transformation caused by estrogens[499].

Treatment for Gliomas: Kaempferol causes human glioma cell death by increasing reactive oxygen species, depolarizing mitochondrial membrane potential, down regulating anti-apoptotic proteins, reducing phosphorylation of extracellular signal- regulated kinase and AKT, while inducing activation of caspase-3[500], among other effects in brain cancer cells.

Treatment for Pancreatic Cancer: As the primary constituent in a methanolic fraction of fruit extract of Trema orientalis (charcoal tree), kaempferol effectively stopped pancreatic cancer cell migration by inhibiting ERK1/2 (extracellular signal-regulated kinase), EGFR (estimated glomerular filtration rate)-related SRC (protein tyrosine kinases), and AKT pathways (involved with inhibiting cell apoptosis and stimulating cell proliferation) by scavenging ROS[501].

Treatment for Endometrial cancer: In HSD17B1 endometrial cancer cells, kaempferol was found to suppress proliferation, promote apoptosis, and limit the tumor-forming, scratch healing, invasion, and migration capacities of the cells, inhibiting tumor growth and promoting apoptosis in a human endometrial cancer xenograft mouse model, with no toxicity to normal cells[502].

Treatment for Liver Cancer: In Huh-7 and SK-Hep-1 liver cancer cell lines, kaempferol reduced the invasion and migration of the cells, dramatically downregulated matrix metalloproteinase-9, and sufficiently suppressed the phosphorylation of Akt expression[503].

[496] Yao, Shihua, Xiaowei Wang, Chunguang Li, Tiejun Zhao, Hai Jin, and Wentao Fang. Kaempferol inhibits cell proliferation and glycolysis in esophagus squamous cell carcinoma via targeting EGFR signaling pathway. Tumor Biology 37.8 (2016): 10247-10256.

[497] Yao, Ke et al. Kaempferol targets RSK2 and MSK1 to suppress UV radiation-induced skin cancer. Cancer prevention research (Philadelphia, Pa.) vol. 7,9 (2014): 958-967.

[498] Kim, Tae, Seon Lee, Mia Kim, Chunhoo Cheon, and Seong-Gyu Ko. Kaempferol induces autophagic cell death via IRE1-JNK-CHOP pathway and inhibition of G9a in gastric cancer cells. Cell Death & Disease 9.9 (2018): 1-14.

[499] Oh, Seung, Yeon Kim, and Kyu Chung. Biphasic effects of kaempferol on the estrogenicity in human breast cancer cells. Archives of Pharmacal Research 29.5 (2006): 354-362.

[500] Jeong, Ji, Min Kim, Thae Kim, and Yong Kim. Kaempferol Induces Cell Death Through ERK and Akt- Dependent Down-Regulation of XIAP and Survivin in Human Glioma Cells. Neurochemical Research 34.5 (2008): 991-1001.

[501] Agrawal S, Das R, Singh AK, Kumar P, Shukla PK, Bhattacharya I, Tripathi AK, Mishra SK, Tiwari KN. Network pharmacology-based anti-pancreatic cancer potential of kaempferol and catechin of Trema orientalis L. through computational approach. Med Oncol. 2023 Apr 3;40(5):133.

[502] Ruan GY, Ye LX, Lin JS, Lin HY, Yu LR, Wang CY, Mao XD, Zhang SH, Sun PM. An integrated approach of network pharmacology, molecular docking, and experimental verification uncovers kaempferol as the effective modulator of HSD17B1 for treatment of endometrial cancer. J Transl Med. 2023 Mar 17;21(1):204.

[503] Ju PC, Ho YC, Chen PN, Lee HL, Lai SY, Yang SF, Yeh CB. Kaempferol inhibits the cell migration of human hepatocellular carcinoma cells by suppressing MMP-9 and Akt signaling. Environ Toxicol. 2021 Oct;36(10):1981-1989.

Treatment for Chemotherapeutic Agent Toxicity: Kaempferol has been shown to ameliorate toxicity caused by the administration of common cancer treatment drugs. This includes protective effects in the vascular endothelium against doxorubicin-induced damage by regulating several key pathways and by inhibiting oxidative stress and apoptosis, improving mitochondrial function[504]. Kaempferol has also been shown to modulate oxidative stress, inflammation, and apoptosis in kidney tissues in a mouse model of cisplatin mediated nephrotoxicity[505]. Kaempferol-induced expression of heme oxygenase-1 also inhibits cisplatin-induced apoptosis in HEI-OC1 cells[506] (mouse cochlear hair cells), which has significant implications for cancer patients who suffer from cisplatin-induced hearing loss. In male BALB/c mice with induced doxorubicin toxicity, treatment with kaempferol reduced renal tubular injury, protected renal tubular epithelial cells, and increased tumor susceptibility to doxorubicin[507]. Finally, in AKI mice, treatment with kaempferol either before or after cisplatin injection markedly improved renal function and ameliorated renal tissue damage[508].

Treatment for Cardiovascular Disease: Kaempferol has been shown to decrease osteopontin (a pro-inflammatory cytokine) and nuclear factor-KB expressions while also decreasing reactive oxygen species generation in human umbilical vein endothelial cells treated with aldosterone[509] - a steroid hormone that affects blood pressure. Other research has demonstrated that kaempferol significantly suppressed the release of tumor necrosis factor alpha, interleukin-1B, interleukin-6, and interleukin-18 while also inhibiting activation of nuclear factor-KB and protein kinase B in lipopolysaccharide plus ATP-induced cardiac fibroblasts[510] (heart tissue). In porcine coronary artery rings, kaempferol has been shown to enhance the action of both endothelium-dependent and endothelium-independent relaxing agents at low concentrations, and to relax the arteries at high concentrations[511]. Kaempferol has also been shown to attenuate lipid profile, infarcted area, and oxidative stress in soprenaline-induced myocardial injury in rats[512], with researchers theorizing that the flavonol likely improves oxygen demand in the infarcted heart tissue. Kaempferol has been shown to inhibit vascular smooth muscle cell proliferation and migration[513] - important because these cells are present during the development of atherosclerotic plaque. Finally, in isoproterenol (ISO)-induced cardiotoxicity in rats, treatment with kaempferol reduced lipid peroxidation, vascular O2, and NoX generation, enhanced the activities of antioxidant enzymes, relieving the rats from hemodynamic depression and further protecting them from free radical–mediated cardiac injury[514].

[504] Wu, Weiqi, Bin Yang, Yang Qiao, Qing Zhou, Huan He, and Ming He. Kaempferol protects mitochondria and alleviates damages against endotheliotoxicity induced by doxorubicin. Biomedicine & pharmacotherapy 126 (2020): 110040-110040.

[505] Wang, Zhu, Wansen Sun, Xi Sun, Ye Wang, and Meilan Zhou. Kaempferol ameliorates Cisplatin induced nephrotoxicity by modulating oxidative stress, inflammation and apoptosis via ERK and NF-KB pathways. AMB Express 10.1 (2020).

[506] Gao, Shang, Byung-Min Choi, Xiao Chen, Ri Zhu, Youngho Kim, HongSeob So, Raekil Park, Meesook Sung, and Bok-Ryang Kim. Kaempferol Suppresses Cisplatin-Induced Apoptosis Via Inductions of Heme Oxygenase-1 and Glutamate-Cysteine Ligase Catalytic Subunit in HEI-OC1 cells. Pharmaceutical Research 27.2 (2009): 235-245.

[507] Wu Q, Chen J, Zheng X, Song J, Yin L, Guo H, Chen Q, Liu Y, Ma Q, Zhang H, Yang Q. Kaempferol attenuates doxorubicin-induced renal tubular injury by inhibiting ROS/ASK1-mediated activation of the MAPK signaling pathway. Biomed Pharmacother. 2023 Jan;157:114087.

[508] Shao, Yf., Tang, Bb., Ding, Yh. et al. Kaempferide ameliorates cisplatin-induced nephrotoxicity via inhibiting oxidative stress and inducing autophagy. Acta Pharmacol Sin 44, 1442–1454 (2023).

[509] Xiao, Hong-Bo, Xiang-Yang Lu, Zi-Kui Liu, and Zhi-Feng Luo. Kaempferol inhibits the production of ROS to modulate OPN-avB3 integrin pathway in HUVECs. Journal of Physiology and Biochemistry 72.2 (2016): 303-313.

[510] Tang, Xi-lan, Jian-xun Liu, Wei Dong, Peng Li, Lei Li, Jin-cai Hou, Yong-qiu Zheng, Cheng-ren Lin, and Jun- guo Ren. Protective Effect of Kaempferol on LPS plus ATP-Induced Inflammatory Response in Cardiac Fibroblasts. Inflammation 38.1 (2014): 94-101.

[511] Xu, Y., D. Yeung, R. Man, and S. Leung. Kaempferol enhances endothelium-independent and dependent relaxation in the porcine coronary artery. Molecular and Cellular Biochemistry 287.2 (2006): 61-67.

[512] Vishwakarma, Anamika, Thakur Singh, Soya Rungsung, Tarun Kumar, Arunvikram Kandasamy, Subhashree Parida, Madhu Lingaraju, Ajay Kumar, Asok Kumar, and Dinesh Kumar. Effect of Kaempferol Pretreatment on Myocardial Injury in Rats. Cardiovascular Toxicology 18.4 (2018): 312-328.

[513] Kim, K., Kim, S., Moh, S.H. et al. Kaempferol inhibits vascular smooth muscle cell migration by modulating BMP-mediated miR-21 expression. Mol Cell Biochem 407, 143-149 (2015).

[514] Krishna, P.S., N, R.K., Swathi et al. Amaranthus viridis methanolic extract and its active compound kaempferol ameliorate myocardial infarction induced by

Treatment Aid for Obesity: Kaempferol has been shown to significantly decrease lipid accumulation in adipocytes (fat storage cell) while also increasing the transcriptional gene profile involved in lipid metabolism, inhibiting adipogenesis (the formation of adipocytes) by up to 62%[515].

Treatment for Diabetes: As an isolated compound from cuscuta pedicellata (a parasitic plant), kaempferol was shown to reduce homeostasis model assessment-insulin resistance and thiobarbituric acid reactive substances levels in rats with high fat diet-induced obesity, leading to a reduction in insulin resistance and oxidative stress, while increasing energy expenditure[516]. As the primary flavonoid constituent in an ethanolic extract made from Mongolian oak cups, kaempferol significantly decreased fasting blood glucose level, indices of heart and liver, and levels of cholesterol and triglyceride in serum and malondialdehyde in heart tissue of alloxan-induced diabetic rats, while also significantly improving HDL levels and superoxide dismutase activity[517].

Treatment for Arthritis: Oral administration of kaempferol has been shown to produce strong anti-arthritis activity, rebalancing the intestinal microbiota, significantly reversing the agitation of metabolites in gut content, and decreasing inflammatory cytokine levels in collagen-induced arthritic mice[518]. Kaempferol has also been shown to prevent osteoclast formation and inhibit bone loss in bone marrow cells harvested from mice[519]. Finally, kaempferol has been shown to exert therapeutic effects in gouty arthritis, partly by regulating the IL-17, AGE-RAGE, p53, TNF, and FoxO signaling pathways[520], among other effects in monosodium urate-induced rats and IL-6-induced peripheral blood mononuclear cells.

Treatment for Depression: In a murine model of chronic social defeat stress, administration of kaempferol decreased oxidative stress markers and inflammatory mediators, and increased activity in the prefrontal cortex, with researchers suggesting the administration of this flavonol in food to treat depression[521].

Treatment for Ulcerative Colitis: Kaempferol was shown to significantly decrease nitric oxide and prostaglandin E2 (a hormone that plays a role in inflammation) levels and profoundly decrease levels of leukotriene B4 (also plays a role in inflammation), suppress colonic mucosa myeloperoxidase, and up-regulate markers for goblet cell function in dextran sulfate sodium-induced colitis in mice[522].

isoproterenol through decreasing oxidative stress and cell death via Nrf-2/HO-1 and MMP/Bax/Bcl-2/TLR-4 pathways in rats. Comp Clin Pathol 32, 661–670 (2023).

[515] Torres-Villarreal, D., A. Camacho, H. Castro, R. Ortiz-Lopez, and A. de la Garza. Anti-obesity effects of kaempferol by inhibiting adipogenesis and increasing lipolysis in 3T3-L1 cells. Journal of Physiology and Biochemistry 75.1 (2018): 83-88.

[516] Mehanna, Eman T, Norhan M El-Sayed, Amany K Ibrahim, Safwat A Ahmed, and Dina M Abo-Elmatty. Isolated compounds from Cuscuta pedicellata ameliorate oxidative stress and upregulate expression of some energy regulatory genes in high fat diet induced obesity in rats. Biomedicine & pharmacotherapy 108 (2019): 1253-1258.

[517] Yin, Peipei, Yu Wang, Lingguang Yang, Jinling Sui, and Yujun Liu. "Hypoglycemic Effects in Alloxan-Induced Diabetic Rats of the Phenolic Extract from Mongolian Oak Cups Enriched in Ellagic Acid, Kaempferol and Their Derivatives." Molecules: A Journal of Synthetic Chemistry and Natural Product Chemistry 23.5 (2018).

[518] Aa, Li-xiang, Fei Fei, Qi Qi, Run-bin Sun, Sheng-hua Gu, Zi-zhen Di, Ji-ye Aa, Guang-ji Wang, and Chang-xiao Liu. "Rebalancing of the gut flora and microbial metabolism is responsible for the anti-arthritis effect of kaempferol." Acta Pharmacologica Sinica 41.1 (2019): 73-81.

[519] Lee, W., Lee, E., Sung, M. et al. Kaempferol Inhibits IL-1B-Stimulated, RANKL-mediated Osteoclastogenesis via Downregulation of MAPKs, c-Fos, and NFATc1. Inflammation 37, 1221-1230 (2014).

[520] Li N, Chen S, Deng W, Gong Z, Guo Y, Zeng S, Xu Q. Kaempferol Attenuates Gouty Arthritis by Regulating the Balance of Th17/Treg Cells and Secretion of IL-17. Inflammation. 2023 Oct;46(5):1901-1916.

[521] Gao, Wenqi, Wei Wang, Yan Peng, and Zhifang Deng. "Antidepressive effects of kaempferol mediated by reduction of oxidative stress, proinflammatory cytokines and up-regulation of AKT/catenin cascade." Metabolic Brain Disease 34.2 (2019): 485-494.

[522] Park, Mi-Young, Geun Ji, and Mi-Kyung Sung. "Dietary Kaempferol Suppresses Inflammation of Dextran Sulfate Sodium-Induced Colitis in Mice." Digestive Diseases and Sciences 57.2 (2011): 355-363.

Treatment for Liver Injury: In induced toxic liver injury in mice, kaempferol increased the expression of the tight junction proteins ZO-1 and occludin, butyrate receptors, and butyrate transporters in the ileum and proximal colon, while also markedly decreasing alanine aminotransferase and aspartate aminotransferase levels[523], which are both markers of liver injury. As isolated from Jindai soybean leaves, kaempferol prevented increases in serum aspartate aminotransferase and serum alanine aminotransferase while also significantly decreasing thiobarbituric acid reactive substances and tumor necrosis factor-alpha levels in mice subjected to carbon tetrachloride-induced liver injury[524], with researchers suggesting that kaempferol as a food additive may help reduce oxidation-related diseases. Kaempferol also significantly decreased propacetamol-induced oxidative stress, serum ALT and AST activities, DNA fragmentation, and early hepatic apoptosis, while also attenuating production of tumor necrosis factor-alpha and interleukin-6[525], and other effects in a model of acetaminophen overdose. Finally, in male Sprague-Dawley rats with carbon tetrachloride (CCl_4)-induced acute liver damage, oral administration of kaempferol at doses of 5 and 10 mg/kg body weight resulted in the amelioration of CCl_4-induced abnormalities in hepatic histology and serum parameters by decreasing the levels of pro-inflammatory mediators, suppressing nuclear factor-kappa B (NF-κB) p65 activation and the phosphorylation of Akt and mitogen-activated protein kinase members[526].

Treatment for Lung Injury: As an isolated compound, kaempferol attenuated lipopolysaccharide (LPS)-mediated production of cytokines and activation of NF-KB, while also modulating TRAF6 polyubiquitination[527] (the flagging of a protein for degradation) in LPS-induced acute lung injury in mice.

Treatment for Psoriasis: In a murine model of imiquimod-induced psoriasis, kaempferol reduced CD3+ T (a protein complex) cell infiltration and gene expression of major proinflammatory cytokines, down-regulated proinflammatory nuclear factor kappa beta signaling in the skin, lowered the percentage of pro-inflammatory cytokines and protein complexes in the spleen and lymph nodes, and suppressed the proliferation of T-cells[528].

Treatment for Allergic Disorders: Kaempferol has been shown to attenuate paw swelling and paw skin mast cell activation, rehabilitate hypothermia, reduce serum concentrations of histamine, TNF-alpha, interleukin-8, and monocyte chemo-attractant protein-1 (chemokines that regulate monocyte migration and infiltration) in mice[529]. In cancerous rat basophil cells, kaempferol upregulated and increased heme oxygenase enzymatic activity and level of expression[530], exerting significant anti-allergic activity.

---

[523] Chen, Jing, Yan-Han Xuan, Ming-Xiao Luo, Xiang-Gui Ni, Li-Qian Ling, Shi-Jia Hu, Jing-Qiao Chen, Jia-Yi Xu, Li-Ya Jiang, Wen-Zhang Si, Lin Xu, Hui Gao, Zheng Liu, and Haiyu Li. Kaempferol alleviates acute alcoholic liver injury in mice by regulating intestinal tight junction proteins and butyrate receptors and transporters. Toxicology 429 (2020): 152338-152338.

[524] Zang, Y., Hashimoto, S., Yu, C. et al. Protective effects of dietary kaempferol glycoside components from unripe soybean (Edamame, Glycine max L. Merrill. 'Jindai') leaves and their serous metabolite on carbon tetrachloride- induced liver injury mice. J Food Sci Technol 55, 4515-4521 (2018).

[525] Tsai, Ming-Shiun, Ying-Han Wang, Yan-Yun Lai, Hsi-Kai Tsou, Gan-Guang Liou, Jiunn-Liang Ko, and Sue- Hong Wang. Kaempferol protects against propacetamol-induced acute liver injury through CYP2E1 inactivation, UGT1A1 activation, and attenuation of oxidative stress, inflammation, and apoptosis in mice. Toxicology letters 290 (2018): 97-109.

[526] Lee C, Yoon S, Moon JO. Kaempferol Suppresses Carbon Tetrachloride-Induced Liver Damage in Rats via the MAPKs/NF-κB and AMPK/Nrf2 Signaling Pathways. Int J Mol Sci. 2023 Apr 7;24(8):6900.

[527] Qian, Jianchang, Xuemei Chen, Xiaojun Chen, Chuchu Sun, Yuchen Jiang, Yuanyuan Qian, Yali Zhang, Zia Khan, Jianmin Zhou, Guang Liang, and Chao Zheng. Kaempferol reduces K63-linked polyubiquitination to inhibit nuclear factor-KB and inflammatory responses in acute lung injury in mice. Toxicology letters 306 (2019): 53-60.

[528] Liu, C., H. Liu, C. Lu, J. Deng, Y. Yan, H. Chen, Y. Wang, C.-L. Liang, J. Wei, L. Han, and Z. Dai. Kaempferol attenuates imiquimod-induced psoriatic skin inflammation in a mouse model. Clinical & Experimental Immunology 198.3 (2019): 403-415.

[529] Cao, Jiao, Chaomei Li, Pengyu Ma, Yuanyuan Ding, Jiapan Gao, Qianqian Jia, Jing Zhu, and Tao Zhang. Effect of kaempferol on IgE-mediated anaphylaxis in C57BL/6 mice and LAD2 cells. Phytomedicine: international journal of phytotherapy and phytopharmacology 79 (2020): 153346-153346.

[530] Hirose, Etsuko, Miyoko Matsushima, Kenzo Takagi, Yui Ota, Keiko Ishigami, Tatsuya Hirayama, Yuta Hayashi, Toshinobu Nakamura, Naozumi Hashimoto, Kazuyoshi Imaizumi, Kenji Baba, Yoshinori Hasegawa, and Tsutomu Kawabe. Involvement of Heme Oxygenase-1 in Kaempferol-Induced Anti-Allergic Actions in RBL-2H3 Cells. Inflammation 32.2 (2009): 99-108.

Treatment for Neurodegenerative Disorders: In cadmium chloride-induced hippocampal damage and memory deficit in rats, kaempferol improved the behavioral scores, preserved hippocampus structure, reduced hippocampal levels of reactive oxygen species and levels of Bax (apoptosis regulator) and cleaved caspase-3, while significantly increasing hippocampal glutathione levels and synaptic proteins[531], among other effects that mitigated synaptic plasticity impairment and memory deficits in rats.

Treatment for Ocular Diseases: Kaempferol has been shown to significantly improve in vitro parallel artificial membrane permeability, in vitro cellular uptake, and ex vivo corneal permeation, and improve the treatment efficacy of corneal alkali burns in mice[532]. Kaempferol has also been shown to increase viability of retinal ganglion cells (RGC) and protect RGC from high-glucose-induced injury[533]. In a murine model of retinal ischemia reperfusion, kaempferol prevented retina thickness change and RGC death while also preventing pro-inflammatory cytokine production and oxidative stress[534].

Treatment for Mastitis: Kaempferol has been shown to prevent mastitis development, decrease myeloperoxidase production, interleukin-6 levels, tumor necrosis factor alpha concentration, and ANGPTL2 (a glycoprotein) expression in murine mastitis[535]. As isolated from the roots of zingiberaceae (a ginger-like spice), kaempferol markedly reduced infiltration of neutrophilic granulocyte, activation of myeloperoxidase, expression of tumor necrosis factor-alpha, interleukin-6, and interleukin-1B, while also suppressing the phosphorylation of nuclear factor-KB in lipopolysaccharide-induced mastitis in mice[536].

Antibacterial Agent: Kaempferol has been shown to exhibit antibacterial activity against helicobacter pylori[537] - a gram-negative bacteria that infects the human digestive tract. Kaempferol has also been shown to inhibit the biofilm formation of Staphylococcus aureus by up to 80%[538].

Treatment for Pulmonary Hypertension: In hypertensive rats, treatment with kaempferol reduced pulmonary artery pressure and pulmonary vascular remodeling, and alleviated right ventricular hypertrophy[539].

Anti-inflammatory Agent: In a rat intestinal microvascular endothelial cell (RIMVEC) line induced to gut-vascular barrier injury via lipopolysaccharide and TNF-alpha, treatment with kaempferol prevented intestinal

[531] El-kott, Attalla Farag, Mashael Mohammed Bin-Meferij, Samy M. Eleawa, and Majed M. Alshehri. Kaempferol Protects Against Cadmium Chloride-Induced Memory Loss and Hippocampal Apoptosis by Increased Intracellular Glutathione Stores and Activation of PTEN/AMPK Induced Inhibition of Akt/mTOR Signaling. Neurochemical Research 45.2 (2020): 295-309.

[532] Zhang, Fan, Rong Li, Meixin Yan, Qiqi Li, Yaru Li, and Xianggen Wu. Ultra-small nanocomplexes based on polyvinylpyrrolidone K-17PF: A potential nanoplatform for the ocular delivery of kaempferol. European journal of pharmaceutical sciences: official journal of the European Federation for Pharmaceutical Sciences 147 (2020): 105289-105289.

[533] Zhao, Lu, Junbo Sun, Suqin Shi, Xiao Qin, Keke Zhang, and Jiangyan Xu. Kaempferol protects retinal ganglion cells from high-glucose-induced injury by regulating vasohibin-1. Neuroscience letters 716 (2020): 134633-134633.

[534] Lin, Chaobin, Fujin Wu, Tongmei Zheng, Xiuchun Wang, Yiwei Chen, and Xiaomin Wu. Kaempferol attenuates retinal ganglion cell death by suppressing NLRP1/NLRP3 inflammasomes and caspase-8 via JNK and NF-KB pathways in acute glaucoma. Eye 33.5 (2018): 777-784.

[535] Hong-Bo Xiao, Guo-Guang Sui, Xiang-Yang Lu, Zhi-Liang Sun Kaempferol modulates Angiopoietin-like protein 2 expression to lessen the mastitis in mice Pharmacological Reports Volume 70, Issue 3, June 2018, Pages 439-445.

[536] Cao, Rongfeng, Kaiqiang Fu, Xiaopei Lv, Weishi Li, and Naisheng Zhang. Protective Effects of Kaempferol on Lipopolysaccharide-Induced Mastitis in Mice. Inflammation 37.5 (2014): 1453-1458.

[537] Escandon, R.A., del Campo, M., Lopez-Solis, R. et al. Antibacterial effect of kaempferol and (-)-epicatechin on Helicobacter pylori . Eur Food Res Technol 242, 1495-1502 (2016).

[538] Ming, Di, Dacheng Wang, Fengjiao Cao, Hua Xiang, Dan Mu, Junjie Cao, Bangbang Li, Ling Zhong, Xiaoyun Dong, Xiaobo Zhong, Lin Wang, and Tiedong Wang. Kaempferol Inhibits the Primary Attachment Phase of Biofilm Formation in Staphylococcus aureus. Frontiers in Microbiology 8 (2017).

[539] Zhang X, Yang Z, Su S, Nan X, Xie X, Li Z, Lu D. Kaempferol ameliorates pulmonary vascular remodeling in chronic hypoxia-induced pulmonary hypertension rats via regulating Akt-GSK3β-cyclin axis. Toxicol Appl Pharmacol. 2023 May 1;466:116478.

angiogenesis by impeding the tube formation and migration of RIMVECs and suppressed the expression of angiogenesis-related signals[540].

IMPLICATIONS FOR HUMAN HEALTH & NUTRITION

Because there are many edible plants, roots, fruits, and vegetables that contain kaempferol, this flavonol is quite common in the average human diet. In fact, kaempferol represents approximately 17% of the total daily flavonoid intake of Americans[541].

To enhance or include dietary kaempferol, consider eating fruits or nuts of almonds, hazelnuts, walnuts, or apples, apricots, bananas, black berries, black currants, cherries, figs, grapes, guava, mangoes, peaches, persimmon, plums, pomegranates, raspberries, or strawberries. Regular consumption of asparagus, beets, broccoli, Brussels sprouts, cabbage, capers, celery, chives, cucumbers, endives, fava beans, green beans, kale, leeks, lettuce, mung beans, olives, onions, parsnips, peas, potatoes, soybean, spinach, squash, or tomatoes will also help improve kaempferol intake. When preparing, cooking, or serving food, consider working with spices that contain kaempferol: basil, bay laurel, black pepper, caraway, chamomile, clove, dill, fennel, ginger, horseradish, nutmeg, oregano, parsley, rosemary, saffron, tarragon, and thyme.

Kaempferol is also found in red wines, aloe vera, cacao, Ginkgo biloba, St. John's Wort, and tea. Cannabis strains that are known to contain kaempferol include Northern Lights, Purple Moose, and White Cookies. Future editions of this text will be updated with further cannabis flavonoid testing results. It should be noted that common methods of cannabis consumption – smoking and vaping – are likely to alter or destroy flavonoid compounds including kaempferol. Eating raw cannabis is the most likely way to obtain some amount of this flavonoid, as well as consumption of specifically prepared oils, and other whole-plant, full-spectrum preparations.

---

[540] Yu R, Zhong J, Zhou Q, Ren W, Liu Z, Bian Y. Kaempferol prevents angiogenesis of rat intestinal microvascular endothelial cells induced by LPS and TNF-α via inhibiting VEGF/Akt/p38 signaling pathways and maintaining gut-vascular barrier integrity. Chem Biol Interact. 2022 Oct 1;366:110135.

[541] Liu, Rui Hai. Health-Promoting Components of Fruits and Vegetables in the Diet. Advances in Nutrition 4.3 (2013): 384S-392S.

Kaempferol Review

Answer the following questions to test your knowledge of this flavonoid:

Question #1: What type of flavonoid is kaempferol?

        a.   Flavane
        b.   Flavone
        c.   Flavonol
        d.   Flavoned

Question #2: The base component in the biosynthesis of kaempferol is:

        a.   Cannabisin
        b.   Rutinoside
        c.   Apigenin
        d.   Naringenin

Question #3: How common is kaempferol in cannabis?

        a.   Top three
        b.   Top five
        c.   Top ten
        d.   Secondary

Question #4: What is the chemical formula for kaempferol?

        a. $C_{15}H_{10}O_6$
        b. $C_{24}H_{10}O_{16}$
        c. $C_{15}H_{10}O_{60}$
        d. $C_{27}H_{30}O_{16}$

Question #5: Name two biological roles of kaempferol in plants?

1 _____ 2 _____

Question #6: Name two potential medical uses of kaempferol:

1 _____ 2 _____

For the answer key to Kaempferol, please visit www.cannabischemistry.org

WOGONIN

Type: Flavone
Chemical Formula: $C_{16}H_{12}O_5$
Molecular Weight: 284.267 g/mol
Boiling Point: 518.80 °C @ 760.00 mm Hg (estimated)[542]
Flash Point: 198.40 °C (estimated by TGSC)
Melting Point: 203 to 206 °C
Solubility: Soluble in water, ethanol, DMS0, dimethyl formamide
Oral LD50: 2.19 g/kg (mouse)[543]
Biological Role: Unknown
Therapeutic Role: Anti-inflammatory agent, anti-cancer agent, neuroprotective agent
Commercial Use: For research purposes only
Occurrence in Cannabis: Top twenty

Occurs in Cannabis Strains: Huckleberry, Purple Moose, Purple Punch 2.0, Skywalker, Strawberry Cough

INTRODUCTION

Wogonin is well-represented in the scientific literature covering flavonoids, despite the fact that this compound is only known to occur in one plant - skullcap (scutellaria baicalensis georgi and scutellaria ocmulgee). Wogonin has been hypothesized to occur in cannabis, and laboratory tests the author conducted found the flavone to be present in a sample of Strawberry Cough grown outside in Maine, United States, in 2019. Interestingly, the same strain grown by the same grower under similar conditions, except indoors, showed no detectable wogonin. Flavonoid testing also confirmed this flavone in Huckleberry, Purple Moose, Purple Punch 2.0, and Skywalker cannabis varieties, confirming for the first time that the compound occurs in a plant outside of the scutellaria genus.

Wogonin has been heavily researched because of its anti-inflammatory and anti-tumor properties, among dozens of other therapeutic benefits, and has been quantified as a primary active ingredient in Sho-Saiko-To, a Japanese herbal supplement used to treat a variety of conditions as part of kampo medicine. Similar formulations have been in use in China for many centuries.

---

[542] The Good Scents Company Data Sheet for Wogonin, from: https://www.thegoodscentscompany.com/data/rw1698441.html. Accessed November 25, 2023.

[543] Cayman Chemical Safety Data Sheet for Wogonin, from: https://cdn.caymanchem.com/cdn/msds/14248m.pdf. Accessed December 4, 2023.

CHEMICAL STRUCTURE

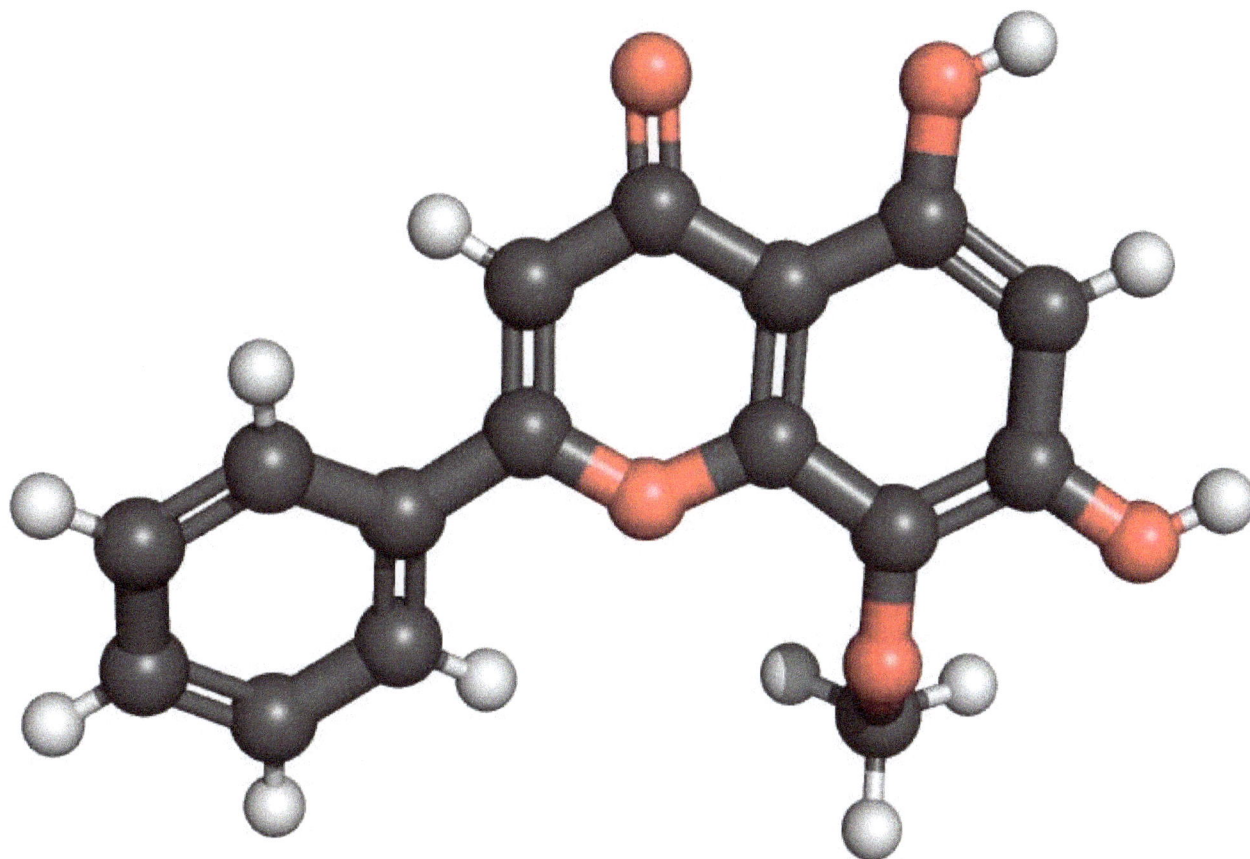

Wogonin is a flavone consisting of sixteen carbon atoms, twelve hydrogen atoms, and five oxygen atoms notated as $C_{16}H_{12}O_5$. The carbon skeleton of wogonin contains eight double bonds: seven that are endocyclic, and one that is an exocyclic oxygenated double bond.

Wogonin is formed in a specialized flavone biosynthetic pathway that not only differs from the typical flavonoid biosynthesis pathways of other plants, but it also differs from that of other plants in its genus[544]. Unlike other flavonoids, wogonin is derived from chrysin[545], and lacks a 4'-hydroxyl group on the B ring; instead, the hydroxy groups are positioned at C-5 and C-7, with a methoxy group at C-8. Other known variations of this flavone include:

Wogonin(1-): a variation of wogonin that has gained an electron.

Wogonin 7-O-glucoside: a glucose molecule is attached at the 7th position.

Wogonin 7-O-beta-D-glucuronate: a glucuronic acid moiety is attached at the 7th position.

[544] Qing Zhao, Jun Yang, Meng-Ying Cui, Jie Liu, Yumin Fang, Mengxiao Yan, Wenqing Qiu, Huiwen Shang, Zhicheng Xu, Reheman Yidiresi, Jing-Ke Weng, Tomas Pluskal, Marielle Vigouroux, Burkhard Steuernagel, Yukun Wei, Lei Yang, Yonghong Hu, Xiao-Ya Chen, Cathie Martin, The Reference Genome Sequence of Scutellaria baicalensis Provides Insights into the Evolution of Wogonin Biosynthesis, Molecular Plant, Volume 12, Issue 7, 2019, Pages 935-950.

[545] Zhao Q, Zhang Y, Wang G, Hill L, Weng JK, Chen XY, Xue H, Martin C. A specialized flavone biosynthetic pathway has evolved in the medicinal plant, Scutellaria baicalensis. Sci Adv. 2016 Apr 8;2(4).

Norwogonin 5,7,8-trihydroxyflavone: a demethylated derivative of wogonin, lacking a methoxy group.

OCCURRENCE IN PLANTS

As of early 2025, wogonin is only reported to occur in skullcap (scutellaria baicalensis and scutellaria ocmulgee), and several chemovars of cannabis.

BIOLOGICAL ACTIVITY IN PLANTS

Despite extensive research, the author found no information regarding biological roles that wogonin might serve in plants.

POTENTIAL USES IN MEDICINE

Purified wogonin is available for purchase for research purposes only; it is well known to researchers and medical professionals as a potent anti-inflammatory agent, and as an anti-tumor agent in the treatment of numerous cancers. There is also potential for this flavone to be used in the treatment of neurological conditions, diabetes, herpes, and cardiovascular diseases, among other potential therapeutic uses.

Treatment for Lung Cancer: Wogonin has been shown to significantly increase the activation of several caspases, ROS production, and apoptosis, and to reduce the viability and tumor development of A549 and A427 lung cancer cells by up to 69%[546], while exhibiting no cytotoxic effects in normal BEAS-2B lung cells. Other research has shown that wogonin can repress the growth and metastatic potential and promote cell apoptosis by repressing MMP1 expression and modulating the PI3K/AKT signaling pathway in A549 and H460 lung cancer cells[547].

Treatment for Breast Cancer: In MCF-7 human breast cancer cells, wogonin has been shown to increase production of ROS, activate caspases, induce PARP cleavage, and change the ratios of antiapoptotic/proapoptotic Bcl-2 (protein that regulates cell death) ratios, inducing mitochondria and death-receptor-mediated apoptotic cell death[548]. In triple-negative breast cancer cells, wogonin has been shown to induce cellular senescence by increasing P16 expression (a protein that slows cell division) and lactase activity, among other effects, inducing permanent proliferation inhibition[549].

Treatment for Cervical Cancer: In HPV (human papilloma virus)-related cervical cancer cells, wogonin suppressed the expression of the E6 and E7 viral oncogenes, modulated the mitochondrial membrane potential and the expression of both pro- and anti-apoptotic factors, and induced the cleavage of caspase-3, -9, and PARP, inducing intrinsic apoptosis[550].

---

[546] Wang, Chengyang, and Chuangcheng Cui. Inhibition of Lung Cancer Proliferation by Wogonin is Associated with Activation of Apoptosis and Generation of Reactive Oxygen Species. Balkan Medical Journal 37.1 (2019): 29- 33.

[547] Guo J, Jin G, Hu Y, Zhao Z, Nan F, Hu X, Hu Y, Han Q. Wogonin Restrains the Malignant Progression of Lung Cancer Through Modulating MMP1 and PI3K/AKT Signaling Pathway. Protein Pept Lett. 2023;30(1):25-34.

[548] Yu, Ji Sun, and An Keun Kim. Wogonin induces apoptosis by activation of ERK and p38 MAPKs signaling pathways and generation of reactive oxygen species in human breast cancer cells. Molecules and cells 31.4 (2011): 327-335.

[549] Yang, Dawei, Qinglong Guo, Yin Liang, Yue Zhao, Xiaoyu Tian, Yuchen Ye, Jieyi Tian, Tao Wu, and Na Lu. Wogonin induces cellular senescence in breast cancer via suppressing TXNRD2 expression. Archives of toxicology 94.10 (2020): 3433-3447.

[550] Kim, Man, Yesol Bak, Yun Park, Dong Lee, Jung Kim, Jeong Kang, Hyuk-Hwan Song, Sei-Ryang Oh, and Do Yoon. Wogonin induces apoptosis by suppressing E6 and E7 expressions and activating intrinsic signaling pathways in HPV-16 cervical cancer cells. Cell Biology and Toxicology 29.4 (2013): 259-272.

Treatment for Brain & Spine Cancer: Wogonin has been shown to activate the AMP-activated protein kinase and p53 signaling pathways in glioblastoma multiforme (GBM) cells, blocking cell cycle progression at the G1 phase, and inducing apoptosis[551]. As the primary constituent in an extract made from the leaves of scutellaria ocmulgee (skullcap), wogonin was shown to delay the growth of F98 intracranial and subcutaneous glioma tumors in rats by inhibiting Akt (protein kinase B), GSK-3 (glycogen synthase kinase 3), and NF-KB signaling[552].

Treatment for Nasopharyngeal Carcinoma: Wogonin has been found to inactivate the Akt signaling pathway and thereby cross-regulate autophagy and apoptosis in human nasopharyngeal carcinoma cells[553]. Research has shown that wogonin induces sub-G1-phase cells, PARP cleavage, and downregulation of delta-Np63 (a protein isoform involved in tumor invasiveness), while also inactivating GSK-3-beta[554] in nasopharyngeal carcinoma cells.

Treatment for Colon Cancer: As an isolated compound, wogonin was shown to induce DNA fragmentation and chromatin condensation, increase cell cycle arrest in the G1 phase, and inactivate the Akt signaling pathway, leading to apoptosis in HT-29 human colorectal cancer cells[555]. Wogonin has also been shown to inhibit YAP1 (yes-associated protein 1, involved in cellular proliferation and suppressing apoptotic genes) expression in vivo and in vitro, suppressing epithelial-mesenchymal transition development and the carcinogenic process of colon cancer through the IRF3-mediated (interferon regulatory factor 3) Hippo (involved in the regulation of cell proliferation, apoptosis, and stem cell self-renewal) signaling pathway[556].

Treatment for Ovarian Cancer: In combination with oridonin (a diterpenoid), wogonin contributed to up-regulation of p53 protein, and the down regulation of Akt, leading to apoptosis and cell cycle modulation in chemo-resistant epithelial ovarian cancer cells[557].

Treatment for Liver Cancer: Wogonin has been shown to cause DNA fragmentation, induce cellular swelling, activate caspase-3, and induce the p53 signaling pathway in SK-HEP-1 hepatocellular carcinoma cells[558]. This flavone has also been shown to induce cell cycle arrest and apoptosis in SMMC7721 and HCCLM3 hepatocellular carcinoma cell lines[559], and inhibit hepatoma cell proliferation by upregulating miR-27b-5p (mediates apoptosis) and downregulating YWHAZ (a gene involved in multiple cellular functions)[560].

[551] Lee, Dae-Hee, Tae Hwa Lee, Chang Hwa Jung, and Young-Ho Kim. Wogonin induces apoptosis by activating the AMPK and p53 signaling pathways in human glioblastoma cells. Cellular signaling 24.11 (2013): 2216-25.

[552] Parajuli, Prahlad, N. Joshee, S. Chinni, A. Rimando, S. Mittal, S. Sethi, and A. Yadav. Delayed growth of glioma by Scutellaria flavonoids involve inhibition of Akt, GSK-3 and NF-KB signaling. Journal of Neuro-Oncology 101.1 (2010): 15-24.

[553] Chow, Shu-Er, Yu-Wen Chen, Chi-Ang Liang, Yao-Kuan Huang, and Jong-Shyan Wang. Wogonin induces cross-regulation between autophagy and apoptosis via a variety of Akt pathway in human nasopharyngeal carcinoma cells. Journal of Cellular Biochemistry 113.11 (2012): 3476-3485.

[554] Chow, Shu-Er, Ying-Ling Chang, Sun-Fa Chuang, and Jong-Shyan Wang. Wogonin induced apoptosis in human nasopharyngeal carcinoma cells by targeting GSK-3B and delta-Np63. Cancer Chemotherapy and Pharmacology 68.4 (2011): 835-845.

[555] Kim, So-Jung, Hyeong-Jin Kim, Hye-Ri Kim, Seung-Ho Lee, Sung-Dae Cho, Chang-Sun Choi, Jeong-Seok Nam, and Ji-Youn Jung. Antitumor actions of baicalein and wogonin in HT-29 human colorectal cancer cells. Molecular Medicine Reports 6.6 (2012): 1443-1449.

[556] You W, Di A, Zhang L, Zhao G. Effects of wogonin on the growth and metastasis of colon cancer through the Hippo signaling pathway. Bioengineered. 2022 Feb;13(2):2586-2597.

[557] Chen, Sophie, Matt Cooper, Matt Jones, Thumuluru Madhuri, Julie Wade, Ashleigh Bachelor, and Simon Butler Manuel. Combined activity of oridonin and wogonin in advanced-stage ovarian cancer cells. Cell Biology and Toxicology 27.2 (2010): 133-147.

[558] Chen, Y., Shen, S., Lee, W. et al. Wogonin and fisetin induction of apoptosis through activation of caspase-3 cascade and alternative expression of p21 protein in hepatocellular carcinoma cells SK-HEP-1. Arch Toxicol 76, 351-359 (2002).

[559] Wu K, Teng M, Zhou W, Lu F, Zhou Y, Zeng J, Yang J, Liu X, Zhang Y, Ding Y, Shen W. Wogonin Induces Cell Cycle Arrest and Apoptosis of Hepatocellular Carcinoma Cells by Activating Hippo Signaling. Anticancer Agents Med Chem. 2022;22(8):1551-1560.

[560] Ma MY, Wang Q, Wang SM, Feng XJ, Xian ZH, Zhang SH. Wogonin inhibits hepatoma cell proliferation by targeting miR-27b-5p/YWHAZ axis. J Biochem Mol Toxicol. 2023 Aug 25:e23508.

Treatment for Neuroblastoma: Treatment with wogonin induced the release of cytochrome c, altered the expression of Bcl-2, Bax, and Bid, and increased the activation of caspase-3, -4, -8, -9, -12, and PARP-1, while significantly promoting apoptosis in two malignant neuroblastoma cell lines[561]. Neuroblastomas generally only affect young children.

Treatment for Bone Cancer: In human osteosarcoma cancer stem cells, wogonin was shown to induce apoptosis, inhibit migration and mobility, and repress the renewal ability of cells by downregulating matrix metallopeptidase-9 expression[562].

Treatment for Lymphoma: Wogonin has been shown to induce apoptosis, regulate the expression of NF-KB, down regulate protein levels of p65 and PU.1, and inhibit the growth of EBV (Epstein-Barr)-positive lymphoma cells[563].

Treatment for Melanoma: In HT144 melanoma cells, treatment with wogonin inhibited the proliferation, colony formation and tumor growth, and exhibited strong anti-inflammatory effects evidenced by the decreased levels of pro-inflammatory factors, the increased level of anti-inflammatory factor, and the decreased expression of inflammatory cytokines[564], among other effects.

Chemotherapy Toxicity Treatment: Wogonin has been shown to inhibit the binding of p65 to Nrf2 (a protein that regulates the expression of antioxidant proteins) by suppression of the KB-binding activity in adriamycin (ADR)-induced resistant human chronic myelogenous leukemia[565]. Wogonin has also been shown to potentiate the anticancer activity of etoposide (a common cancer treatment drug) by suppressing the excretion of calcein (a fluorescent dye) and decreasing the excretion of radiolabeled etoposide[566], while other researchers have shown that wogonin significantly attenuates etoposide-induced oxidative DNA damage and apoptosis[567]. In rats, wogonin ameliorated the nephrotoxicity indices, oxidative stress, inflammation, and apoptosis induced by cisplatin[568] - another common cancer treatment drug. Other research has demonstrated that wogonin induces cell death in aggressive disease phenotype head and neck cancer cells with cisplatin resistance via marked ROS accumulation and the activation of cell death pathways[569].

[561] Ge, Wenliang, Qiyou Yin, and Hua Xian. Wogonin Induced Mitochondrial Dysfunction and Endoplasmic Reticulum Stress in Human Malignant Neuroblastoma Cells Via IRE1a-Dependent Pathway. Journal of Molecular Neuroscience 56.3 (2015): 652-662.

[562] Huynh, Do Luong, Taeho Kwon, Jiao Zhang, Neelesh Sharma, Meeta Gera, Mrinmoy Ghosh, Nameun Kim, Somi Kim Cho, Dong Lee, Yang Park, and Dong Jeong. Wogonin suppresses stem cell-like traits of CD133 positive osteosarcoma cell via inhibiting matrix metallopeptidase-9 expression. BMC Complementary and Alternative Medicine 17.1 (2017): 1-8.

[563] Wu, X., Liu, P., Zhang, H. et al. Wogonin as a targeted therapeutic agent for EBV (+) lymphoma cells involved in LMP1/NF-KB/miR-155/PU.1 pathway. BMC Cancer 17, 147 (2017).

[564] Li L, Ji Y, Zhang L, Cai H, Ji Z, Gu L, Yang S. Wogonin inhibits the growth of HT144 melanoma via regulating hedgehog signaling-mediated inflammation and glycolysis. Int Immunopharmacol. 2021 Dec;101(Pt B):108222.

[565] Xu, Xuefen, Xiaobo Zhang, Yi Zhang, Lin Yang, Yicheng Liu, Shaoliang Huang, Lu Lu, Lingyi Kong, Zhiyu, Qinglong Guo, and Li Zhao. Wogonin reversed resistant human myelogenous leukemia cells via inhibiting Nrf2 signaling by Stat3/NF-KB inactivation. Scientific Reports 7.1 (2017): 1-16.

[566] Enomoto, Riyo, Chika Koshiba, Chie Suzuki, and Eibai Lee. Wogonin potentiates the antitumor action of etoposide and ameliorates its adverse effects. Cancer Chemotherapy and Pharmacology 67.5 (2011): 1063-1072.

[567] Attia, Sabry M, Sheikh Fayaz Ahmad, Gamaleldin I Harisa, Ahmed M Mansour, El Sayed M El Sayed, and Saleh A Bakheet. Wogonin attenuates etoposide-induced oxidative DNA damage and apoptosis via suppression of oxidative DNA stress and modulation of OGG1 expression. Food and chemical toxicology : an international journal published for the British Industrial Biological Research Association 59 (2014): 724-730.

[568] Badawy, Alaa M, Reem N El-Naga, Amany M Gad, Mariane G Tadros, and Hala M Fawzy. Wogonin pre- treatment attenuates cisplatin-induced nephrotoxicity in rats: Impact on PPAR-y, inflammation, apoptosis and Wnt/beta-catenin pathway. Chemico-biological interactions (2019)1.

[569] Kim, Eun, Hyejin Jang, Daiha Shin, Seung Baek, and Jong-Lyel Roh. Targeting Nrf2 with wogonin overcomes cisplatin resistance in head and neck cancer. Apoptosis 21.11 (2016): 1265-1278.

Cardioprotective Agent: In isoproterenol-induced myocardial infarction in rats, wogonin-loaded nanoparticles significantly reduced the cardiac infarct size, serum cardiac markers, lipid peroxidation product, and inflammatory markers, as well as markedly upregulated the protein expression of nuclear Nrf2 and heme oxygenase-1[570] (a gene involved in the prevention of vascular inflammation).

Treatment for Lung Injury: As isolated from the roots of skullcap (scutellaria baicalensis georgi), wogonin reduced the expression of inducible nitric oxide synthase and cyclooxygenase, inhibited the phosphorylation of mitogen-activated protein kinase, and inhibited the phosphorylation of p38 MAPK and JNK, resulting in the inhibition of lung edema and protein leakage in lipopolysaccharide-induced acute lung injury in mice[571]. In the A549 cell line, treatment with wogonin alleviated the inflammation, oxidative stress, and apoptosis in lipopolysaccharide-induced A549 cells by SIRT1 (NAD-dependent deacetylase sirtuin-1)-mediated HMGB1 (High mobility group box 1 protein) deacetylation[572].

Treatment for Testicular Dysfunction: In cadmium-induced testicular toxicity in rats, wogonin significantly improved the reduction in sperm quality and quantity and markedly improved the decrease in body and organ weight[573], among other effects that attenuated the effects of cadmium toxicity.

Treatment for Epilepsy: Wogonin has been shown to improve cognitive function, ameliorate hippocampus damage, and significantly reduce neuroinflammation in induced temporal lobe epilepsy in rats[574].

Neuroprotective Agent: As isolated from the root of scutellaria baicalensis georgi, wogonin reduced levels of malondialdehyde, tumor necrosis factor-a, interleukin-1, and interleukin-6, while also reducing the expression of NF-KB mRNA and protein expression, and ameliorated vacuolization and nuclear pyknosis in the neuronal cells and focal gliosis in irradiated rats[575]. Additionally, research has shown that wogonin can prevent the death of dorsal root ganglion neurons in rats subjected to tunicamycin-induced endoplasmic reticulum stress[576], likely by modulating stress-responsive genes[577].

Treatment for Alzheimer's Disease: Wogonin has been shown to potently promote the clearance of beta-amyloid in primary neural astrocytes while also significantly decreasing beta-amyloid secretion and inhibiting the activity of the enzyme glycogen synthase kinase 3 beta[578], with researchers suggesting wogonin as an mTOR inhibitor to treat Alzheimer's disease.

[570] Bei, Wan, Li Jing, and Nie Chen. Cardio protective role of wogonin loaded nanoparticle against isoproterenol induced myocardial infarction by moderating oxidative stress and inflammation. Colloids and surfaces. B, Biointerfaces 185 (2020): 110635-110635.

[571] Wei, Cheng-Yu, Hai-Lun Sun, Ming-Ling Yang, Ching-Ping Yang, Li-You Chen, Yi-Ching Li, Chien-Ying Lee, and Yu-Hsiang Kuan. Protective effect of wogonin on endotoxin-induced acute lung injury via reduction of p38 MAPK and JNK phosphorylation. Environmental Toxicology 32.2 (2017): 397-403.

[572] Ge J, Yang H, Zeng Y, Liu Y. Protective effects of wogonin on lipopolysaccharide-induced inflammation and apoptosis of lung epithelial cells and its possible mechanisms. Biomed Eng Online. 2021 Dec 14;20(1):125.

[573] Yu, Wen, Zhipeng Xu, Qingqiang Gao, Yang Xu, Bing Wang, and Yutian Dai. Protective role of wogonin against cadmium induced testicular toxicity: Involvement of antioxidant, anti-inflammatory and anti-apoptotic pathways. Life sciences 258 (2020): 118192-118192.

[574] Guo, Xiangyang, Jieying Wang, Nana Wang, Anurag Mishra, Hongyan Li, Hong Liu, Yingli Fan, Na Liu, and Zhongliang Wu. Wogonin preventive impact on hippocampal neurodegeneration, inflammation and cognitive defects in temporal lobe epilepsy. Saudi Journal of Biological Sciences 27.8 (2020): 2149-2156.

[575] Wang, Liying, Chenyu Li, Nagaraja Sreeharsha, Anurag Mishra, Vipin Shrotriya, and Ajay Sharma. Neuroprotective effect of Wogonin on Rat's brain exposed to gamma irradiation. Journal of photochemistry and photobiology. B, Biology 204 (2020): 111775-111775.

[576] Xu, Shujuan, Xin Zhao, Quanlai Zhao, Quan Zheng, Zhen Fang, Xiaoming Yang, Hong Wang, Ping Liu, and Hongguang Xu. Wogonin Prevents Rat Dorsal Root Ganglion Neurons Death via Inhibiting Tunicamycin-Induced ER Stress In Vitro. Cellular and Molecular Neurobiology 35.3 (2014): 389-398.

[577] Chen, Fangyi, Rongbo Wu, Zhu Zhu, Wangping Yin, Min Xiong, Jianwei Sun, Miaozhong Ni, Guoping Cai, and Xinchao Zhang. Wogonin Protects Rat Dorsal Root Ganglion Neurons Against Tunicamycin-Induced ER Stress Through the PERK-eIF2a-ATF4 Signaling Pathway. Journal of Molecular Neuroscience 55.4 (2014): 995-1005.

[578] Zhu, Yuyou, and Juan Wang. Wogonin increases B-amyloid clearance and inhibits tau phosphorylation via inhibition of mammalian target of rapamycin: potential drug to treat Alzheimer's disease. Neurological Sciences 36.7 (2015): 1181-1188.

Treatment for Intracerebral Hemorrhage: As isolated from scutellaria radix, wogonin was shown to promote hematoma clearance, improve neurological recovery, and dramatically reduce inflammatory and oxidative stress responses in a murine model of intracerebral hemorrhage[579].

Treatment for Cerebral Ischemia: Wogonin has been shown to markedly reduce the infarct volume, decrease the production of nitric oxide and inflammatory cytokines, and reduce the activity of NF-KB in a focal cerebral ischemia rat model, while also significantly preventing the reduction of neuronal cell counts in CA1 regions (the first region in the hippocampal circuit) after global ischemic insult in gerbils[580].

Treatment for Diabetes: In a streptozotocin-induced diabetic mouse model, wogonin downregulated monocyte chemotactic protein-1 (MCP-1), tumor necrosis factor-a (TNF-a), interleukin-1B (IL-1B), and NF-KB, while also reducing the expression of extracellular matrix, including fibronectin, collagen IV, a-smooth muscle actin, and transforming growth factor-B1 in the kidneys of diabetic mice[581].

Treatment for Gastric Injury: In ethanol-induced gastric mucosal damage in rats, macroscopic damage was significantly attenuated by pretreatment with wogonin, inhibiting ethanol-induced gastric mucosal injury by up to 64%[582], significantly attenuating gastric hemorrhages and edema.

Treatment for Inflammatory Bowel Disease: In epithelial cell lines derived from colon carcinoma, wogonin suppressed down regulation of tight junction proteins and induced up regulation of inflammatory mediators, suppressing the inflammatory response and maintaining the intestinal barrier function in vitro[583].

Treatment for Inflammatory Skin Disorders: When applied topically to mice with contact dermatitis, wogonin inhibited an edematic response as well as proinflammatory gene expression of cyclooxygenase-2, interleukin-1 B, interferon-y, intercellular adhesion molecule-1, and inducible nitric oxide synthase[584], exhibiting marked inhibition of contact dermatitis.

Treatment for Retinal Disease: In H_2O_2-induced oxidative stress in retinal pigment epithelial (RPE) cells, pre-treatment with wogonin significantly improved the cell viability and markedly reduced the H_2O_2-induced RPE cell death rate[585], likely by modulating the Akt pathway.

Treatment for Herpes: Wogonin has been shown to inhibit herpes simplex virus types 1 and 2 by significantly reducing HSY-induced NF-KB and MAPK pathway activation, reducing viral mRNA transcription, viral protein synthesis, and infectious virion particle titers in vitro[586].

[579] Zhuang, J., Peng, Y., Gu, C. et al. Wogonin Accelerates Hematoma Clearance and Improves Neurological Outcome via the PPAR-y Pathway After Intracerebral Hemorrhage. Transl. Stroke Res. (2020).

[580] Piao, Hua, Shun Jin, Hyang Chun, Jae-Chul Lee, and Won-Ki Kim. Neuroprotective effect of wogonin: Potential roles of inflammatory cytokines. Archives of Pharmacal Research 27.9 (2004): 930-936.

[581] Zheng, Zhi-chao, Wei Zhu, Lei Lei, Xue-qi Liu, and Yong-gui Wu. Wogonin Ameliorates Renal Inflammation and Fibrosis by Inhibiting NF-KB and TGF-B1/Smad3 Signaling Pathways in Diabetic Nephropathy. Drug Design, Development and Therapy 14 (2020): 4135-4148.

[582] Park, Soojin, Ki-Baik Hahm, Tae-Young Oh, Joo-Hyun Jin, and Ryowon Choue. Preventive Effect of the Flavonoid, Wogonin, Against Ethanol-Induced Gastric Mucosal Damage in Rats. Digestive Diseases and Sciences 49.3 (2004): 384-394.

[583] Wang, Wenping, Tingsong Xia, and Xinpu Yu. Wogonin suppresses inflammatory response and maintains intestinal barrier function via TLR4-MyD88-TAK1-mediated NF-KB pathway in vitro. Inflammation Research 64.6 (2015): 423-431.

[584] Lim, Hyun, Haeil Park, and Hyun Kim. Inhibition of contact dermatitis in animal models and suppression of proinflammatory gene expression by topically applied Flavonoid, Wogonin. Archives of Pharmacal Research 27.4 (2004): 442-448.

[585] Yan, Tingqin, Hongsheng Bi, and Yun Wang. Wogonin modulates hydroperoxide-induced apoptosis via PI3K/Akt pathway in retinal pigment epithelium cells. Diagnostic Pathology 9 (2014).

[586] Chu, Y., Lv, X., Zhang, L. et al. Wogonin inhibits in vitro herpes simplex virus type 1 and 2 infection by modulating cellular NF-KB and MAPK pathways. BMC Microbiol 20, 227 (2020).

Antifungal Agent: A study that sought to determine the antifungal activity of a popular Asian extract called ou-gon found that wogonin exhibited potent antifungal activity against three common human pathogenic fungi; trichophyton rubrum, trichophyton mentagrophytes, and aspergillus fumigatus[587], likely by inducing apoptosis as a result of the hyperproduction of reactive oxygen species.

Treatment for HIV: Recent research has shown that wogonin is a novel latency-promoting agent that can inhibit HIV-1 transcription by HIV-1 epigenetic silencing, exhibiting low cytotoxicity and long-lasting inhibition of HIV-1 transcription[588].

Treatment for Lower Back Pain: As a key bioactive ingredient of Huangqi Guizhi formula (a traditional Chinese herbal medicine), treatment with wogonin significantly ameliorated puncture-induced intervertebral disc degeneration and lower back pain by suppressing the upregulated nerve growth factor in rats[589].

Treatment for Spinal Cord Injury: In rats with induced spinal cord injury, treatment with wogonin promoted the recovery of motor function, improved the histopathological morphology, inhibited the activation of the STAT3 signal pathway (a hub of several major signaling pathways involved in inflammatory responses), and reduced neuronal inflammation and apoptosis[590].

Treatment for Colitis: In dextran sulfate sodium (DSS)-induced colitis in mice, treatment with wogonin ameliorated the condition by preventing colon shortening and inhibiting pathological damage, and increasing the expression of PPARγ, restoring intestinal epithelial hypoxia and inhibiting iNOS protein, thereby reducing intestinal nitrite levels[591].

Traumatic Brain Injury Protective Agent: In a murine model of traumatic brain injury, treatment with wogonin improved neurological deficits and learning and memory abilities and relieved cerebral edema by modulation of the PI3K/Akt/Nrf2/HO-1 pathway, and by increasing the levels of antioxidant factors glutathione, superoxide dismutase and catalase in the CA1 region of the hippocampus, and also significantly inhibited the production of malondialdehyde and reactive oxygen species[592].

Treatment for Mastitis: In rat mammary gland tissues induced to inflammation via lipopolysaccharide (LPS), treatment with wogonin relieved oxidative stress in mammary epithelial cells by decreasing ROS generation and MDA levels, and increasing GSH and SOD levels, while also repressing LPS-induced activation of the Akt/NF-κB pathway and increased Nrf2/HO-1 signaling activation[593].

[587] Da, X., Nishiyama, Y., Tie, D. et al. Antifungal activity and mechanism of action of Ou-gon (Scutellaria ro ot extract) components against pathogenic fungi. Sci Rep 9, 1683 (2019).

[588] Zhang H, Cai J, Li C, Deng L, Zhu H, Huang T, Zhao J, Zhou J, Deng K, Hong Z, Xia J. Wogonin inhibits latent HIV-1 reactivation by downregulating histone crotonylation. Phytomedicine. 2023 Jul 25;116:154855.

[589] Li J, Duan W, Chai S, Luo Y, Ma Y, Yang N, Liu M, He W. Wogonin, a Bioactive Ingredient from Huangqi Guizhi Formula, Alleviates Discogenic Low Back Pain via Suppressing the Overexpressed NGF in Intervertebral Discs. Mediators Inflamm. 2023 Feb 20;2023:4436587.

[590] Shao W, Zhang C, Li K, Lu Z, Zhao Z, Gao K, Lv C. Wogonin inhibits inflammation and apoptosis through STAT3 signal pathway to promote the recovery of spinal cord injury. Brain Res. 2022 May 1;1782:147843.

[591] Su Y, Liang J, Zhang M, Zhao M, Xie X, Wang X, Pan Z, Huang S, Yan R, Wang Q, Zhou L, Luo X. Wogonin regulates colonocyte metabolism via PPARγ to inhibit Enterobacteriaceae against dextran sulfate sodium-induced colitis in mice. Phytother Res. 2023 Mar;37(3):872-884.

[592] Feng Y, Ju Y, Yan Z, Ji M, Yang M, Wu Q, Wang L, Sun G. Protective role of wogonin following traumatic brain injury by reducing oxidative stress and apoptosis via the PI3K/Nrf2/HO-1 pathway. Int J Mol Med. 2022 Apr;49(4):53.

[593] He, X., Wang, J., Sun, L. et al. Wogonin attenuates inflammation and oxidative stress in lipopolysaccharide-induced mastitis by inhibiting Akt/NF-κB pathway and activating the Nrf2/HO-1 signaling. Cell Stress and Chaperones (2023).

IMPLICATIONS FOR HUMAN HEALTH & NUTRITION

Dietary consumption of wogonin is currently only possible by making extractions of scutellaria baicalensis or scutellaria ocmulgee, and there is currently no literature to support or inform on the deliberate dietary consumption of this flavone. However, it might be possible to obtain small amounts of wogonin by consuming cannabis strains like Huckleberry, Purple Moose, Purple Punch 2.0, Skywalker, Strawberry Cough, or other chemovars where one of these varieties is a parent. While consumption of intact kaempferol via smoking or vaporizing these strains is unlikely, using whole spectrum cannabis products and/or eating raw cannabis might permit users to obtain some small amount of this compound.

Wogonin Review

Answer the following questions to test your knowledge of this flavonoid:

Question #1: What type of flavonoid is wogonin?

     a. Flavane
     b. Flavone
     c. Flavonol
     d. Flavoned

Question #2: The base component in the biosynthesis of wogonin is:

     a. Cannabisin
     b. Apigenin
     c. Chrysin
     d. Naringenin

Question #3: How common is wogonin in cannabis?

     a. Top three
     b. Top five
     c. Top ten
     d. Top twenty

Question #4: What is the chemical formula for wogonin?

     a. $C_{16}H_{12}O_5$
     b. $C_{26}H_{10}O_{15}$
     c. $C_{16}H_{10}O_{65}$
     d. $C_{26}H_{30}O_{15}$

Question #5: Name the only two plants wogonin is known to occur in:

1 _____ 2 _____

Question #6: Name two potential medical uses of wogonin:

1 _____ 2 _____

For the answer key to Wogonin, please visit www.cannabischemistry.org

SILYMARIN

Type: Flavonoid Complex

Like beta-sitosterol, silymarin is being included in this text primarily as an example of how misinformation can spread in the cannabis industry, particularly where related to scientific aspects of the plant and its use. The author originally selected this compound for inclusion in this work because silymarin has been reported as a flavonoid that occurs in cannabis by several sources. For instance, in an article for US company CBDFx, author Clay Steakley writes "Of the 20 flavonoids found in cannabis, there are a few major ones[594]," and then goes on to list silymarin as one of those major flavonoids. However, no source is provided to substantiate this claim. Similarly, author Patrick Bennett included the compound in a list of cannabis flavonoids in an article for cannabis media giant Leafly, writing; "Other highly active flavonoids found in cannabis include Orientin, Quercetin, Silymarin, and Kaempferol[595]." But once again, no citation or source was given. In one case, researchers searched the scientific literature for flavonoids known to occur in cannabis, and subsequently included silymarin in their results for setting up analyte validation methods, however, when they then tested hemp flower using these methods, no silymarin was quantified[596]. Finally, in an article titled "Anti-Cancer Potential of Cannabinoids, Terpenes, and Flavonoids Present in Cannabis," researchers wrote "Silymarin is a flavonoid derived from milk thistle, but is also present in artichokes, cilantro, coriander, and cannabis[597]," only to offer zero supporting evidence of this claim.

It appears that the only published work available as of early 2025 that quantifies silymarin in cannabis is a study conducted in Nigeria. In this case, researchers tested an ethanolic extraction of dried cannabis leaves provided by the National Drug Law Enforcement Agency, oddly finding seemingly equal amounts of silymarin, quercetin, and kaempferol[598]. Upon querying several cannabis industry researchers regarding this work, skepticism was expressed to the author that these results are accurate.

In general, silymarin is the name given to a flavonoid complex extracted from milk thistle. This complex is well-studied and, in most cases, consists of several flavonoids; silybin, isosilibinin, silicristin, and silidianin, where silybin is widely considered to be the principal active ingredient in the mixture referred to as silymarin. It does not appear that the literature supports including silymarin in this text for two primary reasons:

1. Silymarin is not an individual compound, it is a complex of flavonoids.
2. The likelihood that silymarin has been reliably quantified in cannabis is small.

However, future editions of this book will be updated with new information and emerging research, and, if any of the individual components of silymarin are found to occur in cannabis – most notably silybin – this flavonoid textbook will be updated accordingly with a chapter to support that compound.

[594] Steakley, Clay. What Are Flavonoids in Cannabis? CBDFx Company. Septmeber 21, 2021. From: https://cbdfx.com/what-are-flavonoids-in-cannabis/. Accessed December 1, 2023.

[595] Bennett, Patrick. What are cannabis flavonoids and what do they do? Leafly. February 8, 2018. From: https://www.leafly.com/news/cannabis-101/what-are-marijuana-flavonoids. Accessed December 1, 2023.

[596] York, Jamie L, PhD. Flavonoids...The Next Routine Cannabis Test? Restek Chromatography. 17 June 2021. From: https://www.restek.com/chromablography/flavonoids-the-next-routine-cannabis-test. Accessed December 1, 2023.

[597] Tomko AM, Whynot EG, Ellis LD, Dupré DJ. Anti-Cancer Potential of Cannabinoids, Terpenes, and Flavonoids Present in Cannabis. Cancers (Basel). 2020 Jul 21;12(7):1985.

[598] Nwonuma CO, Nwatu VC, Mostafa-Hedeab G, Adeyemi OS, Alejolowo OO, Ojo OA, Adah SA, Awakan OJ, Okolie CE, Asogwa NT, Udofia IA, Egharevba GO, Aljarba NH, Alkahtani S, Batiha GE. Experimental validation and molecular docking to explore the active components of cannabis in testicular function and sperm quality modulations in rats. BMC Complement Med Ther. 2022 Aug 26;22(1):227.

ISOVITEXIN

Type: Flavone
Chemical Formula: $C_{21}H_{20}O_{10}$
Molecular Weight: 432.38 g/mol
Boiling Point: 807.00 °C @ 760.00 mm Hg (estimated)[599]
Flash Point: 287.10 °C (estimated by TGSC)
Melting Point: 257-258 °C[600]
Solubility: Soluble in DMSO, dimethyl formamide[601], water
Oral TDLO: 15mg/kg (rat)[602]
Biological Role: Antioxidant, allelopathic agent
Therapeutic Role: Anticancer, antibacterial, neurological agent
Commercial Use: Research chemical
Occurrence in Cannabis: Probably top 20

Occurs in Cannabis Strains: Most laboratories are not testing cannabis samples for this flavone as of early 2025, so there are few reports of its occurrence outside of scientific research. One group of researchers quantified isovitexin in three cannabis chemovars in 2020, with one chemovar – CBD Mango Haze – showing much higher levels than the other two varieties[603]. A year later the same research team showed that isovitexin occurred in CBD-dominant chemovars, also noting that the flavonoid content was highest in leaves, and lower or negligent in flowers, bark, and roots[604]. Earlier research that examined two drug type varieties (Skunk and Fourway) and two fiber-type varieties (Kompolti and Fasamo) found that isovitexin occurred at 2.5 times lower concentration in the drug types than in the fiber types[605], again suggesting an association with hemp varieties.

INTRODUCTION

Isoxitexin is in the flavone subclass of flavonoids, and is known to occur in a moderate number of plants as detailed below. This compound is not well-known in the cannabis industry, and at the time of publication of this text, there were no publicly available laboratory testing reports of consumer cannabis products that sought to quantify isovitexin. Some research indicates this flavonoid may be used by plants as an antioxidant, and possibly as an allelopathic agent, while potential medical or therapeutic uses for this compound include in the treatment of cancer and chemotherapeutic agent toxicity, in neurodegenerative diseases, as an antibacterial agent, and many other possible uses in human medicine.

[599] The Good Scents Company Data Sheet for Isovitexin, from: https://www.thegoodscentscompany.com/data/rw1700051.html. Accessed December 2, 2023.

[600] Chemical Book Product Description for Isovitexin, from: https://www.chemicalbook.com/ChemicalProductProperty_US_CB5708574.aspx. Accessed December 2, 2023.

[601] Cayman Chemical Product Information Sheet for Isovitexin, from: https://cdn.caymanchem.com/cdn/insert/31212.pdf. Accessed December 2, 2023.

[602] Cayman Chemical Safety Data Sheet for Isovitexin, from: https://cdn.caymanchem.com/cdn/msds/31212m.pdf. Accessed December 2, 2023.

[603] Jin D, Dai K, Xie Z, Chen J. Secondary Metabolites Profiled in Cannabis Inflorescences, Leaves, Stem Barks, and Roots for Medicinal Purposes. Sci Rep. 2020 Feb 24;10(1):3309.

[604] Jin D, Henry P, Shan J, Chen J. Identification of Chemotypic Markers in Three Chemotype Categories of Cannabis Using Secondary Metabolites Profiled in Inflorescences, Leaves, Stem Bark, and Roots. Front Plant Sci. 2021 Jul 1;12:699530.

[605] Flores-Sanchez IJ, Verpoorte R. PKS activities and biosynthesis of cannabinoids and flavonoids in Cannabis sativa L. plants. Plant Cell Physiol. 2008 Dec;49(12):1767-82.

CHEMICAL STRUCTURE

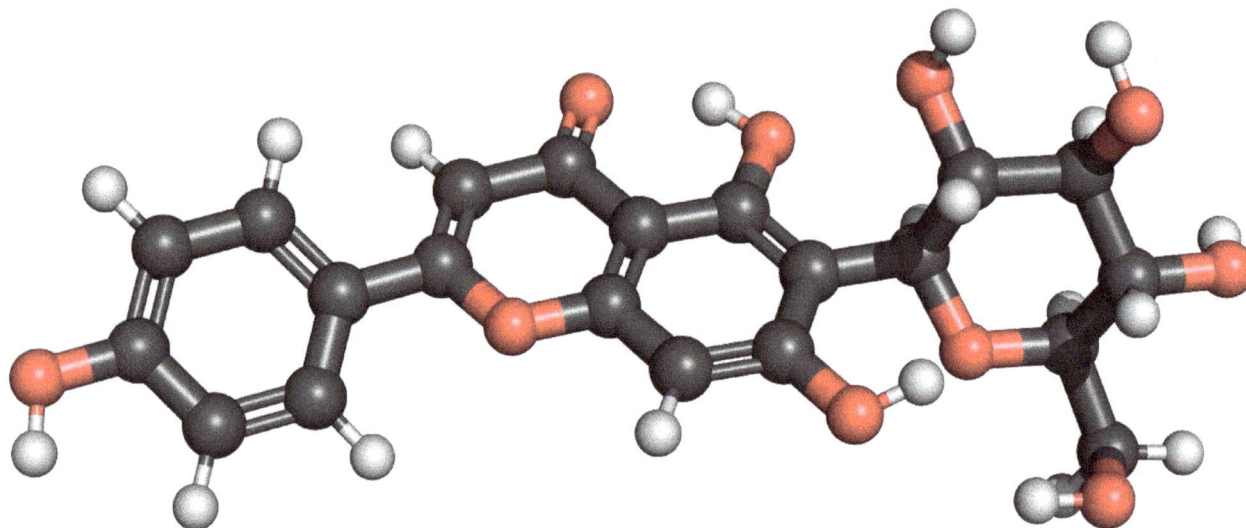

Isovitexin is a flavone consisting of twenty-one carbon atoms, twenty hydrogen atoms, and ten oxygen atoms built around four rings containing seven endocyclic double bonds, and one exocyclic double-bonded oxygen atom. It is important to note that this compound is not an isoflavonoid, which is generally determined by the position of the B-ring on the C-ring, but instead is dubbed with the 'iso' prefix in relation to the position of the glucose functional group.

Isovitexin is generally biosynthesized in the phenylpropanoid metabolic pathway via the production of apigenin. This trihydroxyflavone can be modified in a variety of ways, including:

Isovitexin: the basic structure of isovitexin consists of an apigenin molecule (a flavone) connected to a glucose molecule via a C-glycosidic bond.

Isovitexin 2' '-O-beta-D-glucoside: a second glucose molecule is attached to the 2nd position (2' ') of the primary glucose molecule.

Isovitexin 7-(6' ' '-sinapoylglucoside) 4'-glucoside: a sinapoyl group (derived from sinapic acid) is attached at position 7, and a glucose group at position 4.

Isovitexin 2' '-O-(6' ' '-(E)-p-coumaroyl)glucoside 4': a (E)-p-coumaroyl group (derived from p-coumaric acid) is attached at position 6' ' ' on the glucose molecule at position 2' ', and a glucose molecule is attached at position 4.

Isovitexin 7-O-glucoside 2' '-O-arabinoside: a glucose group is attached at position 7, and an arabinose group at position 2' '.

Isovitexin 7,2' '-di-O-glucoside: two glucose molecules are attached, one at position 7, and the other at position 2.

Isovitexin 7-O-glucoside-2' '-O-rhamnoside: a glucose group is attached at position 7, and a rhamnose group at position 2' '.

Isovitexin 7-O-xyloside-2' '-O-glucoside: a xylose group is attached at position 7, and a glucose group at position 2' '.

Isovitexin 7-O-xyloside: a xylose group is attached at position 7.

Isovitexin-7-olate: isovitexin deprotonated (a proton is removed from the molecule, leaving behind an anion with a negative charge) at position 7.

Isovitexin 4'-O-glucoside 2''-O-arabinoside: a glucose group is attached at position 4, and an arabinose group at position 2''.

Isovitexin 7-O-galactoside-2''-O-glucoside: a galactose group is attached at position 7, and a glucose group at position 2''.

Isovitexin 8-C-beta-glucoside: a glucose molecule is attached at position 8.

7-O-[6-(6-methoxycaffeoyl)glucosyl]isovitexin: a glucose group and a 6-methoxycaffeoyl group is attached at position 7.

2''-O-(beta-D-glucosyl)isovitexin: a glucose group is attached at position 2''.

Each of these variations, however slight, can lead to different functionality and properties, with a virtually unlimited number of variations possible.

OCCURRENCE IN PLANTS

Isovitexin occurs in a moderate number of plants, including some that are common food items in many countries;

Acai palm	Anise	Bamboo
Barley	Black gram	Buckwheat
Cacao	Cannabis	Cucumber
Fenugreek	Flax	Indocalamus latifolius
Lemon	Maize	Mung bean
Oat	Oregano	Passion flower
Patrinia villosa	Pea	Salsify
Tamarind	Tarragon	Tea
Turnip	Winged bean	

BIOLOGICAL ACTIVITY IN PLANTS

Scientific literature examining the biological activity of isovitexin in plants is scarce, and mostly focused on rice plants, with researchers suggesting that the flavone may protect cells from oxidative stress[606]. Other research has shown that a variation of isovitexin isolated from rice plants, isovitexin-2"-O-beta-[6-O-E-p-coumaroylglucopyranoside], dramatically reduced the number of fertile eggs laid by Helicoverpa armigera (African bollworm) to just 7% of control insects when added to insect diets[607], indicating a potential allelopathic effect.

USES IN INDUSTRY

Isovitexin is generally only available commercially as a research chemical.

---

[606] Lin CM, Chen CT, Lee HH, Lin JK. Prevention of cellular ROS damage by isovitexin and related flavonoids. Planta Med. 2002 Apr;68(4):365-7.

[607] Caasi-Lit MT, Tanner GJ, Nayudu M, Whitecross MI. Isovitexin-2'-O-beta-[6-O-E-p-coumaroylglucopyranoside] from UV-B irradiated leaves of rice, Oryza sativa L. inhibits fertility of Helicoverpa armigera. Photochem Photobiol. 2007 Sep-Oct;83(5):1167-73.

POTENTIAL USES IN MEDICINE

While not as well-studied as some of the major flavonoids in this text, isovitexin still offers significant potential in human medicine, with more than a dozen therapeutic uses researched to date:

Treatment for Liver Cancer: As extracted from rice hulls of Oryza sativa, isovitexin has been shown to induce apoptosis via the mitochondrial apoptotic pathway, induce endoplasmic reticulum stress, induce autophagy by enhancing LC3II and other autophagy-related proteins including Beclin1 (a protein that has a central role in autophagy), significantly suppressing the growth of human liver cancer cells[608]. Isovitexin has also been shown to upregulate miR-34a (micro-RNA 34, a tumor suppressing gene), inducing apoptosis and suppressing the stemness (essentially refers to cell plasticity) of hepatocellular carcinoma cells[609].

Treatment for Lung Cancer: Isovitexin has been shown to suppress the stemness of lung cancer stem-like cells by arresting manganese superoxide dismutase (a metalloprotein that prevents mitochondrial dysfunction), Ca 2+ /calmodulin-dependent protein kinase II (mediates signaling cascades), and adenosine monophosphate-activated protein kinase (an enzyme that mediates cellular energy homeostasis) signaling and glycolysis inhibition[610].

Treatment for Colon Cancer: In human colon cancer cells, treatment with isovitexin significantly attenuated cell proliferation, migration, invasion, epithelial-mesenchymal transition, and induced cell apoptosis[611], with researchers suggesting the compound as a potential new treatment for colon cancer.

Treatment for Chemotherapeutic Agent Toxicity: Isovitexin could be a valuable new tool in the fight against toxicity caused by cancer treatment drugs. In one study, researchers found that treatment with isovitexin reduced cisplatin-induced hepatotoxicity and nephrotoxicity in a mouse model of non-small cell lung cancer[612]. As isolated from rice hulls of Oryza sativa, isovitexin inhibited cisplatin-induced inflammation by inhibiting TNF-α, IL-1ß and IL-6 production, inhibited MDA and ROS production, inhibited NF-κB activation, and increased Nrf2 and HO-1 expression in the kidneys of mice[613].

Antioxidant: In lipopolysaccharide (LPS)-activated RAW264.7 macrophages (murine leukemia cells), treatment with isovitexin reduced the production of hydrogen peroxide, markedly reduced LPS-stimulated nitric oxide production, inhibited the expression of iNOS, and inhibited IKK (inhibitor of nuclear factor-κB (IκB) kinase (IKK) complex) activity[614].

---

[608] Lv SX, Qiao X. Isovitexin (IV) induces apoptosis and autophagy in liver cancer cells through endoplasmic reticulum stress. Biochem Biophys Res Commun. 2018 Feb 19;496(4):1047-1054.

[609] Xu C, Cao X, Cao X, Liu L, Qiu Y, Li X, Zhou L, Ning Y, Ren K, Cao J. Isovitexin Inhibits Stemness and Induces Apoptosis in Hepatocellular Carcinoma SK-Hep-1 Spheroids by Upregulating miR-34a Expression. Anticancer Agents Med Chem. 2020;20(14):1654-1663.

[610] Liu F, Yuan Q, Cao X, Zhang J, Cao J, Zhang J, Xia L. Isovitexin Suppresses Stemness of Lung Cancer Stem-Like Cells through Blockage of MnSOD/CaMKII/AMPK Signaling and Glycolysis Inhibition. Biomed Res Int. 2021 May 24;2021:9972057.

[611] Zhu H, Zhao N, Jiang M. Isovitexin attenuates tumor growth in human colon cancer cells through the modulation of apoptosis and epithelial-mesenchymal transition via PI3K/Akt/mTOR signaling pathway. Biochem Cell Biol. 2021 Dec;99(6):741-749.

[612] Chen RL, Wang Z, Huang P, Sun CH, Yu WY, Zhang HH, Yu CH, He JQ. Isovitexin potentiated the antitumor activity of cisplatin by inhibiting the glucose metabolism of lung cancer cells and reduced cisplatin-induced immunotoxicity in mice. Int Immunopharmacol. 2021 May;94:107357.

[613] Liu S, Zhang X, Wang J. Isovitexin protects against cisplatin-induced kidney injury in mice through inhibiting inflammatory and oxidative responses. Int Immunopharmacol. 2020 Jun;83:106437.

[614] Lin CM, Huang ST, Liang YC, Lin MS, Shih CM, Chang YC, Chen TY, Chen CT. Isovitexin suppresses lipopolysaccharide-mediated inducible nitric oxide synthase through inhibition of NF-kappa B in mouse macrophages. Planta Med. 2005 Aug;71(8):748-53.

Antihyperglycemic Agent: As isolated from Wilbrandia ebracteate, isovitexin was shown to exhibit strong antihyperglycemic effects in rats, stimulating insulin secretion resembling oral sulphonylureas[615] (medicines for type 2 diabetes) activity.

Treatment for Alzheimer's Disease: As isolated from Serjania erecta Radlk (Sapindaceae) leaves, isovitexin protected PC12 cells (catecholamine cells) against Aβ25-35 peptide-induced toxicity[616].

Neuroprotective Agent: In glutamate-induced neurotoxicity in hippocampal brain slices of mice, dietary supplementation of an extract high in isovitexin provided significant protective effects against cell damage[617].

Treatment for Lung Injury: In a murine model of acute lung injury, treatment with isovitexin reduced lipopolysaccharide (LPS)-induced pro-inflammatory cytokine secretion, iNOS and COX-2 expression, and decreased the generation of ROS both in vitro and in vivo, effectively protecting against LPS-induced damage, oxidative stress, and inflammation[618].

Treatment for Liver Injury: As isolated from rice hulls of Oryza sativa, isovitexin inhibited inflammatory and oxidative responses induced via treatment with lipopolysaccharide/d-galactosamine or carbon tetrachloride in a murine model of induced liver injury[619].

Antibacterial Agent: Isovitexin has been shown to reduce Staphylococcus aureus adhesion to fibrinogen, reduce biofilm formation, and reduce SpA display on the surface of S. aureus by inhibiting the activity of sortase A[620]. Other researchers have echoed this work, finding that isovitexin can reversibly inhibit sortase A activity in mice with methicillin-resistant Staphylococcus aureus-induced pneumonia[621].

Treatment for Gout: In Sprague-Dawley rats with induced gouty arthritis, treatment with isovitexin alleviated the infiltration of inflammatory cells, ameliorated the proliferation of synovial cells, significantly reduced levels of TNF-α, IL-1β, and IL-6, and remarkably decreased expression levels of TLR4 (toll-like receptor 4), MyD88 (myeloid differentiation primary response 88), and p-NF-κB-p65[622].

Osteoanabolic Agent: In ovariectomized osteopenic mice, treatment with isovitexin completely restored bone strength at L5 (compressive strength) and femur (bending strength), while also inhibiting bone and serum sclerostin

[615] Folador P, Cazarolli LH, Gazola AC, Reginatto FH, Schenkel EP, Silva FR. Potential insulin secretagogue effects of isovitexin and swertisin isolated from Wilbrandia ebracteata roots in non-diabetic rats. Fitoterapia. 2010 Dec;81(8):1180-7.

[616] Guimarães CC, Oliveira DD, Valdevite M, Saltoratto AL, Pereira SI, França Sde C, Pereira AM, Pereira PS. The glycosylated flavonoids vitexin, isovitexin, and quercetrin isolated from Serjania erecta Radlk (Sapindaceae) leaves protect PC12 cells against amyloid-β25-35 peptide-induced toxicity. Food Chem Toxicol. 2015 Dec;86:88-94.

[617] Dos Santos KC, Borges TV, Olescowicz G, Ludka FK, Santos CA, Molz S. Passiflora actinia hydroalcoholic extract and its major constituent, isovitexin, are neuroprotective against glutamate-induced cell damage in mice hippocampal slices. J Pharm Pharmacol. 2016 Feb;68(2):282-91.

[618] Lv H, Yu Z, Zheng Y, Wang L, Qin X, Cheng G, Ci X. Isovitexin Exerts Anti-Inflammatory and Anti-Oxidant Activities on Lipopolysaccharide-Induced Acute Lung Injury by Inhibiting MAPK and NF-κB and Activating HO-1/Nrf2 Pathways. Int J Biol Sci. 2016 Jan 1;12(1):72-86.

[619] Hu JJ, Wang H, Pan CW, Lin MX. Isovitexin alleviates liver injury induced by lipopolysaccharide/d-galactosamine by activating Nrf2 and inhibiting NF-κB activation. Microb Pathog. 2018 Jun;119:86-92.

[620] Mu D, Xiang H, Dong H, Wang D, Wang T. Isovitexin, a Potential Candidate Inhibitor of Sortase A of Staphylococcus aureus USA300. J Microbiol Biotechnol. 2018 Sep 28;28(9):1426-1432.

[621] Tian L, Wu X, Yu H, Yang F, Sun J, Zhou T, Jiang H. Isovitexin Protects Mice from Methicillin-Resistant Staphylococcus aureus-Induced Pneumonia by Targeting Sortase A. J Microbiol Biotechnol. 2022 Oct 28;32(10):1284-1291.

[622] Sun X, Li P, Qu X, Liu W. Isovitexin alleviates acute gouty arthritis in rats by inhibiting inflammation via the TLR4/MyD88/NF-κB pathway. Pharm Biol. 2021 Dec;59(1):1326-1333.

as well as the serum type I collagen cross-linked C-telopeptide, providing an osteoanabolic effect equivalent to teriparatide[623] (a drug used in the treatment of osteoporosis).

Treatment for Kidney Disease: In a murine model of lipopolysaccharide (LPS)-induced renal injury, isovitexin prevented ROS production, increased cell viability, ameliorated mitochondrial membrane potential, and downregulated inflammation and pyroptosis factors in vitro, while in vivo tests showed that isovitexin decreased LPS-induced glomerular atrophy and reduced inflammation-related cytokine release[624].

Treatment for Colitis: In a murine model of dextran sodium sulfate (DSS)-induced colitis, treatment with isovitexin inhibited body weight loss, colonic histological changes, and levels of TNF-α and IL-1β in vivo, while similar treatment in vitro resulted in the inhibition of IL-6 and IL-1β production, and an increase in the expression of tight junction proteins ZO-1 and occludin[625].

IMPLICATIONS FOR HUMAN HEALTH AND NUTRITION

Dietary consumption of isovitexin is not as easy as that of some other flavonoids that are found in many different plants. For cannabis varieties that might contain some amount of this flavone, consider CBD-dominant varieties, including CBD Mango Haze, which is known to produce isovitexin.

When preparing, cooking, or serving foods, consider working with the herbs and spices anise, oregano, or tarragon. Making foods from grains and seeds such as barley, flax, oats, buckwheat, and fenugreek might also allow some consumption of isovitexin. Consider adding lemon to beverages or eating raw lemons, snacking on tamarind products, or consuming teas – particularly those that contain anise. Supplement meals with vegetables and legumes that are known to produce isovitexin, including peas, black gram, cucumbers, maize (corn), mung beans, turnip, winged beans, and salsify.

Readers are urged to use caution and obtain professional counsel on the use of purified isovitexin.

[623] Pal S, Sharma S, Porwal K, Riyazuddin M, Kulkarni C, Chattopadhyay S, Sanyal S, Gayen JR, Chattopadhyay N. Oral Administration of Isovitexin, a Naturally Occurring Apigenin Derivative Showed Osteoanabolic Effect in Ovariectomized Mice: A Comparative Study with Teriparatide. Calcif Tissue Int. 2022 Aug;111(2):196-210.

[624] Tseng CY, Yu PR, Hsu CC, Lin HH, Chen JH. The effect of isovitexin on lipopolysaccharide-induced renal injury and inflammation by induction of protective autophagy. Food Chem Toxicol. 2023 Feb;172:113581.

[625] Mu J, Song J, Li R, Xue T, Wang D, Yu J. Isovitexin prevents DSS-induced colitis through inhibiting inflammation and preserving intestinal barrier integrity through activating AhR. Chem Biol Interact. 2023 Sep 1;382:110583.

Isovitexin Review

Answer the following questions to test your knowledge of this flavonoid:

Question #1: What type of flavonoid is isovitexin?

        a.   Flavane
        b.   Flavonol
        c.   Flavone
        d.   Flavoned

Question #2: The base component in the biosynthesis of isovitexin is:

        a.   Cannabisin
        b.   Apigetrin
        c.   Chrysin
        d.   Apigenin

Question #3: How common is isovitexin in cannabis?

        a.   Top three
        b.   Top five
        c.   Top ten
        d.   Top twenty
        e.   Unknown

Question #4: What is the chemical formula for isovitexin?

        a. $C_{21}H_{20}O_{10}$
        b. $C_{26}H_{10}O_{15}$
        c. $C_{16}H_{10}O_{65}$
        d. $C_{26}H_{30}O_{15}$

Question #5: Name four plants isovitexin is known to occur in:

1 _____   2 _____

3 _____   4 _____

Question #6: Name two potential medical uses of isovitexin:

1 _____   2 _____

For the answer key to Isovitexin, please visit www.cannabischemistry.org

FISETIN

Type: Flavonol
Chemical Formula: $C_{15}H_{10}O_6$
Molecular Weight: 286.24 g/mol
Boiling Point: 599.41 °C @ 760.00 mm Hg (estimated)[626]
Flash Point: 233 °C (estimated by TGSC)
Melting Point: 330 °C @ 760.00 mm Hg
Solubility: Soluble in water (probably)
Intravenous LD50: 180 mg/kg (mouse)[627]
Biological Role: Unknown
Therapeutic Role: Anticancer, senolytic, renal protective, neurological agent
Commercial Use: Anti-inflammatory, anti-aging agent
Occurrence in Cannabis: Unknown

Occurs in Cannabis Strains: Fisetin is theorized to occur in cannabis, although as of early 2025 there were no publicly available laboratory or other quantified tests of cannabis or cannabis products that contained or reported the presence of fisetin. However, ACS Laboratory writes that "The flavonoid fisetin, a yellow plant pigment, occurs in various plants, fruits and vegetables, such as cannabis[628]" but provides no reference or citation for this statement. Additionally, on a web page for Krackeler Scientific, the company writes that their analytical reference standards and other materials can be used to "identify and quantitatively measure the components found in Cannabis products to aid the Cannabis industry in potency testing, terpene profiling, and analysis of contaminants[629]" and then goes on to list fisetin as one of those cannabis constituents. Again, no reference or citation was provided to substantiate the presence of fisetin in cannabis.

INTRODUCTION

Fisetin is a well-known anti-aging and anti-cancer agent, and is one of the most-studied phytochemicals described in this text. This flavonol is used as an anti-aging and anti-inflammatory agent in many products such as skin care and hair care products, and it is used as an additive in some food items. Although the compound is not known to have been quantified in cannabis as of the publication of this text, fisetin likely occurs in at least some varieties. This text will be updated as more information becomes available about fisetin specifically as it occurs in cannabis.

---

[626] The Good Scents Company Data Sheet for Fisetin, from: https://www.thegoodscentscompany.com/data/rw1247011.html. Accessed December 28, 2023.

[627] U.S. Army Armament Research & Development Command, Chemical Systems Laboratory.

[628] ACS Laboratory. Flavonoid Friday: Everything You Need to Know About Fisetin Flavor, Fragrance, and Benefits. From: https://www.acslab.com/flavonoids/flavonoid-friday-everything-you-need-to-know-about-fisetin-flavor-fragrance-and-benefits. Accessed December 28, 2023.

[629] Krackeler Scientific. Cannabis Analysis Standards and Chemicals. Cayman Chemical. From: https://www.krackeler.com/catalog/product/11136/Cannabis-Analysis-Standards-and-Chemicals. Accessed December 28, 2023.

CHEMICAL STRUCTURE

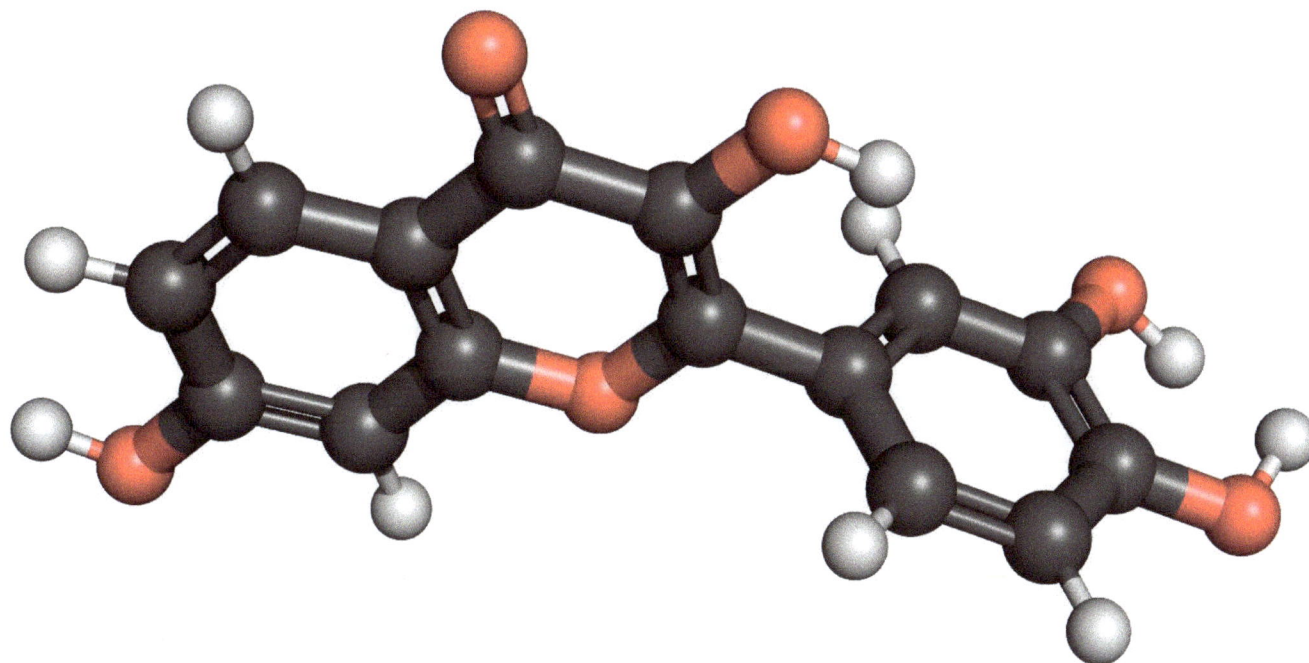

Formed from naringenin in the phenylpropanoid pathway, fisetin is considered a hydroxyflavonol because it has hydroxy groups at positions 3, 3', and 4'. These hydroxy groups are largely responsible for the biological activity of this flavonol. The fisetin molecule contains three rings, seven endocyclic double bonds, one exocyclic carbon to oxygen atom double bond, six oxygen atoms total, fifteen carbon atoms, and ten hydrogen atoms, notated as $C_{15}H_{10}O_6$. Fisetin is structurally similar to quercetin, and in fact hydroxylation at position 5 of the A ring in fisetin converts the molecule to quercetin.

Fisetin can occur as a variety of glucoside or methylation products, including;

Fisetin tetramethyl ether: four methyl groups are attached to different positions of the molecule, often resulting in isomerism.

Fisetin 4'-glucoside: contains a glucoside group at the 4' position.

Fisetin 8-C-glucoside: contains a glucoside group attached at position 8-C.

Fisetin 3-methyl ether: contains a methyl ether group attached to position 3.

Fisetin 3-glucoside: contains a glucoside group at position 3.

These additions and substitutions often result in different functionality and properties of the molecule, including biological activities that may be therapeutically beneficial to humans and animals.

OCCURRENCE IN PLANTS

Fisetin is not known to occur in many plants, however, it does occur in enough sources that most people should be able to obtain dietary fisetin. This flavonol is known to occur and has been studied in Acacia species including A. greggii and A. Berlandieri[630], mulberry trees[631], as well as strawberry, apple, persimmon, grape, onion, and cucumber plants[632].

Acacia (multiple species)	Apple	Cotinus coggygria
Cucumber	Eurasian smoketree	Fabaceae (multiple species)
Grape	Honey locust	Kale
Kiwi	Lotus	Mulberry
Onion	Parrot tree	Peace
Persimmon	Quebracho colorado	Soybean leaf
Strawberry	Sumac	Tomato
Toxicodendron vernicifluum	Yellow cypress	

BIOLOGICAL ACTIVITY IN PLANTS

The biological activities of fisetin in plants is understudied, with some researchers suggesting that the compound possibly extends the life of various organisms that produce this flavonol. Additionally, it is likely that fisetin offers radiation-protective and other beneficial effects attributed to the strong antioxidant properties of this molecule. More research is needed to confirm these and other uses in plants. Interestingly, in mulberry trees, concentrations of fisetin are significantly higher in trees that are twenty or more years old than in younger trees[633].

USES IN INDUSTRY

The potent antioxidant and anti-inflammatory effects of fisetin are used in the food industry to enhance or change the flavor of products, to increase the nutritional value of foods, and to help preserve food products. In some cases, fisetin is used as a yellow or orange pigment as a natural food coloring. Fisetin is Generally Recognized as Safe (GRAS) by the US Food and Drug Administration (FDA), and can be found in various consumer products including cosmetics, skincare and haircare products, and in a variety of supplements including capsules, tablets, powders, drinks, and gummies.

Interestingly, researchers have successfully engineered Escherichia coli as a microbial platform strain for the production of fisetin and related flavonols[634].

---

[630] Forbes TDA, Clement BA. "Chemistry of Acacia's from South Texas". Texas A&M Agricultural Research and Extension Center at Uvalde, 2010. From: https://catbull.com/alamut/Bibliothek/chem%20of%20texas%20acacias.pdf. Accessed January 19, 2024.

[631] Tsurudome N, Minami Y, Kajiya K. Fisetin, a major component derived from mulberry (Morus australis Poir.) leaves, prevents vascular abnormal contraction. Biofactors. 2022 Jan;48(1):56-66.

[632] Kim HJ, Kim SH, Yun JM. Fisetin inhibits hyperglycemia-induced proinflammatory cytokine production by epigenetic mechanisms. Evid Based Complement Alternat Med. 2012;2012:639469.

[633] Tsurudome N, Minami Y, Kajiya K. Fisetin, a major component derived from mulberry (Morus australis Poir.) leaves, prevents vascular abnormal contraction. Biofactors. 2022 Jan;48(1):56-66.

[634] Stahlhut SG, Siedler S, Malla S, Harrison SJ, Maury J, Neves AR, Forster J. Assembly of a novel biosynthetic pathway for production of the plant flavonoid fisetin in Escherichia coli. Metab Eng. 2015 Sep;31:84-93.

POTENTIAL USES IN MEDICINE

Fisetin is a powerful anticancer agent, particularly against breast, oral, and pancreatic cancer, but it is mostly known as one of the world's most potent anti-aging or senolytic compounds. Additionally, this flavonol has potential to be used in the treatment or prevention of many other diseases and conditions, including treatment for Parkinson's disease, heart attack, kidney disease, cognitive disorders, and dozens of other medical and therapeutic uses as detailed below:

Treatment for Breast Cancer: In combination with quercetin and naringenin, fisetin contributed significantly to cell growth reduction, migration suppression, and apoptosis by upregulating miR-1275 and downregulating miR-27a-3p in MCF7 and MDA-MB-231 breast cancer cell lines[635]. Other similar work showed that the combination of fisetin and quercetin synergistically inhibited breast cancer cell proliferation, migration, and colony formation by interacting with the matrix metalloproteinase signaling and apoptotic pathways in mice[636]. Another potent combination – fisetin and kaempferol – exhibited significant cytotoxicity against the triple negative breast cancer MDA-MB-231 cell line, leading to more than 50% cell death at just 20 micrometers[637]. As an individual constituent, fisetin has been shown to induce apoptosis in breast cancer MDA-MB-453 cells through the degradation of HER2/neu and via the PI3K/Akt pathway[638], while other research has demonstrated that fisetin decreases MMP-2 and MMP-9 enzyme activity and gene expression in both protein and mRNA levels in metastatic breast cancer cells[639].

Treatment for Ovarian Cancer: Fisetin has been shown to dose-dependently reduce cell growth and induce apoptosis and necroptosis by mediating the RIP3/MLKL pathway in ovarian carcinoma cells[640].

Treatment for Cervical Cancer: In HeLa cervical cancer cells, treatment with fisetin has been shown to induce morphological changes and inhibit proliferation, change nuclear morphology, induce DNA fragmentation, encourage G2/M arrest and modulate cell cycle regulatory genes, activate extrinsic and intrinsic pathways, modulate expression of pro- and anti-apoptotic proteins, elevate caspase-3, caspase-8, and caspase-9 activity, ameliorate oxidation stress, alleviate inflammation, and change the aberrant MAPK and PI3K/AKT/mTOR activity[641].

Treatment for Colon Cancer: In the colorectal cancer cell line SW-480, fisetin was shown to induce apoptosis by suppressing autophagy and down-regulating nuclear factor erythroid 2-related factor 2 (Nrf2), cleaved caspase-3, and nuclear PARP-1[642]. Treatment of PIK3CA-mutant colorectal cancer cells with fisetin and 5-fluorouracil reduced

[635] Jalalpour Choupanan M, Shahbazi S, Reiisi S. Naringenin in combination with quercetin/fisetin shows synergistic anti-proliferative and migration reduction effects in breast cancer cell lines. Mol Biol Rep. 2023 Sep;50(9):7489-7500.

[636] Hosseini SS, Ebrahimi SO, Haji Ghasem Kashani M, Reiisi S. Study of quercetin and fisetin synergistic effect on breast cancer and potentially involved signaling pathways. Cell Biol Int. 2023 Jan;47(1):98-109.

[637] Afzal M, Alarifi A, Karami AM, Ayub R, Abduh NAY, Saeed WS, Muddassir M. Antiproliferative Mechanisms of a Polyphenolic Combination of Kaempferol and Fisetin in Triple-Negative Breast Cancer Cells. Int J Mol Sci. 2023 Mar 29;24(7):6393.

[638] Guo G, Zhang W, Dang M, Yan M, Chen Z. Fisetin induces apoptosis in breast cancer MDA-MB-453 cells through degradation of HER2/neu and via the PI3K/Akt pathway. J Biochem Mol Toxicol. 2019 Apr;33(4):e22268.

[639] Tsai CF, Chen JH, Chang CN, Lu DY, Chang PC, Wang SL, Yeh WL. Fisetin inhibits cell migration via inducing HO-1 and reducing MMPs expression in breast cancer cell lines. Food Chem Toxicol. 2018 Oct;120:528-535.

[640] Liu Y, Cao H, Zhao Y, Shan L, Lan S. Fisetin-induced cell death in human ovarian cancer cell lines via zbp1-mediated necroptosis. J Ovarian Res. 2022 May 10;15(1):57.

[641] Afroze N, Pramodh S, Shafarin J, Bajbouj K, Hamad M, Sundaram MK, Haque S, Hussain A. Fisetin Deters Cell Proliferation, Induces Apoptosis, Alleviates Oxidative Stress and Inflammation in Human Cancer Cells, HeLa. Int J Mol Sci. 2022 Feb 1;23(3):1707.

[642] Pandey A, Trigun SK. Fisetin induces apoptosis in colorectal cancer cells by suppressing autophagy and down-regulating nuclear factor erythroid 2-related factor 2 (Nrf2). J Cell Biochem. 2023 Sep;124(9):1289-1308.

the expression of PI3K, phosphorylation of AKT, mTOR, its target proteins, and constituents of mTOR signaling complex, and reduced the total number of intestinal tumors in mice[643].

Treatment for Oral Cancer: In SCC-4 human oral cancer cells, treatment with fisetin induced cell death, caused G2/M phase arrest, induced apoptosis, promoted ROS and Ca2+ production, and increased caspase-3, -8, and -9 activities[644]. In the same cell line, fisetin has also been shown remarkable uptake in the cells with significant cytotoxic action with administered via fucoidan/hyaluronic acid cross-linked zein nanoparticles[645]. Finally, fisetin has been shown to markedly inhibit hepatocyte growth factor receptor (c-Met)/Src signaling, downregulate PAK4 expression, inhibit colony formation, inhibit the cell cycle and migration, and induce cleavage of caspase-3 and PARP in human oral squamous cell carcinoma[646].

Treatment for Renal Cell Carcinoma: Fisetin has been shown to inhibit cell viability through cell cycle arrest in the G2/M phase, downregulate cyclin D1 and upregulate p21/p27, inhibit the migration and invasion of human renal carcinoma cells through the downregulation of CTSS, and disintegrin and metalloproteinase 9, and upregulate ERK phosphorylation in 786-O and Caki-1 renal carcinoma cells[647].

Treatment for Non-small Cell Lung Cancer: In combination with cancer treatment drug paclitaxel, fisetin contributed to significant synergistic effects, inducing miotic catastrophe and autophagic cell death in A549 non-small cell lung cancer cells[648]. Fisetin has also been shown to suppress migration, invasion, and stem-cell-like phenotype of human non-small cell lung carcinoma cells via attenuation of epithelial to mesenchymal transition, significantly inhibiting migration and invasion of the cancerous cells[649].

Treatment for Pancreatic Cancer: Fisetin has been shown to markedly inhibit the growth, migration, and infiltration of pancreatic cancer cells by targeting PI3K/AKT/mTOR signaling cascade[650]. In nude mice, treatment with fisetin inhibited cell proliferation, induced DNA damage, induced S-phase arrest in human pancreatic cancer PANC-1-luciferase cells[651].

[643] Khan N, Jajeh F, Eberhardt EL, Miller DD, Albrecht DM, Van Doorn R, Hruby MD, Maresh ME, Clipson L, Mukhtar H, Halberg RB. Fisetin and 5-fluorouracil: Effective combination for PIK3CA-mutant colorectal cancer. Int J Cancer. 2019 Dec 1;145(11):3022-3032.

[644] Su CH, Kuo CL, Lu KW, Yu FS, Ma YS, Yang JL, Chu YL, Chueh FS, Liu KC, Chung JG. Fisetin-induced apoptosis of human oral cancer SCC-4 cells through reactive oxygen species production, endoplasmic reticulum stress, caspase-, and mitochondria-dependent signaling pathways. Environ Toxicol. 2017 Jun;32(6):1725-1741.

[645] Moustafa MA, El-Refaie WM, Elnaggar YSR, El-Mezayen NS, Awaad AK, Abdallah OY. Fucoidan/hyaluronic acid cross-linked zein nanoparticles loaded with fisetin as a novel targeted nanotherapy for oral cancer. Int J Biol Macromol. 2023 Jun 30;241:124528.

[646] Li Y, Jia S, Dai W. Fisetin Modulates Human Oral Squamous Cell Carcinoma Proliferation by Blocking PAK4 Signaling Pathways. Drug Des Devel Ther. 2020 Feb 25;14:773-782.

[647] Hsieh MH, Tsai JP, Yang SF, Chiou HL, Lin CL, Hsieh YH, Chang HR. Fisetin Suppresses the Proliferation and Metastasis of Renal Cell Carcinoma through Upregulation of MEK/ERK-Targeting CTSS and ADAM9. Cells. 2019 Aug 21;8(9):948.

[648] Klimaszewska-Wisniewska A, Halas-Wisniewska M, Tadrowski T, Gagat M, Grzanka D, Grzanka A. Paclitaxel and the dietary flavonoid fisetin: a synergistic combination that induces mitotic catastrophe and autophagic cell death in A549 non-small cell lung cancer cells. Cancer Cell Int. 2016 Feb 16;16:10.

[649] Tabasum S, Singh RP. Fisetin suppresses migration, invasion and stem-cell-like phenotype of human non-small cell lung carcinoma cells via attenuation of epithelial to mesenchymal transition. Chem Biol Interact. 2019 Apr 25;303:14-21.

[650] Xiao Y, Liu Y, Gao Z, Li X, Weng M, Shi C, Wang C, Sun L. Fisetin inhibits the proliferation, migration and invasion of pancreatic cancer by targeting PI3K/AKT/mTOR signaling. Aging (Albany NY). 2021 Nov 25;13(22):24753-24767.

[651] Ding G, Xu X, Li D, Chen Y, Wang W, Ping D, Jia S, Cao L. Fisetin inhibits proliferation of pancreatic adenocarcinoma by inducing DNA damage via RFXAP/KDM4A-dependent histone H3K36 demethylation. Cell Death Dis. 2020 Oct 22;11(10):893.

Treatment for Thyroid Cancer: In human thyroid TPC 1 cancer cells, treatment with fisetin stimulated apoptosis, improved ROS generation, altered MMP and cell cycle phases, upregulated the expression of caspase (-3, -8, and -9), and downregulated JAK 1 and STAT3 expression[652].

Treatment for Chemotherapeutic Agent Toxicity: Fisetin offers significant potential as a treatment for chemotherapeutic agent toxicity, in one case ameliorating cisplatin-induced nephrotoxicity in rats through the modulation of NF-κB activation, and antioxidant defenses[653], while in another case attenuating doxorubicin-induced cardiotoxicity by inhibiting the insulin-like growth factor II receptor apoptotic pathway[654].

Chemotherapeutic Agent Synergist: Fisetin has been shown to synergize the effects of various anticancer drugs, increasing the sensitivity of chemoresistant cells to cisplatin[655], and sensitizing liver cancer[656] and triple negative breast cancer[657] cells to reduce radioresistance (cancer cells that have become resistant to radiation therapy).

Novel Cancer Therapy: In HCT116 human colorectal carcinoma cells induced to senescence via treatment with curcumin, subsequent treatment with quercetin and fisetin induced specific irreversible cell cycle arrest and depletion of cancer cells[658], offering the potential for a novel cancer treatment.

Delayed Postovulatory Oocyte Aging: Fisetin has been shown to delay postovulatory oocyte aging in mice by elevating Sirt1 expression, inhibiting oxidative stress, attenuating abnormal spindle formation, γH2A.X (a marker of DNA double strand breaks and genomic instability) reduction, and apoptosis, attenuating the aging-induced dysfunction of mitochondria, altering the expression of mitochondrial genes, and attenuating the aberrant intensity of H3K9me3[659] (histone 3 lysine 9 trimethylation, which serves key roles in embryonic stem cells).

Treatment for Polycystic Ovary Syndrome: In a murine model of letrozole-induced polycystic ovary syndrome, treatment with fisetin normalized levels of glucose, lipid profile, HOMA-IR (homeostatic model assessment for insulin resistance), testosterone, estradiol, and progesterone, while also increasing expression levels of SIRT1 and AMPK and decreasing expression levels of CYP17A1 (a gene involved in the biosynthesis of androgens) in rat ovaries[660].

[652] Liang, Y., Kong, D., Zhang, Y. et al. Fisetin Inhibits Cell Proliferation and Induces Apoptosis via JAK/STAT3 Signaling Pathways in Human Thyroid TPC 1 Cancer Cells. Biotechnol Bioproc E 25, 197–205 (2020).

[653] Sahu BD, Kalvala AK, Koneru M, Mahesh Kumar J, Kuncha M, Rachamalla SS, Sistla R. Ameliorative effect of fisetin on cisplatin-induced nephrotoxicity in rats via modulation of NF-κB activation and antioxidant defence. PLoS One. 2014 Sep 3;9(9):e105070.

[654] Lin KH, Ramesh S, Agarwal S, Kuo WW, Kuo CH, Chen MY, Lin YM, Ho TJ, Huang PC, Huang CY. Fisetin attenuates doxorubicin-induced cardiotoxicity by inhibiting the insulin-like growth factor II receptor apoptotic pathway through estrogen receptor-α/-β activation. Phytother Res. 2023 Sep;37(9):3964-3981.

[655] Ling J, Zhang L, Wang Y, Chang A, Huang Y, Zhao H, Zhuo X. Fisetin, a dietary flavonoid, increases the sensitivity of chemoresistant head and neck carcinoma cells to cisplatin possibly through HSP90AA1/IL-17 pathway. Phytother Res. 2023 May;37(5):1997-2011.

[656] Kim TW. Fisetin, an Anti-Inflammatory Agent, Overcomes Radioresistance by Activating the PERK-ATF4-CHOP Axis in Liver Cancer. Int J Mol Sci. 2023 May 22;24(10):9076.

[657] Khozooei S, Lettau K, Barletta F, Jost T, Rebholz S, Veerappan S, Franz-Wachtel M, Macek B, Iliakis G, Distel LV, Zips D, Toulany M. Fisetin induces DNA double-strand break and interferes with the repair of radiation-induced damage to radiosensitize triple negative breast cancer cells. J Exp Clin Cancer Res. 2022 Aug 22;41(1):256.

[658] Barra V, Chiavetta RF, Titoli S, Provenzano IM, Carollo PS, Di Leonardo A. Specific Irreversible Cell-Cycle Arrest and Depletion of Cancer Cells Obtained by Combining Curcumin and the Flavonoids Quercetin and Fisetin. Genes (Basel). 2022 Jun 23;13(7):1125.

[659] Xing X, Liang Y, Li Y, Zhao Y, Zhang Y, Li Z, Li Z, Wu Z. Fisetin Delays Postovulatory Oocyte Aging by Regulating Oxidative Stress and Mitochondrial Function through Sirt1 Pathway. Molecules. 2023 Jul 20;28(14):5533.

[660] Mihanfar A, Nouri M, Roshangar L, Khadem-Ansari MH. Ameliorative effects of fisetin in letrozole-induced rat model of polycystic ovary syndrome. J Steroid Biochem Mol Biol. 2021 Oct;213:105954.

Prevention and Treatment of Uterine Leiomyomas: As the primary constituent in an extract made from Rhus verniciflua stokes (the lacquer tree), fisetin exhibited excellent leiomyoma cell cytotoxicity, and induced apoptotic cell death with cell cycle arrest by targeting multiple signaling pathways in uterine leiomyoma cells[661].

Treatment for Parkinson's Disease: In a patient with Parkinson's disease, medical dietary intervention using an enhanced diet rich in fisetin and hexacosanol ameliorated the clinical presentation of cogwheel rigidity, micrographia, bradykinesia, dystonia, constricted arm swing with gait, hypomimia, and retropulsion, with researchers noting that the only worsening of symptoms occurred when the enhanced diet was not followed precisely[662]. In a rat model of rotenone-induced Parkinson's disease, treatment with fisetin improved motor function and reversed the rotenone-induced changes in mitochondrial enzymes, striatal dopamine levels, antioxidant enzyme levels, and histological changes[663]. In the latter study, researchers hypothesized that fisetin might prevent the pathogenesis of Parkinsons's disease. In a similar rat model of rotenone-induced Parkinson's disease study, researchers successfully developed a self-nanoemulsifying drug delivery system of fisetin with improved oral bioavailability and neuroprotective effects[664].

Treatment for Chronic Urticaria: In a murine model of chronic urticaria, treatment with fisetin prevented urticaria-like symptoms in mice, inhibited mast cell activation by suppressing calcium mobilization and degranulation of cytokines and chemokines via binding to MRGPRX2 (regulates mast cell degranulation), and downregulated phosphorylation levels of Akt, P38, NF-κB, and PLCγ in mast cell degranulator compound 48/80-activated LAD2 mast cells[665].

Treatment for Arsenic Toxicity: In male BALB/c mice exposed to arsenic and fluoride, administration of fisetin via drinking water reversed the induced neurobehavioral deficit and restored redox and inflammatory milieu, and cortical and hippocampal neuronal density by regulating TNF-α mediated activation of NLRP3 inflammasome[666] (a multiprotein complex that plays a pivotal role in regulating the innate immune system and inflammatory signaling). In a murine model of arsenic-induced male reproductive toxicity, treatment with fisetin remarkably improved testicular and sperm parameters via antioxidant, anti-lipoperoxidative, anti-apoptotic, and androgenic effects[667].

Treatment for Male Infertility: In a murine model of monosodium glutamate-induced testicular toxicity, administration of fisetin stimulated the hypothalamic-pituitary-gonadal axis, increased plasma sex hormone levels, activated the testicular SIRT1/pAMPK signaling pathway, and inhibited glutamate-induced oxidative stress[668].

[661] Lee JW, Choi HJ, Kim EJ, Hwang WY, Jung MH, Kim KS. Fisetin induces apoptosis in uterine leiomyomas through multiple pathways. Sci Rep. 2020 May 14;10(1):7993.

[662] Renoudet VV, Costa-Mallen P, Hopkins E. A diet low in animal fat and rich in N-hexacosanol and fisetin is effective in reducing symptoms of Parkinson's disease. J Med Food. 2012 Aug;15(8):758-61.

[663] Alikatte K, Palle S, Rajendra Kumar J, Pathakala N. Fisetin Improved Rotenone-Induced Behavioral Deficits, Oxidative Changes, and Mitochondrial Dysfunctions in Rat Model of Parkinson's Disease. J Diet Suppl. 2021;18(1):57-71.

[664] Kumar R, Kumar R, Khurana N, Singh SK, Khurana S, Verma S, Sharma N, Kapoor B, Vyas M, Khursheed R, Awasthi A, Kaur J, Corrie L. Enhanced oral bioavailability and neuroprotective effect of fisetin through its SNEDDS against rotenone-induced Parkinson's disease rat model. Food Chem Toxicol. 2020 Oct;144:111590.

[665] Zhang Y, Huang Y, Dang B, Hu S, Zhao C, Wang Y, Yuan Y, Liu R. Fisetin alleviates chronic urticaria by inhibiting mast cell activation via MRGPRX2. J Pharm Pharmacol. 2023 Oct 5;75(10):1310-1321.

[666] Gopnar VV, Rakshit D, Bandakinda M, Kulhari U, Sahu BD, Mishra A. Fisetin attenuates arsenic and fluoride subacute co-exposure induced neurotoxicity via regulating TNF-α mediated activation of NLRP3 inflammasome. Neurotoxicology. 2023 Jul;97:133-149.

[667] Ijaz MU, Haider S, Tahir A, Afsar T, Almajwal A, Amor H, Razak S. Mechanistic insight into the protective effects of fisetin against arsenic-induced reproductive toxicity in male rats. Sci Rep. 2023 Feb 22;13(1):3080.

[668] Rizk FH, Soliman NA, Abo-Elnasr SE, Mahmoud HA, Abdel Ghafar MT, Elkholy RA, ELshora OA, Mariah RA, Amin Mashal SS, El Saadany AA. Fisetin ameliorates oxidative glutamate testicular toxicity in rats via central and peripheral mechanisms involving SIRT1 activation. Redox Rep. 2022 Dec;27(1):177-185.

Other research has shown that fisetin can protect against infertility caused by long-term testicular exposure to heat[669].

Treatment for Heart Attack: In patulin (a toxic polyketide)-induced apoptosis in H9c2 cardiomyocytes, treatment with fisetin inhibited the protein overexpression of P53, Caspase-9, and Bax, and enhanced the protein expression of Bcl-2[670]. In rats subjected to myocardial infarction surgery, treatment with fisetin improved left atrial expansion, cardiac function, atrial inflammation, fibrosis, and vulnerability to atrial fibrillation[671].

Anti-aging Agent: Fisetin has been shown to reduce ROS production by suppressing NOX1 activation, leading to the inhibition of cellular senescence in vascular smooth muscle cells[672]. Other research showed that treatment with fisetin reduced senescence by increasing PTEN and decreasing mTORC2 protein levels in vascular smooth muscle cells[673]. Fisetin has also been shown to maintain adipose-derived stem cell adipogenic and osteogenic differentiation capacity while also attenuating cellular senescence in human adipose-derived stem cells[674]. Finally, supplementation with fisetin suppressed aging-induced increases in the levels of reactive oxygen species, eryptosis, lipid peroxidation, and protein oxidation in rat erythrocytes[675].

Treatment for Autism: In a valproate-induced rodent model of autism, gestational and post-weaning fisetin treatment significantly improved behavioral impairments by attenuating elevated oxidative stress, ROS, lipid peroxidation, and re-establishing redox homeostasis, while also reinstating the reduced levels of endogenous antioxidants, glutathione, AChE, and ATPases by its antioxidant potential[676].

Treatment for Kidney Disease and Injury: In an adenine diet-induced and unilateral ureteral obstruction-induced chronic kidney disease model in adult male mice, fisetin administration significantly ameliorated tubular injury, inflammation, and tubulointerstitial fibrosis by inhibiting ACSL4-mediated tubular ferroptosis[677]. In similar work, a potassium oxonate- and adenine-induced model of hyperuricemia-induced chronic kidney disease in mice showed that treatment with fisetin improved kidney function, ameliorated renal fibrosis, and restored enteric dysbacteriosis by modulating gut microbiota-mediated tryptophan metabolism and aryl hydrocarbon receptor activation[678]. In streptozotocin-induced diabetic atherosclerosis in low density lipoprotein receptor deficient mice, fisetin treatment effectively attenuated diabetes-exacerbated atherosclerosis by regulating uric acid, urea and creatinine levels in

[669] Pirani M, Novin MG, Abdollahifar MA, Piryaei A, Kuroshli Z, Mofarahe ZS. Protective Effects of Fisetin in the Mice Induced by Long-Term Scrotal Hyperthermia. Reprod Sci. 2021 Nov;28(11):3123-3136.

[670] Xu D, Zhang B, Huang C, Lu J, Li Y, Fu B. Effect and mechanism of Fisetin on myocardial damage induced by Patulin. Mol Biol Rep. 2023 Aug;50(8):6579-6589.

[671] Liu L, Gan S, Li B, Ge X, Yu H, Zhou H. Fisetin Alleviates Atrial Inflammation, Remodeling, and Vulnerability to Atrial Fibrillation after Myocardial Infarction. Int Heart J. 2019 Nov 30;60(6):1398-1406.

[672] Kim SG, Sung JY, Kang YJ, Choi HC. Fisetin alleviates cellular senescence through PTEN mediated inhibition of PKCδ-NOX1 pathway in vascular smooth muscle cells. Arch Gerontol Geriatr. 2023 May;108:104927.

[673] Kim SG, Sung JY, Kim JR, Choi HC. Fisetin-induced PTEN expression reverses cellular senescence by inhibiting the mTORC2-Akt Ser473 phosphorylation pathway in vascular smooth muscle cells. Exp Gerontol. 2021 Dec;156:111598.

[674] Mullen M, Nelson AL, Goff A, Billings J, Kloser H, Huard C, Mitchell J, Hambright WS, Ravuri S, Huard J. Fisetin Attenuates Cellular Senescence Accumulation During Culture Expansion of Human Adipose-Derived Stem Cells. Stem Cells. 2023 Jul 14;41(7):698-710.

[675] Singh S, Garg G, Singh AK, Bissoyi A, Rizvi SI. Fisetin, a potential caloric restriction mimetic, attenuates senescence biomarkers in rat erythrocytes. Biochem Cell Biol. 2019 Aug;97(4):480-487.

[676] Mehra S, Ahsan AU, Sharma M, Budhwar M, Chopra M. Neuroprotective Efficacy of Fisetin Against VPA-Induced Autistic Neurobehavioral Alterations by Targeting Dysregulated Redox Homeostasis. J Mol Neurosci. 2023 Jun;73(6):403-422.

[677] Wang, B., Yang, Ln., Yang, Lt. et al. Fisetin ameliorates fibrotic kidney disease in mice via inhibiting ACSL4-mediated tubular ferroptosis. Acta Pharmacol Sin 45, 150–165 (2024).

[678] Ren Q, Cheng L, Guo F, Tao S, Zhang C, Ma L, Fu P. Fisetin Improves Hyperuricemia-Induced Chronic Kidney Disease via Regulating Gut Microbiota-Mediated Tryptophan Metabolism and Aryl Hydrocarbon Receptor Activation. J Agric Food Chem. 2021 Sep 22;69(37):10932-10942.

urine and serum, ameliorating morphological damages and fibrosis, reducing the production of reactive oxygen species, advanced glycosylation end products, and inflammatory cytokines, and reducing accumulation of extracellular matrix[679]. Finally, in a model of high fat diet-induced nephropathy in mice, treatment with fisetin inhibited inflammation and oxidative stress by blocking iRhom2/NF-κB signaling[680].

Treatment for Ischemic Reperfusion Injury: Fisetin has been shown to exhibit protective, ameliorative, and restorative effects in several types of ischemic reperfusion injury, including renal[681], myocardial[682], and hepatic[683] (kidney, heart, and lung respectively).

Analgesic Agent: In mice with a chronic constriction injury of the sciatic nerve, treatment with fisetin diminished tactile and thermal hypersensitivity, and increased analgesia after combined administration with morphine, buprenorphine, and/or oxycodone[684]. In a murine model of intervertebral disc degeneration and associated lower back pain, treatment with fisetin alleviated H2O2-induced apoptosis, inflammation, and extracellular degradation by suppressing oxidative stress in rat nucleus pulposus mesenchymal stem cells[685].

Treatment for Cognitive Disorders: In pentylenetetrazole kindling-induced cognitive dysfunction in mice, fisetin administration modulated the increased levels of lipid peroxidation and protein carbonyl by increasing the levels of antioxidants in the hippocampus and cortex, upregulating gene expressions of cAMP response element-binding protein and brain-derived neurotrophic factor, and ameliorating altered neurotransmitter levels[686]. Additionally, in a rat model of sepsis-associated encephalopathy, treatment with fisetin significantly improved cognitive dysfunction by reducing ROS, blocking the activation of NLRP3 inflammasome, inhibiting secretion of IL-1β into the central nervous system, and inactivating microglial cells[687].

Preventative and Treatment for Alzheimer's Disease: Fisetin has been shown to inhibit aggregation of the tau fragment K18, disaggregate tau K18 filaments, and prevent the formation of tau aggregates in cells[688], offering significant potential in the treatment of Alzheimer's disease.

[679] Zou TF, Liu ZG, Cao PC, Zheng SH, Guo WT, Wang TX, Chen YL, Duan YJ, Li QS, Liao CZ, Xie ZL, Han JH, Yang XX. Fisetin treatment alleviates kidney injury in mice with diabetes-exacerbated atherosclerosis through inhibiting CD36/fibrosis pathway. Acta Pharmacol Sin. 2023 Oct;44(10):2065-2074.

[680] Chenxu G, Xianling D, Qin K, Linfeng H, Yan S, Mingxin X, Jun T, Minxuan X. Fisetin protects against high fat diet-induced nephropathy by inhibiting inflammation and oxidative stress via the blockage of iRhom2/NF-κB signaling. Int Immunopharmacol. 2021 Mar;92:107353.

[681] Prem PN, Kurian GA. Fisetin attenuates renal ischemia/reperfusion injury by improving mitochondrial quality, reducing apoptosis and oxidative stress. Naunyn Schmiedebergs Arch Pharmacol. 2022 May;395(5):547-561.

[682] Shanmugam K, Prem PN, Boovarahan SR, Sivakumar B, Kurian GA. FIsetin Preserves Interfibrillar Mitochondria to Protect Against Myocardial Ischemia-Reperfusion Injury. Cell Biochem Biophys. 2022 Mar;80(1):123-137.

[683] Pu JL, Huang ZT, Luo YH, Mou T, Li TT, Li ZT, Wei XF, Wu ZJ. Fisetin mitigates hepatic ischemia-reperfusion injury by regulating GSK3β/AMPK/NLRP3 inflammasome pathway. Hepatobiliary Pancreat Dis Int. 2021 Aug;20(4):352-360.

[684] Ciapała K, Rojewska E, Pawlik K, Ciechanowska A, Mika J. Analgesic Effects of Fisetin, Peimine, Astaxanthin, Artemisinin, Bardoxolone Methyl and 740 Y-P and Their Influence on Opioid Analgesia in a Mouse Model of Neuropathic Pain. Int J Mol Sci. 2023 May 19;24(10):9000.

[685] Zhou Q, Zhu C, Xuan A, Zhang J, Zhu Z, Tang L, Ruan D. Fisetin regulates the biological effects of rat nucleus pulposus mesenchymal stem cells under oxidative stress by sirtuin-1 pathway. Immun Inflamm Dis. 2023 May;11(5):e865.

[686] Khatoon S, Samim M, Dahalia M, Nidhi. Fisetin provides neuroprotection in pentylenetetrazole-induced cognition impairment by upregulating CREB/BDNF. Eur J Pharmacol. 2023 Apr 5;944:175583.

[687] Ding H, Li Y, Chen S, Wen Y, Zhang S, Luo E, Li X, Zhong W, Zeng H. Fisetin ameliorates cognitive impairment by activating mitophagy and suppressing neuroinflammation in rats with sepsis-associated encephalopathy. CNS Neurosci Ther. 2022 Feb;28(2):247-258.

[688] Xiao S, Lu Y, Wu Q, Yang J, Chen J, Zhong S, Eliezer D, Tan Q, Wu C. Fisetin inhibits tau aggregation by interacting with the protein and preventing the formation of β-strands. Int J Biol Macromol. 2021 May 1;178:381-393.

Treatment for Psoriasis: Fisetin has been shown to decrease inflammatory responses induced by keratinocytes and CD4+ T lymphocytes, suppress the Akt/mTOR signaling pathway, promote autophagy and differentiation, and improve both the immune and epidermal phenotypes of imiquimod-induced psoriasis-like skin lesions in mice[689], offering potential in the treatment of psoriasis and other inflammation-based skin disorders.

Treatment for Paracetamol-Induced Hepatotoxicity: In paracetamol-induced hepatotoxicity in rats, treatment with fisetin significantly decreased the alanine transaminase, aspartate aminotransferase, and alkaline phosphatase levels, increased SOD activity and GSH levels, decreased malondialdehyde, and mediated CYP2E1 gene expression[690].

Treatment for Lupus: In a murine model of lupus nephritis, treatment with fisetin reduced SPiDER-β-gal (a reagent used to detect β-galactosidase) expression, inhibited TGF-β-induced activation, reduced the expression of p15INK4B (a cyclin-dependent kinase inhibitor that plays a role in hepatic cell cycle arrest), and reduced the number of senescent tubular epithelial cells and myofibroblasts[691].

Treatment for Liver Injury: In rats induced to hepatic fibrosis via administration of thioacetamide, treatment with fisetin restored normal liver functions, increased glutathione, decreased malondialdehyde and inflammatory biomarkers including tumor necrosis factor-alpha and interleukin-6, while also reducing transforming growth factor β1, collagen I, tissue inhibitor of metalloproteinase-1, and matrix metalloproteinase-9 levels[692]. In a murine model of ethanol-induced hepatotoxicity, administration of fisetin promoted SIRT1-mediated autophagy and inhibited Sphk1-mediated endoplasmic reticulum stress, ameliorating ethanol-induced liver injury and fibrosis[693].

Treatment for Cigarette Smoke Exposure and Lung Dysfunction: Fisetin has been shown to effectively treat exposure to cigarette smoke, in one case reversing the expression of epithelial-mesenchymal transition biomarkers, reducing the activity of MMP-2/9, and blocking the migration and invasion potential induced by cigarette smoke exposure[694], while in another case reducing neutrophils and macrophages in bronchoalveolar lavage fluid as well as malondialdehyde, 3-nitrotyrosine, 8-isoprostane, tumor necrosis factor-alpha, interleukin-1beta, granulocyte macrophage-colony stimulating factor, interleukin-4, and interleukin-10 levels in lung tissues, and augmenting lung hemoxinase-1, glutathione peroxidase-2, reduced glutathione, superoxide dismutase, nitric oxide, and nuclear factor erythroid 2-related factor (Nrf2) levels in rats chronically exposed to cigarette smoke[695].

Suppressive Agent Against Atmospheric Particulate Matter-induced Respiratory Injury: Fisetin has potential to prevent or ameliorate particulate matter-induced respiratory injuries, after it was found to increase cell viability and significantly reduce oxidative stress, and upregulate serum- and glucocorticoid-inducible kinase 1 in human pulmonary artery endothelial cells[696].

[689] Roy T, Banang-Mbeumi S, Boateng ST, Ruiz EM, Chamcheu RN, Kang L, King JA, Walker AL, Nagalo BM, Kousoulas KG, Esnault S, Huang S, Chamcheu JC. Dual targeting of mTOR/IL-17A and autophagy by fisetin alleviates psoriasis-like skin inflammation. Front Immunol. 2023 Jan 18;13:1075804.

[690] Ugan RA, Cadirci E, Un H, Cinar I, Gurbuz MA. Fisetin Attenuates Paracetamol-Induced Hepatotoxicity by Regulating CYP2E1 Enzyme. An Acad Bras Cienc. 2023 Apr 3;95(2):e20201408.

[691] Ijima S, Saito Y, Nagaoka K, Yamamoto S, Sato T, Miura N, Iwamoto T, Miyajima M, Chikenji TS. Fisetin reduces the senescent tubular epithelial cell burden and also inhibits proliferative fibroblasts in murine lupus nephritis. Front Immunol. 2022 Nov 17;13:960601.

[692] El-Fadaly AA, Afifi NA, El-Eraky W, Salama A, Abdelhameed MF, El-Rahman SSA, Ramadan A. Fisetin alleviates thioacetamide-induced hepatic fibrosis in rats by inhibiting Wnt/β-catenin signaling pathway. Immunopharmacol Immunotoxicol. 2022 Jun;44(3):355-366.

[693] Zhou ZS, Kong CF, Sun JR, Qu XK, Sun JH, Sun AT. Fisetin Ameliorates Alcohol-Induced Liver Injury through Regulating SIRT1 and SphK1 Pathway. Am J Chin Med. 2022;50(8):2171-2184.

[694] Agraval H, Sharma JR, Prakash N, Yadav UCS. Fisetin suppresses cigarette smoke extract-induced epithelial to mesenchymal transition of airway epithelial cells through regulating COX-2/MMPs/β-catenin pathway. Chem Biol Interact. 2022 Jan 5;351:109771.

[695] Hussain T, Al-Attas OS, Alamery S, Ahmed M, Odeibat HAM, Alrokayan S. The plant flavonoid, fisetin alleviates cigarette smoke-induced oxidative stress, and inflammation in Wistar rat lungs. J Food Biochem. 2019 Aug;43(8):e12962.

[696] Sim, H., Noh, Y., Choo, S. et al. Suppressive Activities of Fisetin on Particulate Matter-induced Oxidative Stress. Biotechnol Bioproc E 26, 568–574 (2021).

Anti-osteoporosis Agent: In a murine model of ovariectomy-induced osteoporosis, treatment with fisetin significantly improved the level of bone mineral content, bone mineral density, and biochemical parameters such as energy, maximum load, stiffness, young modules, and maximum stress, reduced the level of 1,25(OH) 2 D3 and E 2, significantly reduced the level of phosphorus and calcium, increased levels of vitamin D, significantly reduced the malonaldehyde level, and enhanced the glutathione, catalase, superoxide dismutase level in the bone, intestine, and hepatic tissue in rats[697].

Prevention of Preeclampsia: In a model of lipopolysaccharide-induced preeclampsia-like rats, treatment with fisetin reduced hypertension, proteinuria, TNF-α, IL-6, IL-1β, malondialdehyde, and sFlt-1/PlGF ratio, elevated the placental, fetal weight, glutathione, superoxide dismutase, suppressed the TLR4/NF-κB pathway, and promoted the Nrf2/HO-1 pathway in placental tissues[698].

Treatment for Periodontitis: Fisetin offers potential as a novel treatment for periodontitis, after it was shown to reduce the alveolar bone gap, reverse histopathological lesions, inhibit serum inflammatory cytokine concentration in rats, and decrease the inflammatory cytokine contents in the supernatant of lipopolysaccharide-induced human gingival fibroblasts[699].

COVID Preventative Agent: Recent research has shown that fisetin can act as a cell entry inhibitor in SARS CoV-2 infections[700].

Treatment for Sepsis-related Organ Dysfunction: In a sepsis-induced multiple organ dysfunction model in mice, administration of fisetin significantly alleviated lung, liver, and kidney injury, as well as the expression levels of interleukin-6, tumor necrosis factor alpha, and IL-1-beta in bronchoalveolar lavage fluid, while in lipopolysaccharide-treated mouse bone marrow-derived macrophages, application of fisetin again inhibited the expression levels of interleukin-6, tumor necrosis factor alpha, and IL-1-beta, while also inhibiting inducible nitric oxide synthase, and inhibiting the phosphorylation of p38 MAPK, MK2, and transforming growth factor-beta-activated kinase 1[701].

IMPLICATIONS FOR HUMAN HEALTH AND NUTRITION

The best way to obtain fisetin is via the incorporation of a healthy diet that includes vegetables like cucumbers, onions, lotus, kale, soybeans, and tomatoes, and fruits such as grapes, kiwi, strawberries, apples, mulberries, and persimmons. Additionally, fisetin supplements have long been available, although use of these products should be consulted with a qualified physician or doctor.

At this time, it is unknown if fisetin occurs in cannabis. Future editions of this text will be updated with emerging research.

---

[697] Feng P, Shu S, Zhao F. Anti-osteoporosis Effect of Fisetin against Ovariectomy Induced Osteoporosis in Rats: In silico, in vitro and in vivo Activity. J Oleo Sci. 2022;71(1):105-118.

[698] Li Y, Liu Y, Chen J, Hu J. Protective effect of Fisetin on the lipopolysaccharide-induced preeclampsia-like rats. Hypertens Pregnancy. 2022 Feb;41(1):23-30.

[699] Huang X, Shen H, Liu Y, Qiu S, Guo Y. Fisetin attenuates periodontitis through FGFR1/TLR4/NLRP3 inflammasome pathway. Int Immunopharmacol. 2021 Jun;95:107505.

[700] Mishra A, Kaur U, Singh A. Fisetin 8-C-glucoside as entry inhibitor in SARS CoV-2 infection: molecular modelling study. J Biomol Struct Dyn. 2022 Jul;40(11):5128-5137.

[701] Zhang HF, Zhang HB, Wu XP, Guo YL, Cheng WD, Qian F. Fisetin alleviates sepsis-induced multiple organ dysfunction in mice via inhibiting p38 MAPK/MK2 signaling. Acta Pharmacol Sin. 2020 Oct;41(10):1348-1356.

Fisetin Review

Answer the following questions to test your knowledge of this flavonoid:

Question #1: What type of flavonoid is fisetin?

    a. Flavoned
    b. Flavine
    c. Flavone
    d. Flavonol

Question #2: The base component in the biosynthesis of fisetin is:

    a. Cosmoioside
    b. Apigenin
    c. Naringenin
    d. Chrysin

Question #3: How common is fisetin in cannabis?

    a. Top three
    b. Unknown
    c. Top ten
    d. Top twenty

Question #4: What is the chemical formula for fisetin?

    a. $C_{15}H_{10}O_6$
    b. $C_{26}H_{10}O_{15}$
    c. $C_{16}H_{10}O_{65}$
    d. $C_{26}H_{30}O_{15}$

Question #5: Name two plants fisetin is known to occur in:

1 _____ 2 _____

Question #6: Name two specific potential medical uses of fisetin:

1 _____ 2 _____

For the answer key to Fisetin, please visit www.cannabischemistry.org

BAICALIN

Type: Flavone
Chemical Formula: $C_{21}H_{18}O_{11}$
Molecular Weight: 446.365 g/mol
Boiling Point: 836.62 °C @ 760.00 mm Hg (estimated)[702]
Flash Point: 297.20 °C (estimated by TGSC)
Melting Point: 202 to 205 °C
Solubility: Baicalin is soluble in water, while baicalein is poorly soluble in water
Baicalin is also soluble in ethanol and dimethyl sulfoxide
Oral LD50: < 5,000mg/kg (mice)
Biological Role: Plant defense agent
Therapeutic Role: Anti-cancer, anti-depression, antibacterial, lung protective agent
Commercial Use: Research chemical, herbal supplement
Occurrence in Cannabis: Unknown

Occurs in Cannabis Strains:

As of the publication of this text, there was no publicly available information to substantiate that baicalin occurs in cannabis or hemp. ACS Laboratory (Florida, USA) has published an article stating that baicalin occurs in cannabis in small amounts, but they do not cite or prove this claim[703]. Additionally, the Restek Corporation published a paper regarding the development of a method to test hemp and cannabis for flavonoids, which included testing one CBG and one CBD strain for baicalin, whereupon the flavonoid was not found to be present in these samples[704]. However, due to the likelihood that a chemotype of cannabis will be discovered or developed where baicalin is expressed at detectable levels, the author has included this chapter in consideration of how medically important this flavonoid could be.

INTRODUCTION

Baicalin is known primarily to occur in the Skullcap plant, and has been used as an herbal supplement in Chinese culture for centuries. Well-studied for the last 100 years by the scientific community, there are thousands of published papers finding a seemingly vast medical potential for this flavonoid, in particular as an anti-cancer agent. Only *theorized* to occur in cannabis, baicalin has little to no odor, although it does have a bitter flavor common to many purified flavonoids.

---

[702] The Good Scents Company Data Sheet for Baicalin, from: https://www.thegoodscentscompany.com/data/rw1667691.html. Accessed January 31, 2024.

[703] Schmidt, Elena; Flavonoids Friday: Everything You Need to Know about Baicalin Flavor, Fragrance, and Health Benefits. December 28, 2021. From:https://www.acslab.com/flavonoids/marketing-flavonoids-friday-everything-you-need-to-know-about-baicalin-flavor-fragrance-and-health-benefits. Accessed December 7, 2024.

[704] York, Jamie; DeLurio, Dan; The Detection of Flavonoids in Hemp Flower by LC-MS/MS Restek Corporation June 17, 2021. From: https://www.restek.com/chromablography/flavonoids-the-next-routine-cannabis-test. Accessed December 7, 2024.

CHEMICAL STRUCTURE

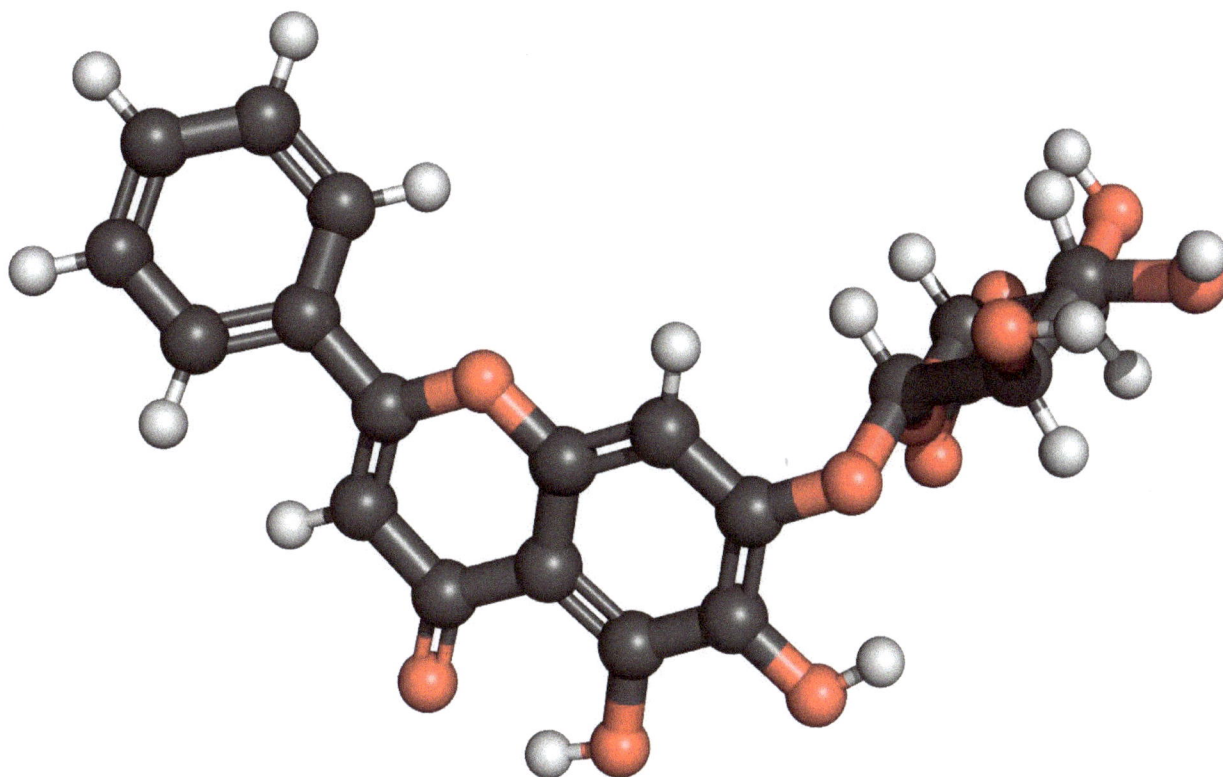

Baicalin is a flavone glycoside comprised of four rings (two of which contain endocyclic oxygen atoms), seven endocyclic double bonds, and two exocyclic double-bonded oxygen atoms, with a total of twenty-one carbon atoms, eighteen hydrogen atoms, and eleven oxygen atoms notated as $C_{21}H_{18}O_{11}$. Further, baicalin features hydroxyl groups at positions 5, 6, and 7, with the group at position 7 also containing a glucuronic acid moiety. Recent work has shown that transcription factor SbMYB12 (a transcription factor belonging to the myeloblastosis family, which plays a pivotal role in regulating gene expression in plants) regulates baicalin biosynthesis in the skullcap plant[705].

Other variations of baicalin can occur due to methylation of hydroxyl groups, acetylation of hydroxyl groups, addition of glucose moieties, replacement of hydroxyl groups with sulfate groups, loss or reduction of double bonds, or other structural changes. For instance, baicalein is the aglycone variation of baicalin, very similar in structure and content to baicalin, but missing the glucuronic acid moiety.

OCCURRENCE IN PLANTS

Baicalin occurs primarily in multiple varieties of Scutellaria, particularly in the roots of Scutellaria baicalensis Georgi, where most commercial and research-grade extractions of this flavonoid are sourced. Interestingly, wogonin is also primarily only known to occur in this same variety of Skullcap.

Baicalin has also been reported to occur in the leaves of Thymus vulgaris and Oroxylum indicum[706].

[705] Wang W, Hu S, Yang J, Zhang C, Zhang T, Wang D, Cao X, Wang Z. A Novel R2R3-MYB Transcription Factor SbMYB12 Positively Regulates Baicalin Biosynthesis in Scutellaria baicalensis Georgi. Int J Mol Sci. 2022 Dec 7;23(24):15452.

[706] Jelić, Dubravko, Lower-Nedza, Agnieszka D., Brantner, Adelheid H., Blažeković, Biljana, Bian, Baolin, Yang, Jian, Brajša, Karmen, Vladimir-Knežević, Sanda, Baicalin and Baicalein Inhibit Src Tyrosine Kinase and Production of IL-6, Journal of Chemistry, 2016, 2510621, 6 pages, 2016.

BIOLOGICAL ACTIVITY IN PLANTS

While publications related to the medical and therapeutic implications of baicalin number in the thousands, there is little information about the biological activity of this flavonoid in plants. Some research indicates that baicalin is likely used as a defense compound, particularly after emissions of this flavone were found to increase significantly in response to chilling stress[707], and exposure to UV-B irradiation[708]. Because baicalin exhibits strong antibacterial properties, it is likely that plants that produce this compound do so to ward off bacterial infections.

USES IN INDUSTRY

Commercial preparations of baicalin are generally for research use, or for use in herbal supplements. Baicalin is a major chemical constituent in at least two widely used herbal supplements: Shuanghuanglian, and Sho-Saiko-To. Purified baicalin appears as a fine and smooth yellow or yellowish-brown crystalline powder.

Interestingly, baicalin and extracts of Scutellaria baicalensis have been successfully used by researchers to impregnate textiles to imbue them with anti-microbial, anti-ultraviolet, and other properties[709]. Researchers noted that the whole plant extract performed far better than baicalin alone, lending further credibility to the concept of synergy among these natural products.

POTENTIAL USES IN MEDICINE

Baicalin has been extensively studied and found to have significant medical and therapeutic value. This flavonoid runs the gambit of medical conditions, offering potential to treat diseases ranging from cancer to depression, obesity to lung injury, and colitis to HIV prevention and much more as detailed below.

Major Antibacterial Agent: Baicalin has been shown to exhibit significant antibacterial properties against a wide range of bacterial strains including Candida albicans[710], Salmonella typhimurium[711, 712], and Staphylococcus aureus[713, 714] including methicillin-resistant strains[715]. In azithromycin-resistant Staphylococcus saprophyticus, treatment with baicalin inhibited biofilm formation by disrupting processes in the primary adhesion and aggregation phases of the bacterium[716]. The latter is particularly important work considering that the formation of biofilm

[707] Yeo HJ, Park CH, Kim JK, Sathasivam R, Jeong JC, Kim CY, Park SU. Effects of Chilling Treatment on Baicalin, Baicalein, and Wogonin Biosynthesis in Scutellaria baicalensis Plantlets. Plants (Basel). 2022 Nov 2;11(21):2958.

[708] Zhang JJ, Li XQ, Sun JW, Jin SH. Nitric oxide functions as a signal in ultraviolet-B-induced baicalin accumulation in Scutellaria baicalensis suspension cultures. Int J Mol Sci. 2014 Mar 18;15(3):4733-46.

[709] Li, H., Li, Z., Liu, Y. et al. Advantages of Scutellaria baicalensis extracts over just baicalin in the ultrasonically assisted multi-functional treatment of linen fabrics. Cellulose 27, 4831–4846 (2020).

[710] Shao J, Xiong L, Zhang MX, Wang TM, Wang CZ. Report: Synergism of sodium bicarbonate and baicalin against clinical Candida albicans isolates via broth microdilution method and checkerboard assay. Pak J Pharm Sci. 2019 May;32(3):1103-1105.

[711] Zhang L, Sun Y, Xu W, Geng Y, Su Y, Wang Q, Wang J. Baicalin inhibits Salmonella typhimurium-induced inflammation and mediates autophagy through TLR4/MAPK/NF-κB signalling pathway. Basic Clin Pharmacol Toxicol. 2021 Feb;128(2):241-255.

[712] Wu SC, Chu XL, Su JQ, Cui ZQ, Zhang LY, Yu ZJ, Wu ZM, Cai ML, Li HX, Zhang ZJ. Baicalin protects mice against Salmonella typhimurium infection via the modulation of both bacterial virulence and host response. Phytomedicine. 2018 Sep 15;48:21-31.

[713] Du Z, Han J, Luo J, Bi G, Liu T, Kong J, Chen Y. Combination effects of baicalin with linezolid against Staphylococcus aureus biofilm-related infections: in vivo animal model. New Microbiol. 2023 Sep;46(3):258-263.

[714] Wang G, Gao Y, Wang H, Niu X, Wang J. Baicalin Weakens Staphylococcus aureus Pathogenicity by Targeting Sortase B. Front Cell Infect Microbiol. 2018 Nov 30;8:418.

[715] Zhang S, Hu B, Xu J, Ren Q, Wang Z, Wang S, Dong Y, Yang G. Baicalin suppress growth and virulence-related factors of methicillin-resistant Staphylococcus aureus in vitro and vivo. Microb Pathog. 2020 Feb;139:103899.

[716] Wang J, Zhu J, Meng J, Qiu T, Wang W, Wang R, Liu J. Baicalin inhibits biofilm formation by influencing primary adhesion and aggregation phases in Staphylococcus saprophyticus. Vet Microbiol. 2021 Nov;262:109242.

prevents antibacterial agents from reaching their target. Baicalin has also been shown to be effective in treating pathogens of the lung, including by reducing Mycoplasma gallisepticum-induced lung inflammation in chickens[717, 718], and attenuating Mycoplasma pneumoniae infection-induced lung injury in mice[719]. Baicalin has also been shown to be effective in the treatment of bacterial pneumonia induced by Escherichia coli in chickens[720], and attenuated Escherichia coli and Staphylococcus aureus-induced endometritis in rabbits[721]. In hypervirulent Klebsiella pneumoniae, combination treatment with baicalin and Levofloxacin (a common quinolone antibiotic) significantly reduced biofilm formation and bacterial burden, likely "by altering the expression of genes regulating efflux pumps and lipopolysaccharide biosynthesis.[722]" Finally, baicalin has been shown to significantly reduce toxin synthesis, sporulation, and spore outgrowth, while also down-regulating genes critical for pathogenesis in Clostridioides difficile[723] (a bacterium that can cause diarrhea, colitis, and other intestinal conditions).

Treatment for Lung Cancer: Baicalin has been shown to suppress lung cancer growth in JB6 Cl41 and H441 lung cancer cells, as well as in a xenograft model of lung cancer in tumor bearing Athymic nude mice by inhibiting PBK/TOPK[724] (T-lymphokine-activated killer cell-originated protein kinase). Baicalin has also been shown to reduce cell proliferation and invasion in A549 and H1299 lung cancer cells by increasing miR-340-5p (microRNA involved in the regulation of gene expression) expression[725].

Treatment for Liver Cancer: In BALB/c mice with induced hepatocellular carcinoma, treatment with baicalin and baicalein decreased STAT3 (signal transducer and activator of transcription protein) activity, downregulated IFN-γ- (interferon) induced PD-L1 (programmed death-ligand 1) expression, thereby restoring T cell sensitivity to kill tumor cells[726]. Other research demonstrated that baicalin can inhibit hepatocellular carcinoma growth primarily by suppressing the expression of ROCK1 (Rho-associated coiled-coil containing protein kinase 1, which is involved in many cellular processes) signaling[727]. Finally, in type 2 diabetes-induced liver cancer, baicalin suppressed tumor progression by regulating the METTL3/m6A/HKDC1 (methyltransferase-like 3/n6-methyladenosine/ hexokinase domain containing 1) axis[728].

[717] Zou M, Yang L, Niu L, Zhao Y, Sun Y, Fu Y, Peng X. Baicalin ameliorates Mycoplasma gallisepticum-induced lung inflammation in chicken by inhibiting TLR6-mediated NF-κB signalling. Br Poult Sci. 2021 Apr;62(2):199-210.

[718] Wu Z, Fan Q, Miao Y, Tian E, Ishfaq M, Li J. Baicalin inhibits inflammation caused by coinfection of Mycoplasma gallisepticum and Escherichia coli involving IL-17 signaling pathway. Poult Sci. 2020 Nov;99(11):5472-5480.

[719] Zhang H, Li X, Wang J, Cheng Q, Shang Y, Wang G. Baicalin relieves Mycoplasma pneumoniae infection-induced lung injury through regulating microRNA-221 to inhibit the TLR4/NF-κB signaling pathway. Mol Med Rep. 2021 Aug;24(2):571.

[720] Peng LY, Yuan M, Song K, Yu JL, Li JH, Huang JN, Yi PF, Fu BD, Shen HQ. Baicalin alleviated APEC-induced acute lung injury in chicken by inhibiting NF-κB pathway activation. Int Immunopharmacol. 2019 Jul;72:467-472.

[721] Miao Y, Ishfaq M, Liu Y, Wu Z, Wang J, Li R, Qian F, Ding L, Li J. Baicalin attenuates endometritis in a rabbit model induced by infection with Escherichia coli and Staphylococcus aureus via NF-κB and JNK signaling pathways. Domest Anim Endocrinol. 2021 Jan;74:106508.

[722] Han J, Luo J, Du Z, Chen Y, Liu T. Synergistic Effects of Baicalin and Levofloxacin Against Hypervirulent Klebsiella pneumoniae Biofilm In Vitro. Curr Microbiol. 2023 Mar 6;80(4):126.

[723] Pellissery AJ, Vinayamohan PG, Venkitanarayanan K. In vitro antivirulence activity of baicalin against Clostridioides difficile. J Med Microbiol. 2020 Apr;69(4):631-639.

[724] Diao X, Yang D, Chen Y, Liu W. Baicalin suppresses lung cancer growth by targeting PDZ-binding kinase/T-LAK cell-originated protein kinase. Biosci Rep. 2019 Apr 9;39(4):BSR20181692.

[725] Zhao F, Zhao Z, Han Y, Li S, Liu C, Jia K. Baicalin suppresses lung cancer growth phenotypes via miR-340-5p/NET1 axis. Bioengineered. 2021 Dec;12(1):1699-1707.

[726] Ke M, Zhang Z, Xu B, Zhao S, Ding Y, Wu X, Wu R, Lv Y, Dong J. Baicalein and baicalin promote antitumor immunity by suppressing PD-L1 expression in hepatocellular carcinoma cells. Int Immunopharmacol. 2019 Oct;75:105824.

[727] Sun J, Yang X, Sun H, Huang S, An H, Xu W, Chen W, Zhao W, He C, Zhong X, Li T, Liu Y, Wen B, Du Q, He S. Baicalin inhibits hepatocellular carcinoma cell growth and metastasis by suppressing ROCK1 signaling. Phytother Res. 2023 Sep;37(9):4117-4132.

[728] Jiang H, Yao Q, An Y, Fan L, Wang J, Li H. Baicalin suppresses the progression of Type 2 diabetes-induced liver tumor through regulating METTL3/m6A/HKDC1 axis and downstream p-JAK2/STAT1/clevaged Capase-3 pathway. Phytomedicine. 2022 Jan;94:153823.

Treatment for Prostate Cancer: After engrafting human prostate cancer cell line LNCaP into nude mice, treatment with baicalin suppressed the cell cycle progression and proliferation of the prostate cancer cells by significantly decreasing the expression of the CDK6/FOXM1 (cyclin-dependent kinase 6/forkhead box protein M1) axis[729].

Treatment for Nasopharyngeal Cancer: In human nasopharyngeal carcinoma CNE-2R cells, treatment with baicalin reversed radioresistance by downregulating radiation-enhanced autophagy[730]. Radioresistance occurs when cancer cells become resistant to treatment with radiation therapy.

Treatment for Mesothelioma: Baicalin has been shown to inhibit the proliferation, migration, and invasion of human mesothelioma (cancer that develops in the thin tissue that lines major organs) cells while increasing apoptosis and the sensitivity of the cells to chemotherapeutic agents including doxorubicin, cisplatin, and pemetrexed[731].

Treatment for Gastric Cancer: In combination with 5-Fluorouracil, baicalin has been shown to inhibit the progression of gastric cancer cells while also increasing intracellular ROS levels, thereby leading to ferroptosis in the cancerous cells[732].

Treatment for Colorectal Cancer: Baicalin has been shown to significantly inhibit viability and proliferation of HCT-116 and CT26 colorectal cancer cells, while strongly triggering mitochondria-mediated apoptosis in both cell lines by inhibiting the NF-κB (Nuclear factor kappa B) signaling pathway[733]. When combined with 5-fluorouracil, baicalin enhanced the antitumor activity of the chemotherapeutic agent in resistant RKO-R10 (colon carcinoma) cells[734]. In xenografted tumor mice, treatment with baicalin blocked the cell cycle and inhibited cell proliferation in colon cancer cells by increasing miR-139-3p expression[735], while similar work in mice showed that baicalin induced apoptosis and suppressed growth in HT-29 cells[736].

Treatment for Bile Duct Cancer: In QBC939 cholangiocarcinoma cells, treatment with baicalin induced apoptosis by inhibiting the mTORC1-p70S6K (mammalian target of rapamycin complex 1 and 70-kDa ribosomal protein S6 kinase, both of which regulate cell functions) signaling pathway after AMPK (AMP-activated protein kinase) activation[737].

[729] Yu Z, Zhan C, Du H, Zhang L, Liang C, Zhang L. Baicalin suppresses the cell cycle progression and proliferation of prostate cancer cells through the CDK6/FOXM1 axis. Mol Cell Biochem. 2020 Jun;469(1-2):169-178.

[730] Wang C, Yang Y, Sun L, Wang J, Jiang Z, Li Y, Liu D, Sun H, Pan Z. Baicalin reverses radioresistance in nasopharyngeal carcinoma by downregulating autophagy. Cancer Cell Int. 2020 Jan 30;20:35.

[731] Xu WF, Liu F, Ma YC, Qian ZR, Shi L, Mu H, Ding F, Fu XQ, Li XH. Baicalin Regulates Proliferation, Apoptosis, Migration, and Invasion in Mesothelioma. Med Sci Monit. 2019 Oct 31;25:8172-8180.

[732] Yuan J, Khan SU, Yan J, Lu J, Yang C, Tong Q. Baicalin enhances the efficacy of 5-Fluorouracil in gastric cancer by promoting ROS-mediated ferroptosis. Biomed Pharmacother. 2023 Aug;164:114986.

[733] Song L, Zhu S, Liu C, Zhang Q, Liang X. Baicalin triggers apoptosis, inhibits migration, and enhances anti-tumor immunity in colorectal cancer via TLR4/NF-κB signaling pathway. J Food Biochem. 2022 Mar;46(3):e13703.

[734] Liu H, Liu H, Zhou Z, Chung J, Zhang G, Chang J, Parise RA, Chu E, Schmitz JC. Scutellaria baicalensis enhances 5-fluorouracil-based chemotherapy via inhibition of proliferative signaling pathways. Cell Commun Signal. 2023 Jun 19;21(1):147.

[735] Cai R, Zhou YP, Li YH, Zhang JJ, Hu ZW. Baicalin Blocks Colon Cancer Cell Cycle and Inhibits Cell Proliferation through miR-139-3p Upregulation by Targeting CDK16. Am J Chin Med. 2023;51(1):189-203.

[736] Tao Y, Zhan S, Wang Y, Zhou G, Liang H, Chen X, Shen H. Baicalin, the major component of traditional Chinese medicine Scutellaria baicalensis induces colon cancer cell apoptosis through inhibition of oncomiRNAs. Sci Rep. 2018 Sep 27;8(1):14477.

[737] Jia M, Yang F, Xu Y, Xu Q, Zeng Y, Dai R, Xiang Y. Baicalin Induced Apoptosis of Human Cholangiocarcinoma Cell through Activating AMPK/mTORC1/p70S6K Signaling Pathway. Bull Exp Biol Med. 2022 Jul;173(3):366-370.

Treatment for Breast Cancer: Baicalin has been shown to significantly suppress cell viability, migration, and invasion of breast cancer cells via the modulation of several key signaling pathways[738]. In MCF-7 and MDA-MB-231 breast cancer cells xenografted into mice, treatment with baicalin suppressed cell invasion, migration, and proliferation, without affecting healthy breast epithelial cells[739]. Baicalin has also been shown to enhance the effects of anti-cancer agent 5-Fluorouracil, inhibiting tumor growth and angiogenesis in Ehrlich solid tumors[740], while other work has shown that baicalin can significantly downregulate the viability of murine and human osteotropic breast cancer cells by inducing apoptosis[741].

Treatment for Chemotherapeutic-induced Cardiotoxicity: In male Swiss albino mice with doxorubicin-induced cardiotoxicity, treatment with baicalin prevented doxorubicin-induced elevation of serum activities of cardiac biomarkers and alterations to the heart, suppressed overexpression of cardiac TLR4 (toll-like receptor 4) and subsequently prevented elevation of both cardiac NF-κB and IL-1β, while also significantly reducing the cardiac levels of DKK1 (Dickkopf-1) and elevated levels of β-catenin (catenin beta-1).[742]

Treatment for Lung Injury: In lipopolysaccharide-induced acute lung injury in male mice, treatment with baicalin suppressed oxidative stress and inflammation by improving the expression of nuclear Nrf2 (nuclear factor erythroid 2-related factor 2) and cytosolic HO-1[743] (heme oxygenase-1), thereby ameliorating acute lung injury. In a murine lung inflammatory injury model induced by inhalation of PM2.5 aerosols (very small particulate matter), treatment with baicalin significantly reduced the expression of inflammatory cytokines and markedly lowered lung injury pathological scores, thereby alleviating pathological damage[744]. In lipopolysaccharide (LPS)-induced acute lung injury in male ICR mice, treatment with a baicalin magnesium salt combination strongly ameliorated LPS-induced inflammatory response and histopathological damages, elevated antioxidant enzyme activity, and downregulated myeloperoxidase and malonaldehyde levels[745]. Similar research working with LPS-induced acute lung injury but in rats found that treatment with baicalin regulated macrophage polarization, thereby alleviating the LPS-induced pulmonary inflammatory response[746]. In an acute lung injury model induced by LPS in male Wistar rats, treatment with baicalin significantly reduced the permeability of the alveolocapillary membrane, alleviated tissue injury and inflammatory infiltration, and inhibited the secretion of inflammatory factors and the infiltration of neutrophils[747]. Finally, interesting research has shown that a novel drug delivery system – baicalin liposome – improved the

[738] Li J, Liu H, Lin Q, Chen H, Liu L, Liao H, Cheng Y, Zhang X, Wang Z, Shen A, Chen G. Baicalin suppresses the migration and invasion of breast cancer cells via the TGF-β/lncRNA-MALAT1/miR-200c signaling pathway. Medicine (Baltimore). 2022 Nov 18;101(46):e29328.

[739] Gao Y, Liu H, Wang H, Hu H, He H, Gu N, Han X, Guo Q, Liu D, Cui S, Shao H, Jin C, Wu Q. Baicalin inhibits breast cancer development via inhibiting IκB kinase activation in vitro and in vivo. Int J Oncol. 2018 Dec;53(6):2727-2736.

[740] Shehatta NH, Okda TM, Omran GA, Abd-Alhaseeb MM. Baicalin; a promising chemopreventive agent, enhances the antitumor effect of 5-FU against breast cancer and inhibits tumor growth and angiogenesis in Ehrlich solid tumor. Biomed Pharmacother. 2022 Feb;146:112599.

[741] Wang B, Huang T, Fang Q, Zhang X, Yuan J, Li M, Ge H. Bone-protective and anti-tumor effect of baicalin in osteotropic breast cancer via induction of apoptosis. Breast Cancer Res Treat. 2020 Dec;184(3):711-721.

[742] El-Ela SRA, Zaghloul RA, Eissa LA. Promising cardioprotective effect of baicalin in doxorubicin-induced cardiotoxicity through targeting toll-like receptor 4/nuclear factor-κB and Wnt/β-catenin pathways. Nutrition. 2022 Oct;102:111732.

[743] Meng X, Hu L, Li W. Baicalin ameliorates lipopolysaccharide-induced acute lung injury in mice by suppressing oxidative stress and inflammation via the activation of the Nrf2-mediated HO-1 signaling pathway. Naunyn Schmiedebergs Arch Pharmacol. 2019 Nov;392(11):1421-1433.

[744] Deng L, Ma M, Li S, Zhou L, Ye S, Wang J, Yang Q, Xiao C. Protective effect and mechanism of baicalin on lung inflammatory injury in BALB/cJ mice induced by PM2.5. Ecotoxicol Environ Saf. 2022 Dec 15;248:114329.

[745] Zhang L, Yang L, Xie X, Zheng H, Zheng H, Zhang L, Liu C, Piao JG, Li F. Baicalin Magnesium Salt Attenuates Lipopolysaccharide-Induced Acute Lung Injury via Inhibiting of TLR4/NF-κB Signaling Pathway. J Immunol Res. 2021 Jun 8;2021:6629531.

[746] Yuan J, Cong R, Xia J, Sun Y, Feng L. [Baicalin alleviates LPS-induced acute lung injury in rats by regulating macrophage polarization]. Xi Bao Yu Fen Zi Mian Yi Xue Za Zhi. 2022 Jan;38(1):9-15. Chinese.

[747] Changle Z, Cuiling F, Feng F, Xiaoqin Y, Guishu W, Liangtian S, Jiakun Z. Baicalin inhibits inflammation of lipopolysaccharide-induced acute lung injury toll like receptor-4/myeloid differentiation primary response 88/nuclear factor-kappa B signaling pathway. J Tradit Chin Med. 2022 Apr;42(2):200-212.

solubility of baicalin, leading to improved treatment outcomes in male KM mice induced to acute lung injury via LPS[748].

Treatment for COPD: In a model of COPD based on exposure to a cigarette smoke extract in 16HBE (human bronchial epithelial) cells, treatment with baicalin reversed inflammation and apoptosis by inhibiting miR-125a[749]. In a similar rat model of COPD that was also based on exposure to cigarette smoke, baicalin markedly reduced inflammation in a dose-dependent manner by inhibiting the NF-kappaB pathway[750]. Other work with cigarette smoke exposure in mice showed that baicalin reduced inflammatory cell infiltration, improve cell viability, and inhibit apoptosis via regulation of HSP72[751] (heat shock protein 72).

Treatment for Liver Injury and Disease: In acetaminophen-induced liver injury in mice, treatment with baicalin promoted liver regeneration by inducing NLRP3 (NOD-, LRR-, and pyrin domain-containing protein 3) inflammasome activation[752]. In a murine model of acute hepatic injury induced by arsenic trioxide, pretreatment with baicalin remarkably improved the oxidative stress response, significantly reduced the accumulation of arsenic in the liver, reduced the production of peroxidation products, and enhanced the antioxidant capacity, thereby reducing inflammation and improving cell apoptosis by modulating the JAK2/STAT3 signaling pathway[753]. In a chlorpyrifos-induced model of liver injury in mice, treatment with baicalin inhibited inflammatory responses, oxidative stress, and cell apoptosis both in vivo and in vitro[754], offering a significant protective effect. Baicalin has also been shown to alleviate alcohol-associated liver disease by increasing the expression of MiR-205[755], and ameliorate alcohol-induced hepatic steatosis by regulating SREBP1c[756] (sterol regulatory element-binding protein 1c). In a high-fat diet-induced mouse model of nonalcoholic fatty liver disease, administration of baicalin exhibited significant protective effects via regulation of the SREBP1/Nrf2/NF-κB signaling pathways[757], while a model of thioacetamide-induced cirrhosis in Sprague-Dawley rats showed that treatment with baicalin also offered significant hepatoprotective effects, in this case via modulation of the NOX4/NF-κB/NLRP3 inflammasome signaling pathways[758].

[748] Long Y, Xiang Y, Liu S, Zhang Y, Wan J, Yang Q, Cui M, Ci Z, Li N, Peng W. Baicalin Liposome Alleviates Lipopolysaccharide-Induced Acute Lung Injury in Mice via Inhibiting TLR4/JNK/ERK/NF-κB Pathway. Mediators Inflamm. 2020 Nov 11;2020:8414062.

[749] Jing X, Huo J, Li L, Wang T, Xu J. Baicalin Relieves Airway Inflammation in COPD by Inhibiting miR-125a. Appl Biochem Biotechnol. 2024 Jun;196(6):3374-3386.

[750] Lixuan Z, Jingcheng D, Wenqin Y, Jianhua H, Baojun L, Xiaotao F. Baicalin attenuates inflammation by inhibiting NF-kappaB activation in cigarette smoke induced inflammatory models. Pulm Pharmacol Ther. 2010 Oct;23(5):411-9.

[751] Hao D, Li Y, Shi J, Jiang J. Baicalin alleviates chronic obstructive pulmonary disease through regulation of HSP72-mediated JNK pathway. Mol Med. 2021 May 30;27(1):53.

[752] Shi L, Zhang S, Huang Z, Hu F, Zhang T, Wei M, Bai Q, Lu B, Ji L. Baicalin promotes liver regeneration after acetaminophen-induced liver injury by inducing NLRP3 inflammasome activation. Free Radic Biol Med. 2020 Nov 20;160:163-177.

[753] He Q, Sun X, Zhang M, Chu L, Zhao Y, Wu Y, Zhang J, Han X, Guan S, Ding C. Protective effect of baicalin against arsenic trioxide-induced acute hepatic injury in mice through JAK2/STAT3 signaling pathway. Int J Immunopathol Pharmacol. 2022 Jan-Dec;36:20587384211073397.

[754] Wang R, Zhang K, Liu K, Pei H, Shi K, He Z, Zong Y, Du R. Protective Effect of Baicalin on Chlorpyrifos-Induced Liver Injury and Its Mechanism. Molecules. 2023 Nov 25;28(23):7771.

[755] Fang L, Wang HF, Chen YM, Bai RX, Du SY. Baicalin confers hepatoprotective effect against alcohol-associated liver disease by upregulating microRNA-205. Int Immunopharmacol. 2022 Jun;107:108553.

[756] Li P, Chen Y, Ke X, Zhang R, Zuo L, Wang M, Chen Z, Luo X, Wang J. Baicalin ameliorates alcohol-induced hepatic steatosis by suppressing SREBP1c elicited PNPLA3 competitive binding to ATGL. Arch Biochem Biophys. 2022 Jun 15;722:109236.

[757] Gao Y, Liu J, Hao Z, Sun N, Guo J, Zheng X, Sun P, Yin W, Fan K, Li H. Baicalin ameliorates high fat diet-induced nonalcoholic fatty liver disease in mice via adenosine monophosphate-activated protein kinase-mediated regulation of SREBP1/Nrf2/NF-κB signaling pathways. Phytother Res. 2023 Jun;37(6):2405-2418.

[758] Zaghloul RA, Zaghloul AM, El-Kashef DH. Hepatoprotective effect of Baicalin against thioacetamide-induced cirrhosis in rats: Targeting NOX4/NF-κB/NLRP3 inflammasome signaling pathways. Life Sci. 2022 Apr 15;295:120410.

Treatment for Depression: In a murine model of chronic mild stress, treatment with baicalin markedly alleviated depression-like behavioral changes via activation of the Rac1-cofilin pathway, thereby improving synaptic plasticity[759] - important findings considering that dysregulation in synaptic plasticity has been implicated in depression. In a similar study of a chronic unpredictable mild stress-induced mouse model of depression, baicalin was shown to promote hippocampal neurogenesis via the Wnt/β-catenin pathway[760]. Increased hippocampal neurogenesis is widely believed to improve depression. Other studies of unpredictable chronic mild stress in mice have shown that treatment with baicalin can ameliorate related depression via modulation of the BDNF/ERK/CREB signaling pathway[761], and by activating the AMPK/PGC-1α pathway and enhancing NIX-mediated mitophagy[762], while the same models in rats have demonstrated that baicalin can exert antidepressant effects through regulating differential metabolites and thereby inhibiting oxidative stress and improving neurogenesis[763]. Some researchers have theorized that impaired neurogenesis in the hippocampus contributes to depression, noting that treatment with baicalin "could promote the differentiation of neurons, which transformation into mature neurons and their survival via the Akt/FOXG1 pathway to exert antidepressant effects[764]." In an olfactory bulbectomized model of depression in rats, administration of baicalin reversed depressive-like behaviors via regulation of the SIRT1-NF-kB signaling pathway[765], offering another target for researchers to pursue in the treatment of depression. Finally, when combined with lithium chloride – a widely-used antidepressant drug – baicalin enhanced neurogenesis via the GSK3β (glycogen synthase kinase-3 beta) pathway[766].

Treatment for Obesity: Treatment with baicalin has been shown to ameliorate obesity, potentially by regulating key inflammatory markers, adipogenesis, and apoptosis[767], and attenuate diet-induced obesity partially through promoting thermogenesis in adipose tissue[768].

Treatment for Atherosclerosis: In an E-deficient mice model of atherosclerosis, treatment with baicalin reduced inflammatory cytokines, prevented the progression of atherosclerotic lesion, attenuated ROS production, inhibited activation of NLRP3 inflammasome, and reduced the expression of the adhesion molecules ICAM-1 and VCAM-1[769]. Baicalin has also been shown to suppress the proliferation and migration of Ox-LDL-VSMCs (vascular smooth

[759] Lu Y, Sun G, Yang F, Guan Z, Zhang Z, Zhao J, Liu Y, Chu L, Pei L. Baicalin regulates depression behavior in mice exposed to chronic mild stress via the Rac/LIMK/cofilin pathway. Biomed Pharmacother. 2019 Aug;116:109054.

[760] Xiao Z, Cao Z, Yang J, Jia Z, Du Y, Sun G, Lu Y, Pei L. Baicalin promotes hippocampal neurogenesis via the Wnt/β-catenin pathway in a chronic unpredictable mild stress-induced mouse model of depression. Biochem Pharmacol. 2021 Aug;190:114594.

[761] Jia Z, Yang J, Cao Z, Zhao J, Zhang J, Lu Y, Chu L, Zhang S, Chen Y, Pei L. Baicalin ameliorates chronic unpredictable mild stress-induced depression through the BDNF/ERK/CREB signaling pathway. Behav Brain Res. 2021 Sep 24;414:113463.

[762] Jin X, Zhu L, Lu S, Li C, Bai M, Xu E, Shen J, Li Y. Baicalin ameliorates CUMS-induced depression-like behaviors through activating AMPK/PGC-1α pathway and enhancing NIX-mediated mitophagy in mice. Eur J Pharmacol. 2023 Jan 5;938:175435.

[763] Ma J, Li X, Yang Z, Liu Q, Liu Y, Liu A. Widely targeted metabolomics unveils baicalin-induced hippocampal metabolic alternations in a rat model of chronic unpredictable mild stress. J Pharm Biomed Anal. 2024 Jan 5;237:115766.

[764]Zhang R, Ma Z, Liu K, Li Y, Liu D, Xu L, Deng X, Qu R, Ma Z, Ma S. Baicalin exerts antidepressant effects through Akt/FOXG1 pathway promoting neuronal differentiation and survival. Life Sci. 2019 Mar 15;221:241-248.

[765] Yu H, Zhang F, Guan X. Baicalin reverse depressive-like behaviors through regulation SIRT1-NF-kB signaling pathway in olfactory bulbectomized rats. Phytother Res. 2019 May;33(5):1480-1489.

[766] Wang Z, Cheng Y, Lu Y, Sun G, Pei L. Baicalin Coadministration with Lithium Chloride Enhanced Neurogenesis via GSK3β Pathway in Corticosterone Induced PC-12 Cells. Biol Pharm Bull. 2022 May 1;45(5):605-613.

[767] Wang ZY, Jiang ZM, Xiao PT, Jiang YQ, Liu WJ, Liu EH. The mechanisms of baicalin ameliorate obesity and hyperlipidemia through a network pharmacology approach. Eur J Pharmacol. 2020 Jul 5;878:173103.

[768] Li H, Tang S. Baicalin attenuates diet-induced obesity partially through promoting thermogenesis in adipose tissue. Obes Res Clin Pract. 2021 Sep-Oct;15(5):485-490.

[769] Zhao J, Wang Z, Yuan Z, Lv S, Su Q. Baicalin ameliorates atherosclerosis by inhibiting NLRP3 inflammasome in apolipoprotein E-deficient mice. Diab Vasc Dis Res. 2020 Nov-Dec;17(6):1479164120977441.

muscle cells that have been exposed to oxidized low-density lipoproteins) in atherosclerosis by upregulating miR-126-5p[770].

Treatment for Glaucoma and Age-Related Vision Loss: Baicalin has been shown to treat both age-related vision loss and glaucoma, by regulating miR-223/NLRP3-mediated pyroptosis in macular degeneration[771], and by regulating the PI3K/AKT pathway signaling in glaucoma[772].

Treatment for Polycystic Ovary Syndrome: In a rat model of polycystic ovary syndrome, treatment with baicalin notably reduced the serum levels of free testosterone, total testosterone, follicle-stimulating hormone, luteinizing hormone, progesterone, and estradiol, while also decreasing body weights, increasing the number of rats with a regular estrous cycle, and ameliorating ovarian histological changes and follicular development[773]. A comparable study with induced polycystic ovary syndrome in rats provided similar results, indicating that treatment with baicalin significantly declined free testosterone and luteinizing hormone levels and alleviated the symptoms of polycystic ovary syndrome by regulating the miR-874-3p/FOXO3 and miR-144/FOXO1 axis[774].

Treatment for Osteoporosis: In dexamethasone-induced osteoporosis in zebrafish, treatment with baicalin enhanced growth and development of larvae, ameliorated mineralization of larvae, regulated the expression of RANKL (receptor activator of nuclear factor kappa beta ligand) and OPG (osteoprotegerin), and modulated expression levels of transcription factors involved in bone remodeling[775].

Treatment for Osteoarthritis: In an experimental model of osteoarthritis in rats, administration of baicalin attenuated muscle dysfunction, decreased muscular ROS, and inhibited nuclear factor erythroid-derived 2-like 2 to attenuate osteoarthritis by inhibiting oxidative stress[776].

Treatment for MRSA related Sepsis: In mice infected with MRSA, treatment with baicalin inhibited the production of IL-6, TNF-α, and other cytokines, inhibited the activation of ERK, JNK MAPK, and NF-κB pathways, reduced the high mortality rate, and, in combination with vancomycin, decreased kidney and liver bacterial loads[777].

Treatment for Tick-Borne Encephalitis Virus: Baicalin has been shown to exhibit an in vitro inhibitory effect against tick-borne encephalitis via direct viricidal activity and direct inhibition of adsorption and viral intracellular replication[778].

---

[770] Chen Z, Pan X, Sheng Z, Yan G, Chen L, Ma G. Baicalin Suppresses the Proliferation and Migration of Ox-LDL-VSMCs in Atherosclerosis through Upregulating miR-126-5p. Biol Pharm Bull. 2019 Sep 1;42(9):1517-1523.

[771] Sun HJ, Jin XM, Xu J, Xiao Q. Baicalin Alleviates Age-Related Macular Degeneration via miR-223/NLRP3-Regulated Pyroptosis. Pharmacology. 2020;105(1-2):28-38.

[772] Zhao N, Shi J, Xu H, Luo Q, Li Q, Liu M. Baicalin suppresses glaucoma pathogenesis by regulating the PI3K/AKT signaling in vitro and in vivo. Bioengineered. 2021 Dec;12(2):10187-10198.

[773] Wang W, Zheng J, Cui N, Jiang L, Zhou H, Zhang D, Hao G. Baicalin ameliorates polycystic ovary syndrome through AMP-activated protein kinase. J Ovarian Res. 2019 Nov 13;12(1):109.

[774] Xu X, Xu X, Wang X, Shen L. Baicalin suppress the development of polycystic ovary syndrome via regulating the miR-874-3p/FOXO3 and miR-144/FOXO1 axis. Pharm Biol. 2023 Dec;61(1):878-885.

[775] Zhao Y, Wang HL, Li TT, Yang F, Tzeng CM. Baicalin Ameliorates Dexamethasone-Induced Osteoporosis by Regulation of the RANK/RANKL/OPG Signaling Pathway. Drug Des Devel Ther. 2020 Jan 15;14:195-206.

[776] Chen DS, Cao JG, Zhu B, Wang ZL, Wang TF, Tang JJ. Baicalin Attenuates Joint Pain and Muscle Dysfunction by Inhibiting Muscular Oxidative Stress in an Experimental Osteoarthritis Rat Model. Arch Immunol Ther Exp (Warsz). 2018 Dec;66(6):453-461.

[777] Shi T, Li T, Jiang X, Jiang X, Zhang Q, Wang Y, Zhang Y, Wang L, Qin X, Zhang W, Zheng Y. Baicalin protects mice from infection with methicillin-resistant Staphylococcus aureus via alleviating inflammatory response. J Leukoc Biol. 2020 Dec;108(6):1829-1839.

[778] Leonova GN, Shutikova AL, Lubova VA, Maistrovskaya OS. Inhibitory Activity of Scutellaria baicalensis Flavonoids against Tick-Borne Encephalitis Virus. Bull Exp Biol Med. 2020 Mar;168(5):665-668.

Treatment for Influenza A: In mice infected with the H1N1 virus (Influenza A), administration of baicalin contributed to the survival of infected lung alveolar epithelial cells in part by inhibiting cell pyroptosis by regulating the caspase-3/GSDME pathway[779].

Treatment for Periodontitis: In periodontal ligament cells (PDLC) challenged with lipopolysaccharides, treatment with baicalein significantly attenuated multiple inflammatory factors, significantly inhibited MAPK signaling, and effectively restored the osteogenic differentiation of LPS-treated PDLC[780].

Treatment for Colitis / Ulcerative Colitis: In dextran sulfate sodium (DSS)-induced colitis in mice, administration of baicalin combined with emodin (an anthraquinone) prevented weight loss, reduced colon shortening, and decreased intestinal damage by inhibiting the activity of NF-κB and exerting anti-inflammatory effects[781]. In a similar DSS-induced colitis model in mice, the administration of baicalin improved intestinal barrier structure and function, regulated the AhR/IL-22 pathway, and induced IL-22 production[782]. Another DSS-induced colitis model in mice showed that baicalin combined with berberine hydrochloride maintained the balance of pro- and anti-inflammatory cytokines and modulated the composition of intestinal microflora by regulating DNA synthesis, replication, and repair of gut microbiota[783].

Anti-Epileptic Agent: Baicalin has been shown to decrease A1 astrocytes in the brain, improve behavioral performance, and reduce levels of interleukin-1α and TNF-α in the cerebral interstitial site in rats induced to epilepsy[784].

Treatment for Autism: In a Wistar rat model of valproic acid (VPA)-induced autism, postnatal administration of baicalin promoted postnatal growth and maturation, improved motor development, and ameliorated repetitive behavior and social deficits while also enhancing neuronal mitochondrial functions[785].

Treatment for Hyperuricemic Nephropathy: In a murine model of potassium oxonate-induced hyperuricemic nephropathy, treatment with baicalin significantly ameliorated the levels of renal function, decreased p-PI3K, p-AKT and p-p65 expression, down-regulated BAX/BCL2 and CASP3, and blunted the mRNA levels of TNF-α, IL-1β, and IL-18[786].

Treatment for Spinal Cord Injury: In a rat model of spinal cord injury, administration of baicalin dramatically decreased the water content of spinal cord tissue, the permeability of blood-spinal cord barrier, oxidant stress,

[779] Wei Z, Gao R, Sun Z, Yang W, He Q, Wang C, Zhang J, Zhang X, Guo L, Wang S. Baicalin inhibits influenza A (H1N1)-induced pyroptosis of lung alveolar epithelial cells via caspase-3/GSDME pathway. J Med Virol. 2023 May;95(5):e28790.

[780] Ren M, Zhao Y, He Z, Lin J, Xu C, Liu F, Hu R, Deng H, Wang Y. Baicalein inhibits inflammatory response and promotes osteogenic activity in periodontal ligament cells challenged with lipopolysaccharides. BMC Complement Med Ther. 2021 Jan 23;21(1):43.

[781] Xu B, Huang S, Chen Y, Wang Q, Luo S, Li Y, Wang X, Chen J, Luo X, Zhou L. Synergistic effect of combined treatment with baicalin and emodin on DSS-induced colitis in mouse. Phytother Res. 2021 Oct;35(10):5708-5719.

[782] Li YY, Wang XJ, Su YL, Wang Q, Huang SW, Pan ZF, Chen YP, Liang JJ, Zhang ML, Xie XQ, Wu ZY, Chen JY, Zhou L, Luo X. Baicalein ameliorates ulcerative colitis by improving intestinal epithelial barrier via AhR/IL-22 pathway in ILC3s. Acta Pharmacol Sin. 2022 Jun;43(6):1495-1507.

[783] Yan Y, Li L, Wu K, Zhang G, Peng L, Liang Y, Wang Z. A Combination of Baicalin and Berberine Hydrochloride Ameliorates Dextran Sulfate Sodium-Induced Colitis by Modulating Colon Gut Microbiota. J Med Food. 2022 Aug;25(8):853-862.

[784] Li G, Zhang S, Cheng Y, Lu Y, Jia Z, Yang X, Zhang S, Guo W, Pei L. Baicalin suppresses neuron autophagy and apoptosis by regulating astrocyte polarization in pentylenetetrazol-induced epileptic rats and PC12 cells. Brain Res. 2022 Jan 1;1774:147723.

[785] Elesawy RO, El-Deeb OS, Eltokhy AK, Arakeep HM, Ali DA, Elkholy SS, Kabel AM. Postnatal baicalin ameliorates behavioral and neurochemical alterations in valproic acid-induced rodent model of autism: The possible implication of sirtuin-1/mitofusin-2/ Bcl-2 pathway. Biomed Pharmacother. 2022 Jun;150:112960.

[786] Liu Z, Xiang H, Deng Q, Fu W, Li Y, Yu Z, Qiu Y, Mei Z, Xu L. Baicalin and baicalein attenuate hyperuricemic nephropathy via inhibiting PI3K/AKT/NF-κB signalling pathway. Nephrology (Carlton). 2023 Jun;28(6):315-327.

proinflammatory cytokine expression, and apoptosis, while also significantly improving the recovery of limb function[787].

Protective Agent Against UVB Radiation: In a model of (UVB)-induced photoaging in the dorsal skin of female C57BL/6 mice, topical application of baicalin reduced epidermal thickening of mouse skin, increased the production of collagen I and III, and decreased the expression of MMP-1 and MMP-3, while the researchers also noted that "baicalin has positive effects on UVB-SIPS human dermal fibroblasts by inducing cell proliferation via decreasing senescence-related proteins, increasing collagen production, and decreasing collagen degradation[788]."

Prevention of HIV: Baicalin has been shown to inhibit HIV antigen expression and P24 antigen production in an H9 cell culture, with low levels of cytotoxicity to healthy cells[789].

IMPLICATIONS FOR HUMAN HEALTH AND NUTRITION

Dietary supplementation or consumption of baicalin is only possible by purchasing herbal supplements such as Shuanghuanglian or Sho-Saiko-To, or by working with purified baicalin. However, despite the high safety profile of this flavone, we recommend taking professional advice and extreme caution when using any purified flavonoid or related compound. Because baicalin has not been quantified in cannabis yet, it seems unlikely that consumption of this plant or products made from it will result in the dietary acquisition of the flavonoid.

---

[787] Cao Y, Li G, Wang YF, Fan ZK, Yu DS, Wang ZD, Bi YL. Neuroprotective effect of baicalin on compression spinal cord injury in rats. Brain Res. 2010 Oct 21;1357:115-23.

[788] Zhang JA, Yin Z, Ma LW, Yin ZQ, Hu YY, Xu Y, Wu D, Permatasari F, Luo D, Zhou BR. The protective effect of baicalin against UVB irradiation induced photoaging: an in vitro and in vivo study. PLoS One. 2014 Jun 20;9(6):e99703.

[789] Zhang X, Tang X, Chen H. Inhibition of HIV replication by baicalin and S. baicalensis extracts in H9 cell culture. Chin Med Sci J. 1991 Dec;6(4):230-2.

Baicalin Review

Answer the following questions to test your knowledge of this flavonoid:

Question #1: What type of flavonoid is baicalin?

  a. Flavone
  b. Flavines
  c. Flavoned
  d. Flavonol

Question #2: What is the primary difference between baicalin and baicalein?

  a. Hydroxyl group methylation
  b. Acetylation of the hydroxyl group
  c. Sulfonation
  d. Glucuronic acid moiety

Question #3: How common is baicalin in cannabis?

  a. Top three
  b. Unknown
  c. Top ten
  d. Top twenty

Question #4: What is the chemical formula for baicalin?

  a. $C_{15}H_{10}O_6$
  b. $C_{21}H_{18}O_{11}$
  c. $C_{16}H_{10}O_{65}$
  d. $C_{26}H_{30}O_{15}$

Question #5: Name the primary plant baicalin is known to occur in:

_____

Question #6: Name two specific potential medical uses of baicalin:

1 _____ 2 _____

For the answer key to Baicalin, please visit www.cannabischemistry.org

CHRYSIN

Type: Flavone
Chemical Formula: $C_{15}H_{10}O_4$
Molecular Weight: 254.24 g/mol
Boiling Point: 491.91 °C @ 760.00 mm Hg (estimated by TGSC)
Flash Point: 192.50 °C (estimated by TGSC)
Melting Point: 285.50 °C @ 760.00 mm Hg[790]
Solubility: Poorly soluble in water, Soluble in ethanol, methanol, dimethyl sulfoxide, acetone, dimethylformamide; Partially soluble in chloroform, ethyl acetate
Oral LD50: > 2,000 mg/kg (rats)
Biological Role: Insecticidal, nematicidal agent
Therapeutic Role: Anticancer agent, chemotherapeutic adjuvant, antidepressant, antiobesity agent
Commercial Use: Research chemical, skin cream ingredient
Occurrence in Cannabis: Unconfirmed, likely only trace amounts

Occurs in Cannabis Strains:

As of the publication of this text, there were no publicly available reports substantiating the presence of chrysin in cannabis. However, due to the likelihood that a chemotype of cannabis will be discovered or developed where chrysin is expressed at detectable levels, the author has included this chapter in consideration of how medically important this flavonoid could be, particularly in the treatment of cancer.

INTRODUCTION

Chrysin is a flavone that has been moderately studied in the treatment of different types of cancer; it is well-known for being found in honey and propolis, and in several types of fungi. Chrysin is found in and sometimes derived from the same plant that is the primary source of wogonin and baicalin - Scutellaria baicalensis, or Skullcap. There are no known organoleptic details or scent/flavor profiles for chrysin, although it is likely that the flavone is bitter like other similar molecules.

---

[790] The Good Scents Company Data Sheet for Chrysin, from: https://www.thegoodscentscompany.com/data/rw1229801.html. Accessed December 23, 2024.

CHEMICAL STRUCTURE

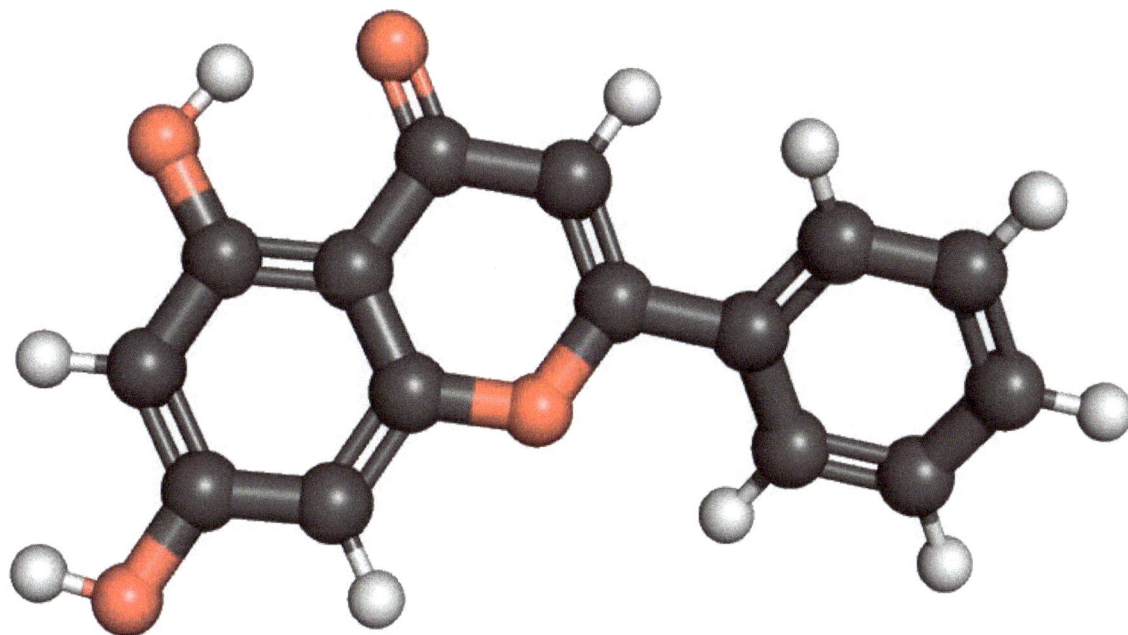

Formed in the phenylpropanoid pathway and derived from naringenin, chrysin is a flavone with the technical name of 5,7-dihydroxyflavone, which indicates that two hydroxy groups are located at positions 5 and 7. The general structure of the chrysin molecule features three rings, seven endocyclic double bonds, and one exocyclic double bonded oxygen atom, with a total of four oxygen atoms, notated as $C_{15}H_{10}O_4$.

Chrysin can undergo several modifications to create variants including some covered in this book such as apigenin (4′,5,7-trihydroxyflavone) and luteolin (3′,4′,5,7-tetrahydroxyflavone), where the former has one additional hydroxy group at the 4 position of the B ring, while the latter has two additional hydroxy groups at the 3 and 4 position. Other variants include methoxylated versions where the hydroxyl groups have been replaced with methoxy groups, such as 5,7-dimethoxyflavone, and chrysin-6-methyl ether (5,7-dihydroxy-6-methoxyflavone). Glycosylated variants include the addition of sugar moieties, such as chrysin-7-glucoside, while other possibilities include acetylated variants such as chrysin-5,7-diacetate, as well as prenylated, geranylated, and halogenated versions of the chrysin molecule.

OCCURRENCE IN PLANTS

Chrysin occurs in just a few plants and foods, including honey, honeycomb, and propolis, as well as several different types of fungi. Recent research quantified chrysin as a major component of a fungi isolated from the marine green alga Chaetomorpha antennina[791]. Other sources of this flavone include:

[791] Parthasarathy R, Chandrika M, Sruthi D, Yashavantha Rao HC, Jayabaskaran C. Clonostachys rosea, a marine algal endophyte, as an alternative source of

Blue passion flower	Honeycomb	Propolis
Carrot	Lactarius deliciosus	Scutellaria baicalensis
Chamomile	Oroxylum indicum	Sour cherry
Dysphania graveolens	Pinus monticola	Suillus bellinii
Honey	Pleurotus ostreatus	Sweet orange

BIOLOGICAL ACTIVITY IN PLANTS

Little is known about the biological activity of chrysin in plants. However, some research has shown that chrysin inhibits the growth of larvae and female oviposition behavior of Zeugodacus cucurbitae (melon fruit fly), a major agricultural pest[792]. Chrysin may also be used by plants as a nematicidal agent, considering that research published in 2020 showed that chrysin exhibited strong nematicidal activity against the root-knot nematode[793], another common agricultural pest.

USES IN INDUSTRY

Commercial preparations of chrysin are primarily for research purposes only and appear as a light yellow solid or yellow powder, typically extracted from the blue passion flower. This flavone sometimes appears in skin creams and other topical applications, such as the Chrysin Plus DIM Cream available on Amazon, a product that bills itself as a Topical Aromatase Inhibitor Cream that allegedly supports estrogen balance, with the description of the product as follows: "Chrysin Plus Dim cream is not intended to be slathered over areas of your body. The cream works systemically, meaning that it helps balance estrogen throughout your whole body. Using more IS NOT better. If you use too much, you may experience unwanted side effects. As this cream affects hormone balance, you may want to check with your doctor to make sure that this product is appropriate for you before you purchase[794]."

POTENTIAL USES IN MEDICINE

Chrysin is hydrophobic and has low bioavailability coupled with rapid excretion, which means that it is generally poorly bioavailable. However, recent scientific research has sought to overcome this problem, with scientists creating a tannic acid coated chrysin nanosuspension that improves intestinal absorption[795], and other work that developed chrysin conjugated with methoxypolyethylene glycols, which improved water solubility[796]. Medical and therapeutic potential of chyrsin includes the following:

Treatment for Breast Cancer: Chrysin has been shown to induce cell cycle arrest and apoptosis in breast cancer cells both as an isolated compound, and in combination with quercetin, with the highest toxicity in the combination

chrysin and its anticancer effect. Arch Microbiol. 2023 Jul 6;205(8):275.

[792] Puri, S., Singh, S. & Sohal, S.K. Inhibitory effect of chrysin on growth, development and oviposition behaviour of melon fruit fly, Zeugodacus cucurbitae (Coquillett) (Diptera: Tephritidae). Phytoparasitica 50, 151–162 (2022).

[793] Bano, S., Iqbal, E.Y., Lubna et al. Nematicidal activity of flavonoids with structure activity relationship (SAR) studies against root knot nematode Meloidogyne incognita. Eur J Plant Pathol 157, 299–309 (2020).

[794] Amazon sales page for Chrysin-Plus-DIM-Cream, from: https://www.amazon.com/Chrysin-Plus-DIM-Cream-Aromatase/dp/B0CSC3HPVZ. Accessed December 23, 2024.

[795]Salama A, Hamed Salama A, Hasanein Asfour M. Tannic acid coated nanosuspension for oral delivery of chrysin intended for anti-schizophrenic effect in mice. Int J Pharm. 2024 May 10;656:124085.

[796] Oggero J, Gasser FB, Zacarías SM, Burns P, Baravalle ME, Renna MS, Ortega HH, Vaillard SE, Vaillard VA. PEGylation of Chrysin Improves Its Water Solubility while Preserving the In Vitro Biological Activity. J Agric Food Chem. 2023 Dec 13;71(49):19817-19831.

of the two flavonoids[797]. Chrysin has also been shown to increase the cytotoxicity of NK-92 and Jurkat-T cells targeting MCF-7 and MDA-MB-231 breast cancer cells[798], while the combination of chrysin and pyrotinib "yielded a potent synergistic effect to induce more evident cell cycle arrest, inhibit the proliferation of BT-474 and SK-BR-3 BC cells, and repress in vivo tumor growth in xenograft mice models[799]." Finally, treatment with co-nanoencapsulated curcumin and chrysin caused synergistic growth inhibition in MDA-MB-231 breast cancer cells by upregulating miR-132 and miR-502c expression[800].

Treatment for Liver Cancer: Combinatorial treatment with diosmetin and chrysin induced apoptosis, enhanced autophagy, reduced inflammatory mediator production, and improved the tumor cell microenvironment by inhibiting the PI3K/AKT/mTOR/NF-κB signaling pathway in hepatocellular carcinoma cells[801].

Treatment for Lung Cancer: In benzo(a)pyrene-induced lung carcinogenesis in Swiss albino mice, administration of chrysin decreased lipid peroxides and carcinoembryonic antigen with concomitant increase in the levels of both enzymatic antioxidants and non-enzymatic antioxidants, downregulated the expression of PCNA, COX-2 and NF-κB, and maintained cellular homeostasis[802].

* *Treatment for Colorectal Cancer:* Apigenin combined with chrysin was shown to significantly reduce cell clone numbers, migration, and invasion ability, and increase cell apoptosis in two colorectal cancer cell lines primarily by suppressing the activity of P38-MAPK/AKT pathway[803]. Another chrysin combination – curcumin and chrysin in the form of co-encapsulated polymeric nanoparticles – was shown to exhibit a synergistic antiproliferative and inhibitory effect on hTERT gene expression in Caco-2 colorectal cancer cells[804].

Treatment for Cervical Cancer: In the immortal HeLa cervical cancer cell line, treatment with chrysin exhibited profound effects by increasing the expression of proapoptotic genes and caspases, while decreasing the expression level of antiapoptotic genes and cell cycle regulatory genes, significantly inhibiting proliferation and inducing apoptosis by modulation of the AKT/MAPK pathways[805].

[797] Ramos PS, Ferreira C, Passos CLA, Silva JL, Fialho E. Effect of quercetin and chrysin and its association on viability and cell cycle progression in MDA-MB-231 and MCF-7 human breast cancer cells. Biomed Pharmacother. 2024 Oct;179:117276.

[798] Durmus E, Ozman Z, Ceyran IH, Pasin O, Kocyigit A. Chrysin Enhances Anti-Cancer Activity of Jurkat T Cell and NK-92 Cells Against Human Breast Cancer Cell Lines. Chem Biodivers. 2024 Oct;21(10):e202400806.

[799] Liu X, Zhang X, Shao Z, Zhong X, Ding X, Wu L, Chen J, He P, Cheng Y, Zhu K, Zheng D, Jing J, Luo T. Pyrotinib and chrysin synergistically potentiate autophagy in HER2-positive breast cancer. Signal Transduct Target Ther. 2023 Dec 18;8(1):463.

[800] Javan N, Khadem Ansari MH, Dadashpour M, Khojastehfard M, Bastami M, Rahmati-Yamchi M, Zarghami N. Synergistic Antiproliferative Effects of Co-nanoencapsulated Curcumin and Chrysin on MDA-MB-231 Breast Cancer Cells Through Upregulating miR-132 and miR-502c. Nutr Cancer. 2019;71(7):1201-1213.

[801] Yu X, Zhang D, Hu C, Yu Z, Li Y, Fang C, Qiu Y, Mei Z, Xu L. Combination of Diosmetin With Chrysin Against Hepatocellular Carcinoma Through Inhibiting PI3K/AKT/mTOR/NF-κB Signaling Pathway: TCGA Analysis, Molecular Docking, Molecular Dynamics, In Vitro Experiment. Chem Biol Drug Des. 2024 Oct;104(4):e70003.

[802] Eshvendar Reddy Kasala, Lakshmi Narendra Bodduluru, Chandan C Barua, Rajaram Mohanrao Madhana, Vicky Dahiya, Mukesh Kumar Budhani, Ramana Reddy Mallugari, Suseela Reddy Maramreddy, Ranadeep Gogoi, Chemopreventive effect of chrysin, a dietary flavone against benzo(a)pyrene induced lung carcinogenesis in Swiss albino mice, Pharmacological Reports, Volume 68, Issue 2, 2016, Pages 310-318, ISSN 1734-1140.

[803] Zhang X, Zhang W, Chen F, Lu Z. Combined effect of chrysin and apigenin on inhibiting the development and progression of colorectal cancer by suppressing the activity of P38-MAPK/AKT pathway. IUBMB Life. 2021 May;73(5):774-783.

[804] Lotfi-Attari J, Pilehvar-Soltanahmadi Y, Dadashpour M, Alipour S, Farajzadeh R, Javidfar S, Zarghami N. Co-Delivery of Curcumin and Chrysin by Polymeric Nanoparticles Inhibit Synergistically Growth and hTERT Gene Expression in Human Colorectal Cancer Cells. Nutr Cancer. 2017 Nov-Dec;69(8):1290-1299.

[805] Raina R, Afroze N, Kedhari Sundaram M, Haque S, Bajbouj K, Hamad M, Hussain A. Chrysin inhibits propagation of HeLa cells by attenuating cell survival and inducing apoptotic pathways. Eur Rev Med Pharmacol Sci. 2021 Mar;25(5):2206-2220.

Radiotherapy Enhancement Agent: In B16-F10 melanoma cells, combination therapy of chrysin with radiation exhibited a synergistic effect, increasing cell apoptosis, inducing immunologic cell death, and intensifying the radiotherapy-induced immunogenicity[806]. In triple negative breast cancer cells, combination treatment of chrysin and radiotherapy increased Bax (pro-apoptotic gene) and p53 levels, decreased the expression of HIF-1α, and reduced the expression of Bcl-2, thereby inducing apoptosis in MDA-MB-231 cells[807].

Chemotherapeutic Adjuvant / Supplemental Therapy Agent: Chrysin has been shown to enhance and supplement various anticancer therapies, including synergizing the effects of the chemotherapeutic agent docetaxel[808], reducing the toxicity and DNA damage cause by the anticancer drug carboplatin[809], and ameliorating bortezomib-induced nephrotoxicity[810].

Treatment for Neurodegenerative Disease: In a mouse model of aluminum chloride-induced neurotoxicity, treatment with chrysin and a glycosylated form of chrysin exerted antioxidant effects in the cytoplasm of SH-SY5Y cells, recovering memory loss and counteracting neuronal death in part due to the increased bioavailability of the combination of molecules[811].

Treatment for Depression: In a murine model of depression, administration of low doses of chrysin produced a rapid antidepressant-like effect, whereas high doses of chrysin produced a delayed but sustained effect, with researchers noting that the effects were similar to allopregnanolone and fluoxetine[812], common medications for depression. In clonidine-induced depression-like behavior in rats, administration of lycopene and chrysin reversed all clonidine-induced alterations and clinical signs of behavioral hopelessness, primarily by mitigating neuroinflammation and oxidative stress[813].

Treatment for SARS / COVID: As isolated from Scutellaria baicalensis Georgi, chrysin 7-O-β-D-glucuronide was shown to potently inhibit SARS-CoV-2 binding and reduce the levels of pro-inflammatory cytokines in Vero E6 (African green monkey) cells[814].

---

[806] Jafari S, Ardakan AK, Aghdam EM, Mesbahi A, Montazersaheb S, Molavi O. Induction of immunogenic cell death and enhancement of the radiation-induced immunogenicity by chrysin in melanoma cancer cells. Sci Rep. 2024 Oct 5;14(1):23231.

[807] Jafari, S., Dabiri, S., Mehdizadeh Aghdam, E. et al. Synergistic effect of chrysin and radiotherapy against triple-negative breast cancer (TNBC) cell lines. Clin Transl Oncol 25, 2559–2568 (2023).

[808] Ghamkhari A, Pouyafar A, Salehi R, Rahbarghazi R. Chrysin and Docetaxel Loaded Biodegradable Micelle for Combination Chemotherapy of Cancer Stem Cell. Pharm Res. 2019 Oct 23;36(12):165.

[809] Jan BL, Ahmad A, Khan A, Rehman MU, Alkharfy KM. Protective effect of chrysin, a flavonoid, on the genotoxic activity of carboplatin in mice. Drug Chem Toxicol. 2022 Sep;45(5):2146-2152.

[810] Kankılıç NA, Şimşek H, Akaras N, Gür C, Küçükler S, İleritürk M, Gencer S, Kandemir FM. The ameliorative effects of chrysin on bortezomib-induced nephrotoxicity in rats: Reduces oxidative stress, endoplasmic reticulum stress, inflammation damage, apoptotic and autophagic death. Food Chem Toxicol. 2024 Aug;190:114791.

[811] Okoh VI, Campos HM, Yasmin de Oliveira Ferreira P, Pereira RM, Souza Silva Y, Arruda EL, Pagliarani B, de Almeida Ribeiro Oliveira G, Lião LM, Franco Dos Santos G, Vaz BG, Sabino JR, Alcantara Dos Santos FC, Costa EA, Tarozzi A, Menegatti R, Ghedini PC. Chrysin bonded to β-d-glucose tetraacetate enhances its protective effects against the neurotoxicity induced by aluminum in Swiss mice. J Pharm Pharmacol. 2024 Apr 3;76(4):368-380.

[812] Guillén-Ruiz G, Bernal-Morales B, Limón-Vázquez AK, Olmos-Vázquez OJ, Rodríguez-Landa JF. Involvement of the GABAA Receptor in the Antidepressant-Like Effects Produced by Low and High Doses of the Flavonoid Chrysin in the Rat: A Longitudinal Study. J Integr Neurosci. 2024 Mar 4;23(3):51.

[813] Abd Al Haleem EN, Ahmed HI, El-Naga RN. Lycopene and Chrysin through Mitigation of Neuroinflammation and Oxidative Stress Exerted Antidepressant Effects in Clonidine-Induced Depression-like Behavior in Rats. J Diet Suppl. 2023;20(3):391-410.

[814] Yi Y, Yu R, Xue H, Jin Z, Zhang M, Bao YO, Wang Z, Wei H, Qiao X, Yang H. Chrysin 7-O-β-D-glucuronide, a dual inhibitor of SARS-CoV-2 3CLpro and PLpro, for the prevention and treatment of COVID-19. Int J Antimicrob Agents. 2024 Jan;63(1):107039.

Treatment for Epilepsy: In a PTZ-induced kindling mouse model of epilepsy, treatment with an ethanolic extract of Pyrus pashia and chrysin reduced seizure severity and prevented memory impairment, increased the levels of BDNF and CREB, and reduced apoptotic biomarkers in the hippocampus[815].

Treatment for Organophosphate Toxicity: In chlorpyrifos-induced dysfunction of the hypothalamic-pituitary-testicular axis in rats, treatment with chrysin nanocrystals exhibited antioxidant and anti-inflammatory properties, effectively counteracting reductions in luteinizing hormone, serum testosterone, follicle-stimulating hormone, and testicular enzyme biomarkers, while also enhancing antioxidant defenses and reducing inflammatory markers[816].

Treatment for Endometrial Hyperplasia: In estradiol-induced endometrial hyperplasia in rats, treatment with chrysin increased PPARα activity, exerting anti-proliferative, antioxidant, and anti-inflammatory effects that attenuated the condition[817].

Treatment for Knee Osteoarthritis: In a murine model of monosodium iodoacetate-induced knee osteoarthritis, treatment with chrysin mitigated neuropathic pain and peripheral sensitization by repressing the RAGE/PI3K/AKT pathway modulated by HMGB1[818] (high mobility group box 1, a protein that plays a role in DNA repair).

Treatment for Diet-related Heart Stress: In high-fat diet-fed rats, administration of chrysin reduced heart endoplasmic reticulum stress-induced apoptosis by inhibiting PERK and Caspase 3-7, providing significant cardioprotective effects[819]. In a similar study involving high fat diet-fed male Sprague Dawley rats, oral administration of chrysin attenuated myocardial oxidative stress by upregulating eNOS and Nrf2 target genes[820].

Treatment for Non-Alcoholic Fatty Liver Disease: In rats induced with non-alcoholic fatty liver disease via administration of a high fructose diet, treatment with chrysin caused a significant decrease in serum fasting glucose and improved insulin resistance, dyslipidemia, and liver enzymes, significantly decreased liver weight and hepatic free fatty acids, triglyceride, and cholesterol content while also exerting antioxidant effects, reducing carbonyl content, advanced glycation end products, collagen, TNF-α, and IL-6 concentrations in the liver, and significantly reducing the hepatic gene expression of SREBP-1c, and increasing that of PPAR-α[821].

Antiobesity Agent: Chrysin has been shown to down-regulate adipogenesis, lipogenesis and ROS, and up-regulate lipolysis and antioxidant enzyme in differentiated 3T3-L1 adipocytes[822]. In diet-induced obese mice, administration of chrysin improved glycemic control and insulin sensitivity by regulating the expression of IKKε/TBK1 in the liver of subject animals[823]. In another study of high fat diet-fed mice, supplementation with chrysin alleviated adiposity

[815] Sharma P, Kumari A, Singh P, Srivas S, Thakur MK, Hemalatha S. Pyrus pashia fruit extract and its major phytometabolite chrysin prevent hippocampal apoptosis and memory impairment in PTZ-kindled mice. Nutr Neurosci. 2024 Aug;27(8):836-848.

[816] Farkhondeh T, Roshanravan B, Samini F, Samarghandian S. Effect of Chrysin and Chrysin Nanocrystals on Chlorpyrifos-Induced Dysfunction of the Hypothalamic-Pituitary-Testicular Axis in Rats. Curr Mol Pharmacol. 2024;17:e18761429305457.

[817] Eid, B.G. Chrysin attenuates estradiol-induced endometrial hyperplasia in rats via enhancing PPARα activity. Environ Sci Pollut Res 29, 54273–54281 (2022).

[818] Xu B, Xu Y, Kong J, Liu Y, Zhang L, Shen F, Wang J, Shen X, Chen H. Chrysin mitigated neuropathic pain and peripheral sensitization in knee osteoarthritis rats by repressing the RAGE/PI3K/AKT pathway regulated by HMGB1. Cytokine. 2024 Aug;180:156635.

[819] Yuvaraj S, Vasudevan V, Puhari SSM, Sasikumar S, Ramprasath T, Selvi MS, Selvam GS. Chrysin reduces heart endoplasmic reticulum stress-induced apoptosis by inhibiting PERK and Caspase 3-7 in high-fat diet-fed rats. Mol Biol Rep. 2024 May 25;51(1):678.

[820] Yuvaraj S, Ramprasath T, Saravanan B, Vasudevan V, Sasikumar S, Selvam GS. Chrysin attenuates high-fat-diet-induced myocardial oxidative stress via upregulating eNOS and Nrf2 target genes in rats. Mol Cell Biochem. 2021 Jul;476(7):2719-2727.

[821] Pai SA, Munshi RP, Panchal FH, Gaur IS, Juvekar AR. Chrysin ameliorates nonalcoholic fatty liver disease in rats. Naunyn Schmiedebergs Arch Pharmacol. 2019 Dec;392(12):1617-1628.

[822] John CM, Arockiasamy S. Enhanced Inhibition of Adipogenesis by Chrysin via Modification in Redox Balance, Lipogenesis, and Transcription Factors in 3T3-L1 Adipocytes in Comparison with Hesperidin. J Am Nutr Assoc. 2022 Nov-Dec;41(8):758-770.

[823] Amir Siddiqui M, Badruddeen, Akhtar J, Uddin S, Chandrashekharan SM, Ahmad M, Khan MI, Khalid M. Chrysin modulates protein kinase IKKε/TBK1,

and insulin resistance by promoting subcutaneous adipocyte browning and systematic energy expenditure via the regulation of PDGFRα and microRNA expressions[824], while in fructose-fed rats, consumption of chrysin attenuated metabolic disease[825].

Treatment for Acute Lung Injury: In mice induced to acute lung injury via lipopolysaccharide, pre-treatment with chrysin attenuated inflammation by reducing the production of myeloperoxidase and pro-inflammatory cytokine levels in the lung and bronchoalveolar lavage fluid, improving lung edema by reducing the vascular permeability, improving the antioxidant capacity by increasing the activities of superoxide dismutase (SOD) and glutathione peroxidase (GSH-Px) in the lung tissue, and suppressing the lipopolysaccharide-induced expression of glucose-regulated protein 78 (GRP78) and phosphorylated inositol-requiring enzyme 1α (p-IRE1α)[826].

Treatment for Brain Aging: Chrysin has been shown to restore antioxidant status and inhibit oxidative stress and the apoptotic pathways in the hippocampus and prefrontal cortex of D-galactose-induced aging rat brains[827].

IMPLICATIONS FOR HUMAN HEALTH AND NUTRITION

Dietary consumption of chrysin is the safest way to obtain this flavone. Consider eating the vegetables carrots or oroxylum, or fruits such as sour cherry or sweet orange. Teas or other products that contain chamomile, and many types of honey and honeycomb are also likely to contain chrysin. Extractions made from blue passion flower or skullcap plants might contain high concentrations of chrysin, so extreme care must be used when working with these compounds. Some skincare products also contain chrysin, although the author cannot vouch for the authenticity or efficacy of any such products.

insulin sensitivity and hepatic fatty infiltration in diet-induced obese mice. Drug Dev Res. 2022 Feb;83(1):194-207.

[824] Wang X, Cai H, Shui S, Lin Y, Wang F, Wang L, Chen J, Liu J. Chrysin Stimulates Subcutaneous Fat Thermogenesis in Mice by Regulating PDGFRα and MicroRNA Expressions. J Agric Food Chem. 2021 Jun 2;69(21):5897-5906.

[825] Andrade N, Andrade S, Silva C, Rodrigues I, Guardão L, Guimarães JT, Keating E, Martel F. Chronic consumption of the dietary polyphenol chrysin attenuates metabolic disease in fructose-fed rats. Eur J Nutr. 2020 Feb;59(1):151-165.

[826] Chen M, Li J, Liu X, Song Z, Han S, Shi R, Zhang X. Chrysin prevents lipopolysaccharide-induced acute lung injury in mice by suppressing the IRE1α/TXNIP/NLRP3 pathway. Pulm Pharmacol Ther. 2021 Jun;68:102018.

[827] Prajit R, Saenno R, Suwannakot K, Kaewngam S, Anosri T, Sritawan N, Aranarochana A, Sirichoat A, Pannangrong W, Wigmore P, Welbat JU. Chrysin mitigates neuronal apoptosis and impaired hippocampal neurogenesis in male rats subjected to D-galactose-induced brain aging. Biogerontology. 2024 Nov;25(6):1275-1284.

Chrysin Review

Answer the following questions to test your knowledge of this flavonoid:

Question #1: What type of flavonoid is chrysin?

     a. Flavane
     b. Flavonene
     c. Flavonol
     d. Flavone

Question #2: The base component in the biosynthesis of chrysin is:

     a. Cannabisin
     b. Naringenin
     c. Chrysinite
     d. Apigenin

Question #3: What two plants is chrysin usually sourced from?

     a. Carrot and Sour Cherry
     b. Sweet Orange and Skullcap
     c. Skullcap and Blue passion flower
     d. Honey and Propolis

Question #4: What is the chemical formula for chrysin?

     a. $C_{15}H_{10}O_4$
     b. $C_{26}H_{10}O_{15}$
     c. $C_{16}H_{10}O_{65}$
     d. $C_{26}H_{30}O_{15}$

Question #5: In what position are this molecule's two hydroxy groups located:

1 _____ 2 _____

Question #6: Name two potential medical uses of chrysin:

1 _____ 2 _____

For the answer key to Chrysin, please visit www.cannabischemistry.org

NARINGENIN

Type: Flavanone
Chemical Formula: $C_{15}H_{12}O_5$
Molecular Weight: 272.25 g/mol
Melting Point: 251 °C
Solubility: Slightly soluble in water, Soluble in ethanol, methanol, acetone, and other organic solvents.
Oral LD50: Likely > 5,000 mg/kg (rats)[828]
Previous studies indicating 330 mg/kg (rats) are unreliable
Biological Role: Allelopathic, antibacterial agent
Therapeutic Role: Anticancer, antidepressant, antinociceptive agent
Commercial Use: Research chemical, skin conditioning agent
Occurrence in Cannabis: Trace

Occurs in Cannabis Strains:

As of early 2025, there were very few publicly available reports of naringenin occurring in cannabis. Concerning commercial or recreational/drug types, naringenin was found to be the least averaged polyphenol compound across six cultivars grown by a licensed producer in Edmonton, Canada: Alien Dawg, Tangerine Dream, Sensi Star, Quadra, Gabriola (Frosty Monster), and Island Honey[829]. In industrial hemp, naringenin was detected in small amounts in the Kompolti, Tiborszallasi, Antal, and Carmagnola Cs varieties[830], while other reports indicate that this flavanone is found in several Italian varieties of industrial hemp, namely the Fedora 17, Ferimon, and Futura 75 cultivars[831].

INTRODUCTION

Naringenin is a flavanone that is often the precursor chemical to many other flavonoids and related compounds. In general, compounds such as *naringin* lose a sugar group to become naringenin in the human digestive tract (conversion via metabolism), which means that in most cases, this molecule is not likely to be found in significant concentration in cannabis varieties, although a chemovar could be developed that specifically expresses naringenin.

There are no reported organoleptic properties for naringenin, which means that when it does occur in cannabis, it probably does not contribute to the aroma or flavor of the variety in question, although purified forms of the compound are likely to taste bitter. In high quantities, naringenin may impart certain medical or therapeutic value as indicated in the "Potential Uses in Medicine" section below, while in smaller quantities it may synergize the effects of other compounds or contribute no effects at all.

Interestingly, naringenin appears to be produced or sequestered by the larvae of the Oriental hornet (Vespa orientalis), and, as extracted from these larvae, the flavanone exhibited significant activity against breast cancer cells[832].

[828] Ortiz-Andrade RR, Sánchez-Salgado JC, Navarrete-Vázquez G, Webster SP, Binnie M, García-Jiménez S, León-Rivera I, Cigarroa-Vázquez P, Villalobos-Molina R, Estrada-Soto S. Antidiabetic and toxicological evaluations of naringenin in normoglycaemic and NIDDM rat models and its implications on extra-pancreatic glucose regulation. Diabetes Obes Metab. 2008 Nov;10(11):1097-104.

[829] Wishart DS, Hiebert-Giesbrecht M, Inchehborouni G, Cao X, Guo AC, LeVatte MA, Torres-Calzada C, Gautam V, Johnson M, Liigand J, Wang F, Zahraei S, Bhumireddy S, Wang Y, Zheng J, Mandal R, Dyck JRB. Chemical Composition of Commercial Cannabis. J Agric Food Chem. 2024 Jun 26;72(25):14099-14113.

[830] Izzo L, Castaldo L, Narváez A, Graziani G, Gaspari A, Rodríguez-Carrasco Y, Ritieni A. Analysis of Phenolic Compounds in Commercial Cannabis sativa L. Inflorescences Using UHPLC-Q-Orbitrap HRMS. Molecules. 2020 Jan 31;25(3):631.

[831] Radwan MM, Chandra S, Gul S, ElSohly MA. Cannabinoids, Phenolics, Terpenes and Alkaloids of Cannabis. Molecules. 2021 May 8;26(9):2774.

[832] Zedan AMG, Sakran MI, Bahattab O, Hawsawi YM, Al-Amer O, Oyouni AAA, Nasr Eldeen SK, El-Magd MA. Oriental Hornet (Vespa orientalis) Larval

CHEMICAL STRUCTURE

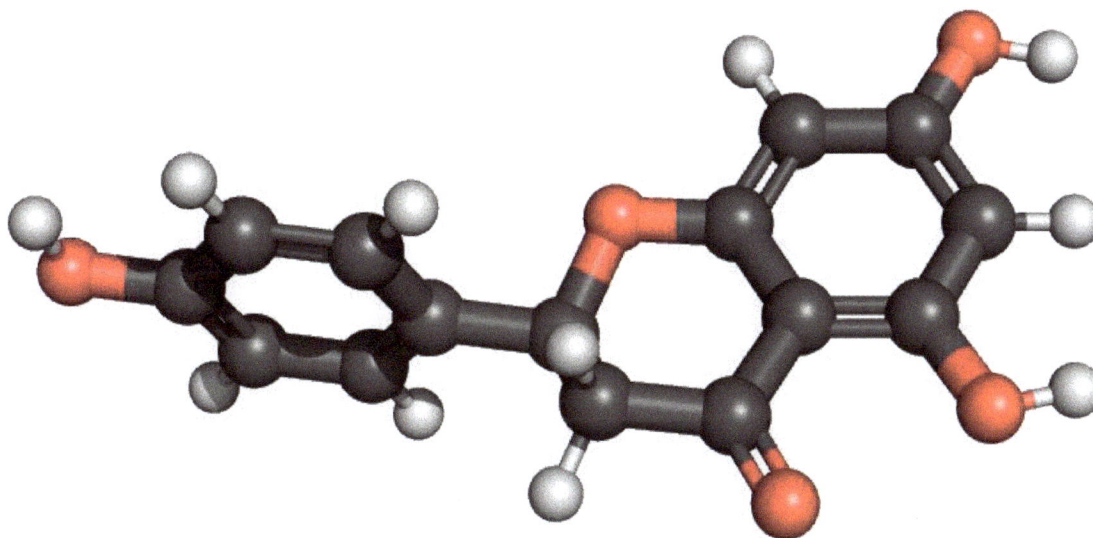

Naringenin is biosynthesized from p-coumaroyl-CoA and malonyl-CoA, with molecules like naringin losing a sugar group to become naringenin. Naringenin is often the base or precursor chemical in the formation of many other types of flavonoids, including flavones, flavonols, anthocyanins, flavan-3-ols, dihydroflavonols, and catechins.

The naringenin molecule is based on three rings; Ring A is comprised of benzene with two hydroxyl groups, Ring B is also a benzene ring, but attached via the heterocyclic pyran Ring C. Overall naringenin contains six endocyclic double bonds and one exocyclic double bonded oxygen atom in the B ring, with five oxygen atoms total, notated as $C_{15}H_{12}O_5$.

Notably, naringenin has one chiral center at the C2 position of Ring C, making it optically active, with the natural form being the (S)-enantiomer.

Like most flavonoids, naringenin variations can occur via glycosylation, methylation, hydroxylation, halogenation, and other modifications. Consider the following variations:

Naringin - naringenin molecule attached to a neohesperidose sugar (a disaccharide of glucose and rhamnose) at the 7-position.

Naringenin-7-O-glucoside - glucose attached to the 7-position hydroxyl group on the A-ring.

7-O-Methyl-Naringenin - the hydroxyl group at the 7-position is replaced with a methoxy (-OCH3) group.

4'-O-Methyl-Naringenin - the hydroxyl group at the 4'-position on the B-ring is replaced with a methoxy group.

Eriodictyol – a hydroxyl group is added at the 3'-position of the B-ring.

Extracts Induce Antiproliferative, Antioxidant, Anti-Inflammatory, and Anti-Migratory Effects on MCF7 Cells. Molecules. 2021 May 31;26(11):3303.

OCCURRENCE IN PLANTS

Naringenin does occur naturally, including in cannabis and hemp, but it is predominantly present as glycosides in fruits and vegetables, often occurring as naringin or other compounds. Other plants and foods found to contain this flavanone include:

Allium	Fig	Orange
Angelica	Flaxseed	Orange mint
Artichoke	Garden cress	Papaya
Asparagus	Garlic	Parsnip
Avocado	Grapefruit	Passion fruit
Barley	Hazelnut	Peanut
Beet	Hemp	Pecan
Black walnut	Hops	Peppermint
Brazil nut	Horseradish	Pineapple
Buckwheat	Kiwi	Poppy
Burdock	Kohlrabi	Pummelo
Capers	Lemon	Quince
Cannabis	Lemongrass	Rape
Caraway	Lentils	Safflower
Cardamom	Lima bean	Saffron
Cauliflower	Loquat	Star anise
Cherry	Lovage	Star fruit
Chestnut	Lupine	Sunflower
Chicory	Mandarin orange	Sweet bay
Chickpea	Mango	Sweet potato
Chives	Mentha	Tomato
Cinnamon	Mugwort	Turmeric
Corn mint	Mulberry	Turnip
Cucumber	Nutmeg	Walnut
Cucurbita	Oat	Watermelon
Date	Olive	Wild carrot
Endive	Onion	Wild celery

BIOLOGICAL ACTIVITY IN PLANTS

Little research exists concerning the biological activities of naringenin in plants, however, some work indicates that this molecule might serve allelopathic and/or antibacterial functions, among others. As described previously, naringenin is often used as a precursor chemical in the biosynthesis of other flavonoid and related compounds.

Allelopathic Agent: In rice seedlings, exposure to naringenin caused root browning, delay of leaf/root development and shoot dwarfing, although the compound did not outright kill the plants[833]. Other research has demonstrated that naringenin strongly inhibits the germination of Arabidopsis thaliana (thale cress) by impairing auxin transport[834].

---

[833] Deng, F., Aoki, M. and Yogo, Y. (2004), Effect of naringenin on the growth and lignin biosynthesis of gramineous plants. Weed Biology and Management, 4: 49-55.

[834] Hernández I, Munné-Bosch S. Naringenin inhibits seed germination and seedling root growth through a salicylic acid-independent mechanism in Arabidopsis thaliana. Plant Physiol Biochem. 2012 Dec;61:24-8.

Increases Stress Tolerance: Naringenin has been shown to increase salt/osmotic stress tolerance in bean plants, likely by regulating nitrogen assimilation pathways, improving expression levels of tolerance-associated genes, and by modulating the antioxidant capacity and AsA-GSH (ascorbate-glutathione) redox-based systems[835].

Antibacterial Agent: Recent work has shown that the roots of bacterial wilt-resistant tobacco (mutant KCB-1) limited the growth and reproduction of Ralstonia solanacearum (bacteria) by increasing the expression of naringenin in root metabolites and secretions[836].

USES IN INDUSTRY

Naringenin is primarily a research chemical, although it is also used in some commercial or consumer products. Purified forms of this compound appear as pale yellow or beige crystalline powders. According to the Flavor and Extract Manufacturers Association, naringenin is sometimes used in food or other products to reduce bitter flavors. Naringenin is also thought to be the active constituent in the traditional Chinese herbal medicine called Chai Hu Shu Gan San, where it is believed to exhibit activity against nonalcoholic steatohepatitis[837]. Experimental research includes using naringenin in contact lenses as a slow-release drug delivery system[838], and the compound has also been proposed as a food industry fat replacer, particularly as a replacer for low-fat cream[839].

A significant portion of the current research into naringenin seeks to produce this compound via means that do not include plant extraction. For instance, both naringenin[840] and (2S)-naringenin[841] have been produced by engineered Saccharomyces cerevisiae (brewer's yeast), while other work has used naringenin as a starting point to create dihydroquercetin[842].

POTENTIAL USES IN MEDICINE

In addition to the research mentioned above regarding the production of naringenin from alternative sources, medical and therapeutic research has focused on making this hydrophobic compound more bioavailable. For instance, recent work has shown that specially developed chitosan oligosaccharide modified bovine serum albumin nanoparticles can help the human body better absorb and use naringenin[843]. Other work shows that this flavanone can be a potent anticancer agent, antidepressant, antibacterial agent, and other medical and therapeutic potential as detailed below:

---

[835] Ozfidan-Konakci C, Yildiztugay E, Alp FN, Kucukoduk M, Turkan I. Naringenin induces tolerance to salt/osmotic stress through the regulation of nitrogen metabolism, cellular redox and ROS scavenging capacity in bean plants. Plant Physiol Biochem. 2020 Dec;157:264-275.

[836] Shi H, Jiang J, Yu W, Cheng Y, Wu S, Zong H, Wang X, Ding A, Wang W, Sun Y. Naringenin restricts the colonization and growth of Ralstonia solanacearum in tobacco mutant KCB-1. Plant Physiol. 2024 Jun 28;195(3):1818-1834.

[837] Ren Y, Xiao K, Lu Y, Chen W, Li L, Zhao J. Deciphering the mechanism of Chaihu Shugan San in the treatment of nonalcoholic steatohepatitis using network pharmacology and molecular docking. J Pharm Pharmacol. 2024 Nov 4;76(11):1521-1533.

[838] Chau Thuy Nguyen D, Dowling J, Ryan R, McLoughlin P, Fitzhenry L. Controlled release of naringenin from soft hydrogel contact lens: An investigation into lens critical properties and in vitro release. Int J Pharm. 2022 Jun 10;621:121793.

[839] Zhang J, Cheng T, Sun M, Li Y, Zhang G, Hu Z, Wang D, Guo Z, Wang Z. Application of soy protein isolate-naringenin complexes as fat replacers in low-fat cream: Based on protein conformational changes, aggregation states and interfacial adsorption behavior. Int J Biol Macromol. 2024 Aug;274(Pt 1):133315.

[840] Koopman F, Beekwilder J, Crimi B, van Houwelingen A, Hall RD, Bosch D, van Maris AJ, Pronk JT, Daran JM. De novo production of the flavonoid naringenin in engineered Saccharomyces cerevisiae. Microb Cell Fact. 2012 Dec 8;11:155.

[841] Gao S, Lyu Y, Zeng W, Du G, Zhou J, Chen J. Efficient Biosynthesis of (2S)-Naringenin from p-Coumaric Acid in Saccharomyces cerevisiae. J Agric Food Chem. 2020 Jan 29;68(4):1015-1021.

[842] Yu S, Li M, Gao S, Zhou J. Engineering Saccharomyces cerevisiae for the production of dihydroquercetin from naringenin. Microb Cell Fact. 2022 Oct 15;21(1):213.

[843] Fang R, Liao Y, Qiu H, Liu Y, Lin S, Chen H. Chitosan Oligosaccharide Modified Bovine Serum Albumin Nanoparticles for Improving Oral Bioavailability of Naringenin. Curr Drug Deliv. 2024;21(8):1142-1150.

Treatment for Breast Cancer: Naringenin has been shown to block the migration and invasion of MDA-MB-231 breast cancer cells by down-regulating MMP-2 and MMP-9 expressions[844], while other work with the same breast cancer cell line showed that treatment with naringenin inhibited the cancer cells by inducing programmed cell death, caspase stimulation, G2/M phase cell cycle arrest, and suppressing metastasis[845]. As the primary constituent in an extract of Vespa orientalis larvae, naringenin contributed to potent antioxidant activities, inhibited cell viability, modulated the expression of apoptosis-related genes, inhibited intracellular ROS and induced antioxidant status, inhibited cell migration, and reduced the expression of inflammation-related genes in MCF7 breast cancer cells[846]. Additionally, naringenin combined with metformin was shown to reduce tumor volume, weight, and multiplicity in two animal models of breast cancer[847]. In one of the hardest to treat breast cancers – triple negative breast cancer – treatment with naringenin and metformin synergized the effects of the anticancer drug doxorubicin, reducing tumor volume and weight, increasing necrosis, and inhibiting cell proliferation[848].

Treatment for Non-Small Cell Lung Cancer: In A549 and H1299 non-small cell lung cancer cells, treatment with apigenin and naringenin caused significant cytotoxicity with cell cycle arrest at G2/M phases, significantly enhanced mitochondria dysfunction, elevated oxidative stress, and activated the apoptotic pathway[849] in both cell lines.

Treatment for Liver Cancer: In HepG2 human hepatocellular carcinoma cells, treatment with naringenin inhibited proliferation, induced apoptosis, and induced cell cycle arrest[850].

Treatment for Oral Cancer: Naringenin has been shown to induce intracellular ROS to trigger programmed cell death and endoplasmic reticulum stress via apoptosis and autophagy in human oral squamous carcinoma cells[851]. In CAL-27 tongue carcinoma cells, treatment with naringenin induced apoptotic cell death by modulating the Bid and Bcl-xl signaling pathways[852].

Treatment for Colorectal Cancer: In azoxymethane/dextran sulfate-induced colorectal cancer in mice, naringenin combined with 5-fluorouracil significantly reduced cardiotoxicity and liver damage and attenuated colorectal injuries by increasing ROS levels and decreasing the mitochondrial membrane potential, thereby stimulating AMPK/mTOR signaling[853].

[844] Sun Y, Gu J. Study on effect of naringenin in inhibiting migration and invasion of breast cancer cells and its molecular mechanism. Zhongguo Zhong Yao Za Zhi. 2015 Mar;40(6):1144-50. Chinese.

[845] Qi Z, Kong S, Zhao S, Tang Q. Naringenin inhibits human breast cancer cells (MDA-MB-231) by inducing programmed cell death, caspase stimulation, G2/M phase cell cycle arrest and suppresses cancer metastasis. Cell Mol Biol (Noisy-le-grand). 2021 Sep 29;67(2):8-13.

[846] Zedan AMG, Sakran MI, Bahattab O, Hawsawi YM, Al-Amer O, Oyouni AAA, Nasr Eldeen SK, El-Magd MA. Oriental Hornet (Vespa orientalis) Larval Extracts Induce Antiproliferative, Antioxidant, Anti-Inflammatory, and Anti-Migratory Effects on MCF7 Cells. Molecules. 2021 May 31;26(11):3303.

[847] Pateliya B, Burade V, Goswami S. Enhanced antitumor activity of doxorubicin by naringenin and metformin in breast carcinoma: an experimental study. Naunyn Schmiedebergs Arch Pharmacol. 2021 Sep;394(9):1949-1961.

[848] Pateliya B, Burade V, Goswami S. Combining naringenin and metformin with doxorubicin enhances anticancer activity against triple-negative breast cancer in vitro and in vivo. Eur J Pharmacol. 2021 Jan 15;891:173725.

[849] Liu X, Zhao T, Shi Z, Hu C, Li Q, Sun C. Synergism Antiproliferative Effects of Apigenin and Naringenin in NSCLC Cells. Molecules. 2023 Jun 23;28(13):4947.

[850] Arul D, Subramanian P. Naringenin (citrus flavonone) induces growth inhibition, cell cycle arrest and apoptosis in human hepatocellular carcinoma cells. Pathol Oncol Res. 2013 Oct;19(4):763-70.

[851] Liu JF, Chang TM, Chen PH, Lin JS, Tsai YJ, Wu HM, Lee CJ. Naringenin induces endoplasmic reticulum stress-mediated cell apoptosis and autophagy in human oral squamous cell carcinoma cells. J Food Biochem. 2022 Nov;46(11):e14221.

[852] Du Y, Lai J, Su J, Li J, Li C, Zhu B, Li Y. Naringenin-induced Oral Cancer Cell Apoptosis Via ROS-mediated Bid and Bcl-xl Signaling Pathway. Curr Cancer Drug Targets. 2024;24(6):668-679.

[853] Wang D, Zhou Y, Hua L, Hu M, Zhu N, Liu Y, Zhou Y. The role of the natural compound naringenin in AMPK-mitochondria modulation and colorectal cancer inhibition. Phytomedicine. 2024 Aug;131:155786.

Prevention of Skin Cancer: In chemically induced skin cancer in Swiss albino mice, oral administration of naringenin reduced the number and size of papillomas and suppressed papillomagenesis[854].

Treatment for Bone Cancer: In the HOS and U2OS osteosarcoma cell lines, naringenin significantly inhibited cell viability and proliferation, induced cell cycle arrest, inhibited cell growth, induced endoplasmic reticulum stress-mediated apoptosis, and increased autophagy[855].

Treatment for Retinoblastoma: In the Y79 retinoblastoma cell line, naringenin combined with resveratrol significantly decreased proliferation and increased apoptosis by modulating expression levels of Bax, Bcl2, and P21 mRNAs[856].

Treatment for Myeloid Leukemia: Naringenin has been shown to act synergistically with dasatinib and ponatinib, two tyrosine kinase inhibitors used to treat chronic myeloid leukemia (CML). Researchers showed that naringenin decreased colony formation, self-renewal, and viability of CML cells, enhanced the inhibitory effects of dasatinib, and upregulated genes related to mitochondrial biogenesis while downregulating antioxidant defense genes[857].

Treatment for Doxorubicin Toxicity: Doxorubicin is a common anticancer drug that often causes cardiac toxicity, thereby limiting the potential of this important chemotherapeutic agent. Researchers are working on ways to limit the damage that doxorubicin causes to healthy cells, and naringenin might play a significant role in this effort. For instance, in male Swiss albino mice with doxorubicin-induced cardiac toxicity, treatment with naringenin ameliorated all biochemical markers of doxorubicin toxicity[858]. In K562 chronic myelogenous leukemia cells treated with doxorubicin, additional treatment with naringenin reduced the levels of ROS and lipid peroxidation while increasing the activity of superoxide dismutase and glutathione peroxidase[859]. Meanwhile, naringenin combined with p-coumaric acid ameliorated all biochemical markers of doxorubicin induced cardiac toxicity in Swiss albino rats[860], and in a Dalton's lymphoma ascites tumor-bearing mouse model, changes in antioxidant enzymes and lipid peroxidation caused by doxorubicin were prevented when the mice were supplemented with naringenin[861].

Treatment for Radiation Induced Intestinal Injury: Radiation treatment can lead to intestinal injury in some patients. In C57BL/6J mice exposed to a single dose of 13 Gy X-ray total abdominal irradiation, administration of naringenin prolonged the survival rate, protected crypts and villi from damage, alleviated the level of radiation-

[854] Kumar R, Bhan Tiku A. Naringenin Suppresses Chemically Induced Skin Cancer in Two-Stage Skin Carcinogenesis Mouse Model. Nutr Cancer. 2020;72(6):976-983.

[855] Lee CW, Huang CC, Chi MC, Lee KH, Peng KT, Fang ML, Chiang YC, Liu JF. Naringenin Induces ROS-Mediated ER Stress, Autophagy, and Apoptosis in Human Osteosarcoma Cell Lines. Molecules. 2022 Jan 7;27(2):373.

[856] Rakhshan R, Atashi HA, Hoseinian M, Jafari A, Haghighi A, Ziyadloo F, Razizadeh N, Ghasemian H, Nia MMK, Sefidi AB, Arani HZ. The Synergistic Cytotoxic and Apoptotic Effect of Resveratrol and Naringenin on Y79 Retinoblastoma Cell Line. Anticancer Agents Med Chem. 2021 Oct 28;21(16):2243-2249.

[857] Wang M, Deng Q, Zhang W, Qi Q. Oxidative lipid damage by naringenin selectively sensitizes chronic myeloid leukemia cell lines and patient samples to Bcr-Abl tyrosine kinase inhibitors. Biochem Biophys Res Commun. 2024 Nov 12;733:150653.

[858] Arafa HM, Abd-Ellah MF, Hafez HF. Abatement by naringenin of doxorubicin-induced cardiac toxicity in rats. J Egypt Natl Canc Inst. 2005 Dec;17(4):291-300.

[859] Feng YQ, Zuo XL, Li RF, Zhang KJ, Chen F, Xiao H. [Protection against doxorubicin-induced oxidative damage in normal blood cells by naringenin]. Zhongguo Shi Yan Xue Ye Xue Za Zhi. 2008 Aug;16(4):790-3. Chinese.

[860] Shiromwar SS, Chidrawar VR. Combined effects of p-coumaric acid and naringenin against doxorubicin-induced cardiotoxicity in rats. Pharmacognosy Res. 2011 Jul;3(3):214-9.

[861] Kathiresan V, Subburaman S, Krishna AV, Natarajan M, Rathinasamy G, Ganesan K, Ramachandran M. Naringenin Ameliorates Doxorubicin Toxicity and Hypoxic Condition in Dalton's Lymphoma Ascites Tumor Mouse Model: Evidence from Electron Paramagnetic Resonance Imaging. J Environ Pathol Toxicol Oncol. 2016;35(3):249-262.

induced inflammation, and mitigated intestinal barrier damage in the irradiated mice, while also reducing immune cell infiltration and intestinal epithelial cell apoptosis[862].

Sensitizes Cancer Cells to Death: Naringenin combined with hesperetin has been shown to act as a tyrosine kinase inhibitor, preferentially targeting and sensitizing HER2 positive breast cancer cells to cell death[863].

Treatment for Depression: In olfactory bulbectomy-induced neuroinflammation, oxidative stress, altered kynurenine pathway, and behavioral deficits in BALB/c mice, treatment with naringenin restored alterations in the kynurenine pathway, exhibiting neuroprotective activity comparable to that of fluoxetine[864] (an SSRI used to treat depression). In rats induced to depression via treatment with streptozotocin, treatment with naringenin-loaded hybridized nanoparticles reduced depressive symptoms by influencing brain neurotransmitters and exhibiting antioxidant and anti-inflammatory effects[865]. Finally, for mice induced with depression via administration of dexamethasone, treatment with naringenin improved neurogenesis, decreased neuroinflammation, and restored monoamines levels, reliving depressive symptoms such as anxiety, anhedonia, and despair[866].

Treatment for Polycystic Ovarian Syndrome: In a rat model of polycystic ovarian syndrome, treatment with naringenin combined with morin (a flavonol) reduced insulin resistance and endometrial hyperplasia, improving the inflammatory and oxidative microenvironment of ovarian tissues[867]. In letrozole-induced polycystic ovary syndrome in Sprague Dawley rats, treatment with naringenin increased ovulation potential and decreased cystic follicles and androgen levels[868].

Treatment for Tinnitus: In sodium salicylate-induced tinnitus in male Wistar rats, treatment with naringenin decreased Bax gene expression, increased Bcl-2 gene expression, and restored the expression levels of both Tfam and Pgc-1α genes, thereby reducing apoptosis and modulating the mitochondrial state[869].

Treatment for Asthma: In a murine model of asthma, administration of naringenin lowered levels of IL4 and IL13, inhibited pulmonary IkappaBalpha degradation and NF-kappaB DNA-binding activity, and significantly reduced the levels of CCL5, CCL11, and iNOS, inhibiting allergen-induced airway inflammation[870]. Naringenin has also been shown to be effective in the treatment of diesel particulate matter-induced airway abnormalities[871].

---

[862] Ling Z, Wang Z, Chen L, Mao J, Ma D, Han X, Tian L, Zhu Q, Lu G, Yan X, Ding Y, Xiao W, Chen Y, Peng A, Yin X. Naringenin Alleviates Radiation-Induced Intestinal Injury by Inhibiting TRPV6 in Mice. Mol Nutr Food Res. 2024 Apr;68(8):e2300745.

[863] Chandrika BB, Steephan M, Kumar TRS, Sabu A, Haridas M. Hesperetin and Naringenin sensitize HER2 positive cancer cells to death by serving as HER2 Tyrosine Kinase inhibitors. Life Sci. 2016 Sep 1;160:47-56.

[864] Bansal Y, Singh R, Saroj P, Sodhi RK, Kuhad A. Naringenin protects against oxido-inflammatory aberrations and altered tryptophan metabolism in olfactory bulbectomized-mice model of depression. Toxicol Appl Pharmacol. 2018 Sep 15;355:257-268.

[865] El-Marasy SA, AbouSamra MM, Moustafa PE, Mabrok HB, Ahmed-Farid OA, Galal AF, Farouk H. Anti-depressant effect of Naringenin-loaded hybridized nanoparticles in diabetic rats via PPARγ/NLRP3 pathway. Sci Rep. 2024 Jun 12;14(1):13559.

[866] Abdelkawy YS, Elharoun M, Sheta E, Abdel-Raheem IT, Nematalla HA. Liraglutide and Naringenin relieve depressive symptoms in mice by enhancing Neurogenesis and reducing inflammation. Eur J Pharmacol. 2024 May 15;971:176525.

[867] Yi Yang, Liu J, Xu W. Naringenin and morin reduces insulin resistance and endometrial hyperplasia in the rat model of polycystic ovarian syndrome through enhancement of inflammation and autophagic apoptosis. Acta Biochim Pol. 2022 Feb 10;69(1):91-100.

[868] Rashid R, Tripathi R, Singh A, Sarkar S, Kawale A, Bader GN, Gupta S, Gupta RK, Jha RK. Naringenin improves ovarian health by reducing the serum androgen and eliminating follicular cysts in letrozole-induced polycystic ovary syndrome in the Sprague Dawley rats. Phytother Res. 2023 Sep;37(9):4018-4041.

[869] Safavi-Naeini SM, Nasehi M, Zarrindast MR, Safavi-Naeini SA. Exploring the effects of naringenin on cell functioning and energy synthesis in the hippocampus of male Wistar rats with chronic tinnitus, by examining genetic indicators such as Bax, Bcl-2, Tfam, and Pgc-1α. Gene. 2025 Jan 15;933:148980.

[870] Shi Y, Dai J, Liu H, Li RR, Sun PL, Du Q, Pang LL, Chen Z, Yin KS. Naringenin inhibits allergen-induced airway inflammation and airway responsiveness and inhibits NF-kappaB activity in a murine model of asthma. Can J Physiol Pharmacol. 2009 Sep;87(9):729-35.

[871] Shi R, Su WW, Zhu ZT, Guan MY, Cheng KL, Fan WY, Wei GY, Li PB, Yang ZY, Yao HL. Regulation effects of naringin on diesel particulate matter-induced abnormal airway surface liquid secretion. Phytomedicine. 2019 Oct;63:153004.

Antinociceptive / Analgesic Agent: Naringenin has been suggested as a potent anti-inflammatory drug that relieves pain effectively and can be used in pain management therapy[872], with some research indicating that one variation of the flavanone – (2S)-naringenin – works by inhibiting NaV1.8 voltage-gated sodium channels[873]. In particular, naringenin has been shown to suppress paclitaxel-induced pain[874], alleviate bone cancer pain[875], and mitigate orofacial pain[876].

Treatment for Diabetic Kidney Disease: In streptozotocin-induced diabetes in male Sprague Dawley rats, treatment with naringenin combined with Lisinopril showed significant improvement in biochemical and urine parameters, attenuating renal oxidative stress and renal damage[877].

Treatment for Hypertension: In pregnancy-induced hypertension in mice, administration of naringenin significantly decreased blood pressure, total urine protein level, plasma levels of VCE, α-ADR and angiotensin by suppressing SHP-1 expression in vascular endothelial cells of the subject animals[878]. In a high fat diet-induced rat model of obesity-associated hypertension, treatment with naringenin reduced body and fat weight, reduced blood pressure, regulated lipid parameters, reduced serum MDA and NO, increased serum SOD and GSH, regulated adipocytokines, and decreased the phosphorylation of STAT3[879].

Treatment for Cigarette Smoke-induced Lung Diseases: Naringenin has been found to suppress BEAS-2B-derived extracellular vesicular cargoes disorder caused by cigarette smoke extract (CSE), thereby inhibiting M1 macrophage polarization[880]. In Sprague Dawley rats exposed to CSE, co-exposure of naringenin attenuated airway cilia structural and functional injury by modulating the IL-17 signaling pathway and activating the cAMP signaling pathway[881].

Treatment for Neurological Diseases: In a rotenone-induced model of Parkinson's disease in rats, administration of naringenin considerably reduced motor and non-motor impairments, increased antioxidant enzyme activities, and

[872] Xue N, Wu X, Wu L, Li L, Wang F. Antinociceptive and anti-inflammatory effect of Naringenin in different nociceptive and inflammatory mice models. Life Sci. 2019 Jan 15;217:148-154.

[873] Zhou Y, Cai S, Moutal A, Yu J, Gómez K, Madura CL, Shan Z, Pham NYN, Serafini MJ, Dorame A, Scott DD, François-Moutal L, Perez-Miller S, Patek M, Khanna M, Khanna R. The Natural Flavonoid Naringenin Elicits Analgesia through Inhibition of NaV1.8 Voltage-Gated Sodium Channels. ACS Chem Neurosci. 2019 Dec 18;10(12):4834-4846.

[874] Pan C, Xu Y, Jiang Z, Fan C, Chi Z, Zhang Y, Miao M, Ren Y, Wu Z, Xu L, Mei C, Chen Q, Xi Y, Chen X. Naringenin relieves paclitaxel-induced pain by suppressing calcitonin gene-related peptide signalling and enhances the anti-tumour action of paclitaxel. Br J Pharmacol. 2024 Sep;181(17):3136-3159.

[875] Li Y, Zheng G, Tang Y, Chen Y, Yang M, Zheng Q, Bao Y. Naringenin alleviates bone cancer pain via NF-κB/uPA/PAR2 pathway in mice. J Orthop Surg (Hong Kong). 2024 May-Aug;32(2):10225536241266671.

[876] Yajima S, Sakata R, Watanuki Y, Sashide Y, Takeda M. Naringenin Suppresses the Hyperexcitability of Trigeminal Nociceptive Neurons Associated with Inflammatory Hyperalgesia: Replacement of NSAIDs with Phytochemicals. Nutrients. 2024 Nov 15;16(22):3895.

[877] Kulkarni YA, Suryavanshi SV. Combination of Naringenin and Lisinopril Ameliorates Nephropathy in Type-1 Diabetic Rats. Endocr Metab Immune Disord Drug Targets. 2021;21(1):173-182.

[878] Duan B, Li Y, Geng H, Ma A, Yang X. Naringenin prevents pregnancy-induced hypertension via suppression of JAK/STAT3 signalling pathway in mice. Int J Clin Pract. 2021 Oct;75(10):e14509.

[879] Liu H, Zhao H, Che J, Yao W. Naringenin Protects against Hypertension by Regulating Lipid Disorder and Oxidative Stress in a Rat Model. Kidney Blood Press Res. 2022;47(6):423-432.

[880] Chen Z, Wu H, Fan W, Zhang J, Yao Y, Su W, Wang Y, Li P. Naringenin suppresses BEAS-2B-derived extracellular vesicular cargoes disorder caused by cigarette smoke extract thereby inhibiting M1 macrophage polarization. Front Immunol. 2022 Jul 18;13:930476.

[881] Zhang J, Fan W, Wu H, Yao Y, Jin L, Chen R, Xu Z, Su W, Wang Y, Li P. Naringenin attenuated airway cilia structural and functional injury induced by cigarette smoke extract via attenuated airway cilia structural and functional injury. Phytomedicine. 2024 Apr;126:155053.

restored detrimental changes in neurotransmitter levels[882]. Research has also shown that treatment with naringenin can improve the learning and memory ability of mice induced to Alzheimer's disease[883].

Treatment for Atopic Dermatitis: In a mouse model of atopic dermatitis, treatment with naringenin decreased dermatitis scores and right ear thickness, alleviated skin lesions, and reduced the number of infiltrated mast cells and eosinophilic granulocytes[884], effectively ameliorating the condition.

Treatment for Hair Loss: Naringenin has been shown to significantly increase hair growth, hair follicle diameter expansion, and hair follicle quantity in albino Wistar mice[885], primarily due to the flavanone's antioxidant and anti-inflammatory activity.

Slows Brain Aging: Recent work has shown that naringenin can improve the lifespan and health of C. elegans (a roundworm), and slow brain aging in mice[886].

Treatment for Nonalcoholic Steatohepatitis: As the primacy active constituent in the traditional Chinese herbal medicine Chaihu Shugan San, naringenin has been shown to ameliorate liver lipid metabolism and inflammatory damage in mice with experimental nonalcoholic steatohepatitis induced by a choline-deficient high-fat diet[887].

IMPLICATIONS FOR HUMAN HEALTH AND NUTRITION

As is the case with many flavonoids, the simplest way to obtain this compound is by eating fruits, vegetables, spices, herbs, nuts, and other foods that are known to contain naringenin. For instance, when cooking, preparing, or serving food, consider working with the spices and herbs caraway, cardamom, chives, cinnamon, corn mint, garlic, nutmeg, peppermint, saffron, sweet bay, and turmeric. You can also consume fruits such as cherries, dates, figs, grapefruit, kiwis, lemons, mandarins, mangoes, mulberries, oranges, papaya, passion fruit, pineapples, star fruit, and watermelon. Vegetables that likely contain naringenin include artichokes, asparagus, avocados, beets, capers, cauliflower, chickpeas, cucumbers, garden cress, horseradish, kohlrabi, lentils, lima beans, olives, onions, parsnip, sweet potatoes, tomatoes, turnips, and wild carrots. Snacking on nuts such as black walnuts, Brazil nuts, chestnuts, flax seed, hazelnuts, peanuts, pecans, sunflower seeds, and walnuts may result in the dietary acquisition of naringenin.

Some cannabis and hemp varieties may contain naringenin, although the content is likely to be considered "trace." It should be noted that, if a cannabis variety does contain naringenin, the best way to preserve and subsequently consume this compound is by eating raw cannabis leaves, stalks, shoots, or flowers. Current known varieties of cannabis that contain naringenin include the recreational or drug types Alien Dawg, Tangerine Dream, Sensi Star, Quadra, Gabriola (Frosty Monster), and Island Honey, and industrial hemp varieties Kompolti, Tiborszallasi, Antal, Carmagnola Cs, Fedora 17, Ferimon, and Futura 75.

[882] Madiha S, Batool Z, Shahzad S, Tabassum S, Liaquat L, Afzal A, Sadir S, Sajid I, Mehdi BJ, Ahmad S, Haider S. Naringenin, a Functional Food Component, Improves Motor and Non-Motor Symptoms in Animal Model of Parkinsonism Induced by Rotenone. Plant Foods Hum Nutr. 2023 Dec;78(4):654-661.

[883] Zhu Y, Guo X, Li S, Wu Y, Zhu F, Qin C, Zhang Q, Yang Y. Naringenin ameliorates amyloid-β pathology and neuroinflammation in Alzheimer's disease. Commun Biol. 2024 Jul 28;7(1):912.

[884] Tian L, Wang M, Wang Y, Li W, Yang Y. Naringenin ameliorates atopic dermatitis by inhibiting inflammation and enhancing immunity through the JAK2/STAT3 pathway. Genes Genomics. 2024 Mar;46(3):333-340.

[885] Khayoon N, Gany S, Hadi NR, Al Mudhafar A. Effect of topical naringenin and its combination with minoxidil on enhancing hair growth in a mouse model. J Med Life. 2023 Nov;16(11):1685-1691.

[886] Piragine E, De Felice M, Germelli L, Brinkmann V, Flori L, Martini C, Calderone V, Ventura N, Da Pozzo E, Testai L. The Citrus flavanone naringenin prolongs the lifespan in C. elegans and slows signs of brain aging in mice. Exp Gerontol. 2024 Sep;194:112495.

Ren Y, Xiao K, Lu Y, Chen W, Li L, Zhao J. Deciphering the mechanism of Chaihu Shugan San in the treatment of nonalcoholic steatohepatitis using network pharmacology and molecular docking. J Pharm Pharmacol. 2024 Nov 4;76(11):1521-1533.
[887]

Naringenin Review

Answer the following questions to test your knowledge of this flavonoid:

Question #1: What type of flavonoid is naringenin?

    a. Flavoned
    b. Flavine
    c. Flavanone
    d. Flavonol

Question #2: The base components in the biosynthesis of naringenin are:

    a. Cosmoioside-delta1 and apigenin
    b. Apigenin and chrysin-3-ol
    c. p-coumaroyl-CoA and malonyl-CoA
    d. Malonoxide Ab and Phenylprobate

Question #3: How common is naringenin in cannabis?

    a. Top three
    b. Unknown
    c. Top ten
    d. Top Twenty

Question #4: What is the chemical formula for naringenin?

    a. $C_{15}H_{12}O_5$
    b. $C_{26}H_{10}O_{15}$
    c. $C_{16}H_{10}O_{65}$
    d. $C_{26}H_{30}O_{15}$

Question #5: Name two plants naringenin is known to occur in:

1 _____ 2 _____

Question #6: Name two specific potential medical uses of naringenin:

1 _____ 2 _____

For the answer key to Naringenin, please visit www.cannabischemistry.org

APIGETRIN

Type: Flavone
Chemical Formula: $C_{21}H_{20}O_{10}$
Molecular Weight: 432.381 g·mol
Boiling Point: 788.9 °C @ 760.00 mm Hg (estimated by TGSC)[888]
Flash Point: 280.70 °C (537.00 °F.) (estimated by TGSC)
Melting Point: 230 - 237°C
Solubility: Soluble in methanol, DMSO, other organic solvents,
slightly soluble in water, insoluble in petroleum ether, chloroform.
Oral LD50: Unknown
Biological Role: Unknown
Therapeutic Role: Anticancer, anti-diabetes, antiobesity agent
Commercial Use: Research Chemical
Occurrence in Cannabis: Unknown

Occurs in Cannabis Strains: At the time of publication of this textbook, there were no publicly available studies, reports, or laboratory tests quantifying the occurrence of apigetrin in cannabis or hemp, however, it is likely that this molecule occurs in the plant in trace amounts. Additionally, laboratory standards for this compound are probably not present during most analytical testing of cannabis – it is recommended to request that a laboratory acquire the standard and test for this compound specifically. Cannabis varieties may be bred to express this flavone, hence the author's choice to include it in this book.

INTRODUCTION

This chapter began as a section to research and discuss cosmosioside, which was anecdotally linked to cannabis. However, cosmosioside is a synonym for apigetrin, which is an enantiomer of apigenin – a compound covered earlier in this book. Because a molecule may have dozens of synonyms, it is important to use the term most widely in scientific circulation at the time whenever possible. In this case, apigetrin is more appropriate than cosmosioside, or cosmosiin, or cosmetin, or any of the other seventy or more synonyms for this molecule.

Apigetrin is a flavone with no known organoleptic properties. This coupled with the likelihood that apigetrin occurs as a trace element – if at all – in cannabis means it is improbable that this compound contributes much to the aroma or flavor profile of cannabis or hemp.

---

[888] The Good Scents Company Data Sheet for Cosmosiin, from: https://www.thegoodscentscompany.com/data/rw1588411.html. Accessed January 13, 2025.

CHEMICAL STRUCTURE

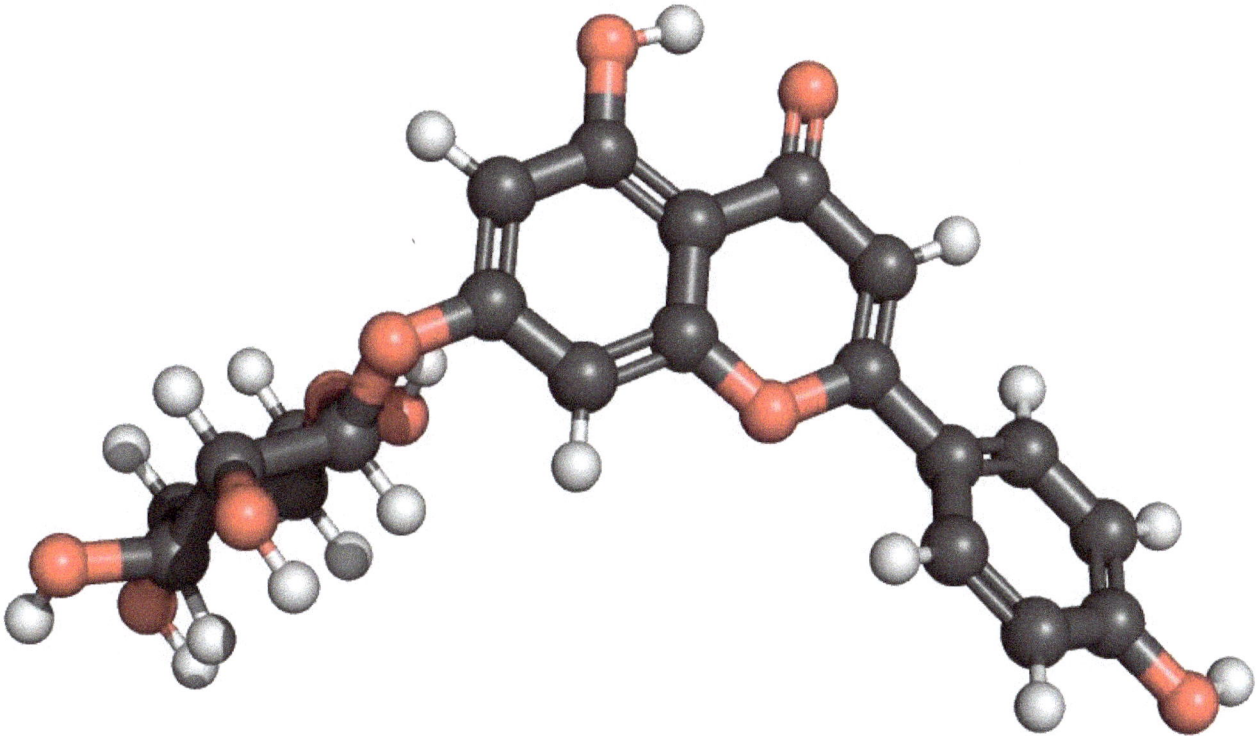

Apigetrin is biosynthesized in the phenylpropanoid pathway, where p-coumaroyl-CoA and malonyl-CoA eventually form naringenin, which is then hydroxylated to form apigenin, which then becomes apigetrin via glycosylation. Structurally, this molecule contains four rings, six endocyclic double bonds, one exocyclic double bonded oxygen atom, with ten oxygen atoms in total, scientifically notated as $C_{21}H_{20}O_{10}$.

Variations of apigetrin, which are structural derivatives of apigenin-7-O-glucoside, commonly arise due to differences in the glycosylation pattern, substitution of the sugar moiety, or modifications to the apigenin aglycone, such as the following examples:

Apigenin-7-O-rhamnoside - contains a rhamnose sugar group

Apigenin-7-O-galactoside - contains a galactose sugar group

Apigenin-7-O-(rhamnosyl-glucoside) - a rhamnose attached to the glucose group

Apigenin-4'-O-glucoside - glucose is attached at the 4'-hydroxy group on the B-ring

Apigenin-5-O-glucoside - glucose is attached at the 5-hydroxy group on the A-ring

Apigenin-7,4'-di-O-glucoside - two glucose units are attached at both the 7-O and 4'-O positions

Luteolin-7-O-glucoside - apigetrin with an extra hydroxyl group at C3'

7-O-methyl-apigenin glucoside - methylation of the hydroxyl group at the 7 position

Apigenin-7-O-(6'-malonyl-glucoside) - sugar moiety is acylated with malonic acid

OCCURRENCE IN PLANTS

Apigetrin is known to occur in a moderate number of plants, although it is likely that with more testing for this particular molecule, it will be discovered in additional species.

Acanthus ebracteatus	Coreopsis lanceolata	Patchouli
Anise	Cumin	Peppermint
Apple heart wood	Dandelion root	Rosemary
Barley	Humulus lupulus	Sage
Broussonetia leaves	Lemon balm	Scutellaria baicalensis Georgi
Cabocla	Lupine	Stachys tibetica Vatke
Cacao	Marjoram	Teucrium gnaphalodes
Carrot	Olive leaf	Thyme
Chamomile	Oregano	Water mint
Chrysanthemum	Parsley	

BIOLOGICAL ACTIVITY IN PLANTS

The author found no scientific or other literature or research that investigated the biological activity of apigetrin in plants. However, based on common biological activity of other flavonoids, it may be reasonable to predict that apigetrin serves UVB protection, antioxidant, allelopathic, and antibacterial roles, among other possible functions.

USES IN INDUSTRY

Commercial preparations of apigetrin appear as white or pale yellow to yellow crystals, and are generally used for research purposes only. Apigetrin is not typically extracted directly from plants in its glycosylated form; it is usually synthesized from apigenin, although some researchers have successfully produced this flavone from engineered Escherichia coli[889].

POTENTIAL USES IN MEDICINE

Apigetrin appears to be a potent anticancer agent for several types of carcinomas, and it also offers potential as an antiobesity agent, exhibits anti-diabetes properties, and may even help in the battle against Influenza A.

Treatment for Liver Cancer: In HepG2 liver cancer cells, treatment with apigetrin suppressed cell growth and caused cell death, induced G2/M phase cell cycle arrest, induced apoptosis and chromatin condensation, and induced death receptor-mediated apoptosis[890]. In Hep3B liver cancer cells, treatment with apigetrin inhibited cell growth and proliferation, induced apoptosis and necroptosis, increased ROS levels, up-regulated TNFα while down-regulating phosphorylation of p-p65 and IκB, and inhibited the expression of Bcl-xl while increasing Bax levels[891].

---

[889] Thuan NH, Chaudhary AK, Van Cuong D, Cuong NX. Engineering co-culture system for production of apigetrin in Escherichia coli. J Ind Microbiol Biotechnol. 2018 Mar;45(3):175-185.

[890] Bhosale PB, Kim HH, Abusaliya A, Jeong SH, Park MY, Kim HW, Seong JK, Ahn M, Park KI, Heo JD, Kim YS, Kim GS. Inhibition of Cell Proliferation and Cell Death by Apigetrin through Death Receptor-Mediated Pathway in Hepatocellular Cancer Cells. Biomolecules. 2023 Jul 14;13(7):1131.

[891] Bhosale PB, Abusaliya A, Kim HH, Ha SE, Park MY, Jeong SH, Vetrivel P, Heo JD, Kim JA, Won CK, Kim HW, Kim GS. Apigetrin Promotes TNFα-Induced Apoptosis, Necroptosis, G2/M Phase Cell Cycle Arrest, and ROS Generation through Inhibition of NF-κB Pathway in Hep3B Liver Cancer Cells. Cells. 2022 Sep 1;11(17):2734.

Treatment for Gastric Cancer: In AGS human gastric cancer cells, treatment with apigetrin reduced cell proliferation, induced G2/M phase cell cycle arrest, induced apoptotic cell death, increased the expression of extrinsic apoptosis pathway proteins and mRNA, and increased autophagic cell death by modulating the PI3K/AKT/mTOR pathway[892]. In vivo studies have shown that apigetrin can significantly inhibit gastric cancer cell xenograft tumorigenesis through inducing apoptosis and inhibiting the STAT3/JAK2 pathways[893].

Treatment for Colorectal Cancer: Referring to apigetrin as cosmosiin, researchers found that the compound may be useful in the treatment of colorectal cancer after it was found to suppress the expression of programmed death-ligand 1and trigger apoptosis in colorectal carcinoma cells[894].

Treatment for Leukemia: In K562 chronic leukemia cells, treatment with apigetrin induced erythroid differentiation and G2/M phase cell cycle arrest while downregulating the expression of several proteins involved in cell cycle regulation, protein synthesis, and nuclear import and export of signaling molecules[895].

Treatment for Doxorubicin Toxicity: In doxorubicin-induced testicular damage in male albino rats, apigetrin supplementation significantly reversed all the induced testicular damages due to its androgenic, anti-apoptotic, anti-inflammatory, and antioxidant properties[896].

Treatment for Influenza A Infection: As the primary constituent in an extract made from Ixeris sonchifolia Hance (an herb used in traditional Chinese medicine), apigetrin significantly improved Influenza A virus (IAV)-induced oxidative stress, inhibited the IAV-induced cytokine storm by suppressing the excessive activation of TLR3/4/7, JNK/p38 MAPK and NF-κB, decreased autophagosome accumulation and promoted degradation of IAV protein, significantly increased the average survival time, and reduced the lung edema and IAV replication[897].

Treatment for Diabetes: In streptozotocin (STZ)-induced cell damage in RINm5F cells, treatment with apigetrin inhibited the elevation of intracellular ROS levels, restored the impairment of antioxidant enzymes, recovered the disruption of redox homeostasis, significantly suppressed the STZ-induced apoptosis, and attenuated STZ-induced endoplasmic reticulum (ER) stress[898].

Treatment for Acute Otitis: In a mouse model of lipopolysaccharide-induced acute otitis, treatment with apigetrin reduced higher mucosa thickness, inhibited the inflammatory response and reduced inflammatory factors including interleukin-1β, tumor necrosis factor α, IL-6, and vascular endothelial growth factor, and reduced ROS levels[899].

[892] Kim SM, Vetrivel P, Ha SE, Kim HH, Kim JA, Kim GS. Apigetrin induces extrinsic apoptosis, autophagy and G2/M phase cell cycle arrest through PI3K/AKT/mTOR pathway in AGS human gastric cancer cell. J Nutr Biochem. 2020 Sep;83:108427.

[893] Sun Q, Lu NN, Feng L. Apigetrin inhibits gastric cancer progression through inducing apoptosis and regulating ROS-modulated STAT3/JAK2 pathway. Biochem Biophys Res Commun. 2018 Mar 25;498(1):164-170.

[894] Han JH, Lee EJ, Park W, Choi JG, Ha KT, Chung HS. Cosmosiin Induces Apoptosis in Colorectal Cancer by Inhibiting PD-L1 Expression and Inducing ROS. Antioxidants (Basel). 2023 Dec 18;12(12):2131.

[895] Tsolmon S, Nakazaki E, Han J, Isoda H. Apigetrin induces erythroid differentiation of human leukemia cells K562: proteomics approach. Mol Nutr Food Res. 2011 May;55 Suppl 1:S93-S102.

[896] Ijaz MU, Yaqoob S, Hamza A, David M, Afsar T, Husain FM, Amor H, Razak S. Apigetrin ameliorates doxorubicin prompted testicular damage: biochemical, spermatological and histological based study. Sci Rep. 2024 Apr 20;14(1):9049.

[897] He M, Ren Z, Goraya MU, Lin Y, Ye J, Li R, Dai J. Anti-influenza drug screening and inhibition of apigetrin on influenza A virus replication via TLR4 and autophagy pathways. Int Immunopharmacol. 2023 Nov;124(Pt B):110943.

[898] Zhang R, Shi J, Wang T, Qiu X, Liu R, Li Y, Gao Q, Wang N. Apigetrin ameliorates streptozotocin-induced pancreatic β-cell damages via attenuating endoplasmic reticulum stress. In Vitro Cell Dev Biol Anim. 2020 Sep;56(8):622-634.

[899] Guo H, Li M, Xu LJ. Apigetrin treatment attenuates LPS-induced acute otitis media though suppressing inflammation and oxidative stress. Biomed Pharmacother. 2019 Jan;109:1978-1987.

Treatment for Obesity: Apigetrin has been shown to inhibit cell proliferation, induce cell cycle delay, suppress the mRNA levels of C/EBP-α, PPAR-γ, SREBP-1c and FAS, and suppress the mRNA level of pro-inflammatory genes, thereby inhibiting adipogenesis (the production of fat cells) of 3T3-L1 preadipocytes[900].

Treatment for Skeletal Muscle Diseases: In lipopolysaccharide-induced inflammation in L6 skeletal muscle cells, treatment with apigetrin inhibited the expression of iNOS and COX-2, inhibited the induced phosphorylation of p65 and IκB-α, and significantly downregulated the phosphorylation of JNK and p38[901].

IMPLICATIONS FOR HUMAN HEALTH AND NUTRITION

Dietary acquisition is the most likely way to consume apigetrin. When preparing, cooking, or serving foods, consider working with herbs or spices of anise, cacao, cumin, marjoram, oregano, parsley, peppermint, rosemary, sage, thyme, water mint, which are all known to contain apigetrin. You can also try drinking chamomile, lemon balm, or peppermint tea, or drinks that contain hops including most beers. Adding the vegetables carrots or dandelion roots to your meals may also allow you to acquire some apigetrin.

Working with purified apigetrin is for experimental purposes only and should be undertaken only with extreme caution and professional guidance, as pure compounds can be dangerous in some situations.

[900] Hadrich F, Sayadi S. Apigetrin inhibits adipogenesis in 3T3-L1 cells by downregulating PPARγ and CEBP-α. Lipids Health Dis. 2018 Apr 25;17(1):95.

[901] Ha SE, Bhagwan Bhosale P, Kim HH, Park MY, Abusaliya A, Kim GS, Kim JA. Apigetrin Abrogates Lipopolysaccharide-Induced Inflammation in L6 Skeletal Muscle Cells through NF-κB/MAPK Signaling Pathways. Curr Issues Mol Biol. 2022 Jun 8;44(6):2635-2645.

Apigetrin Review

Answer the following questions to test your knowledge of this flavonoid:

Question #1: What type of flavonoid is apigetrin?

    a. Flavonol
    b. Flavorall
    c. Chalcone
    d. Flavone

Question #2: The base flavonoid in the biosynthesis of apigetrin is:

    a. Naringeniis
    b. Apigenin
    c. Quercetin
    d. Luteolin

Question #3: What is the primary difference between apigetrin and apigenin?

    a. Position 3' and 5 Hydroxylation
    b. Two double bonded oxygen atoms
    c. Glycosylation
    d. Methylation

Question #4: What is the chemical formula for apigetrin?

    a. $C_{24}H_{26}O$
    b. $C_{15}H_{24}O_6$
    c. $C_{26}H_{28}O_6$
    d. $C_{21}H_{20}O_{10}$

Question #5: Name two synonyms for apigetrin:

1 _____ 2 _____

Question #6: Name two potential medical uses of apigetrin:

1 _____ 2 _____

For the answer key to Apigetrin, please visit www.cannabischemistry.org

CHRYSOERIOL

Type: Flavone
Chemical Formula: $C_{16}H_{12}O_6$
Molecular Weight: 300.26 g-mol
Boiling Point: 574.32 °C @ 760.00 mm Hg (estimated by TGSC[902])
Flash Point: 219.40 °C (427.00 °F.) (estimated by TGSC)
Melting Point: 330 - 331 °C[903]
Solubility: Poorly soluble in water, soluble in ethanol, methanol, dimethyl sulfoxide, and acetone
Oral LD50: Unknown, but likely low toxicity
Biological Role: Insecticidal Agent
Therapeutic Role: Anti-inflammatory, anticancer agent
Commercial Use: Research chemical
Occurrence in Cannabis: Precursor element and trace

Occurs in Cannabis Strains:

As of early 2025, there were no publicly available laboratory tests or other confirmation that chrysoeriol occurs in cannabis plants, although it is likely that the flavone occurs in at least trace amounts in some varieties. However, a 2022 study found that chrysoeriol, kaempferol, and luteolin were detected at high levels in an unnamed variety of hemp seeds[904].

INTRODUCTION

Chrysoeriol is an O-methylated flavone with poor water solubility, and no known organoleptic properties. If this compound occurs in cannabis, it is unlikely to contribute significantly to the aroma or flavor profile of a given variety. However, it is interesting to note that chrysoeriol is a major chemical precursor to cannflavin A and cannflavin B[905], flavonoids covered in earlier chapters in this book and often thought only to occur in cannabis.

Chrysoeriol is understudied compared to many of the other flavonoids in this textbook, although it is known to be a potent anti-inflammatory agent and exhibits some anticancer properties. This flavone is primarily used for research purposes and does not appear to be used as an individual component in commercial products.

---

[902] The Good Scents Company Data Sheet for Chrysoeriol, from: https://www.thegoodscentscompany.com/data/rw1486351.html. Accessed January 14, 2025.

[903] Human Metabolome Database (HMDB), located at: http://www.hmdb.ca/metabolites/HMDB0030667. Accessed January 14, 2025.

[904] Ning K, Hou C, Wei X, Zhou Y, Zhang S, Chen Y, Yu H, Dong L, Chen S. Metabolomics Analysis Revealed the Characteristic Metabolites of Hemp Seeds Varieties and Metabolites Responsible for Antioxidant Properties. Front Plant Sci. 2022 Jun 21;13:904163.

[905] Rea KA, Casaretto JA, Al-Abdul-Wahid MS, Sukumaran A, Geddes-McAlister J, Rothstein SJ, Akhtar TA. Biosynthesis of cannflavins A and B from Cannabis sativa L. Phytochemistry. 2019 Aug;164:162-171.

CHEMICAL STRUCTURE

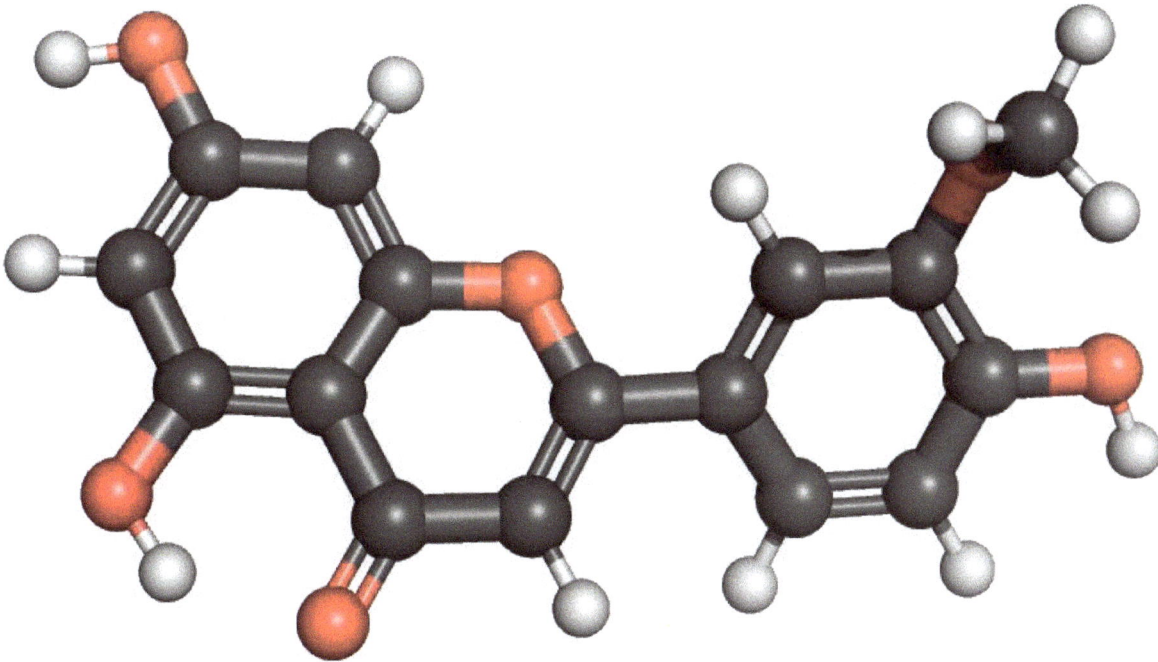

Biosynthesized in the phenylpropanoid pathway, chrysoeriol is the 3'-O-methyl derivative of luteolin (covered in a separate chapter), which means that methylation of the 3'-hydroxyl group on the B-ring of luteolin results in the formation of chrysoeriol. Chrysoeriol's molecular structure includes the classic three ring flavonoid backbone, with seven endocyclic double bonds, one exocyclic double bonded oxygen atom, and six oxygen atoms total, chemically notated as $C_{16}H_{12}O_6$. In this flavone, the A-ring contains hydroxyl groups at the 5 and 7 positions, the B-Ring is substituted with a hydroxyl group at the 4'-position and a methoxy group at the 3'-position, and the C-Ring has a central three-carbon ketone bridge forming part of the chromenone core.

Major variations of chrysoeriol include hydroxylation variants such as apigenin (included as a separate chapter in this book), and methylation variants, which involve adding methyl groups to hydroxyl groups, forming methoxy groups. For instance, chrysoeriol is the 3'-methoxy derivative of luteolin, and another methylated version of chrysoeriol is 7-methoxy-luteolin.

Glycosylation variants also exist, which are formed by attaching sugar moieties to the hydroxyl groups, such as chrysoeriol-7-O-glucoside. Acylation variants occur when acyl groups (including acetyl or p-coumaroyl) are added to glycosides of chrysoeriol, such as chrysoeriol-7-O-(p-coumaroyl)-glucoside. Other variants include sulfation and phosphorylation variants where sulfate or phosphate groups are attached to hydroxyl groups, and metal complex variants that can arise due to chrysoeriol's ability to chelate metal ions through its hydroxyl groups.

Each of the above types of variants changes several attributes of the molecule, including water solubility and bioavailability, function, boiling and melting points, and propensity for further changes in the molecule.

OCCURRENCE IN PLANTS

Alfalfa	Celery	Halocnemum strobilaceum
Black huckleberry	Chamomile	Hazelnut
Buckwheat	Corn	Lemongrass
Cardamom	Fennel	Lippia nodiflora

Loquat	Perilla frutescens	Tarragon
Origanum scabrum	Sage	Tea rooibos
Parsley	Schouwia thebaica	Thalassia testudinum
Pineapple	Soybean	Thyme
Peanut	Strawberry	Turmeric
Petroselinum hortense	Swiss chard	

BIOLOGICAL ACTIVITY IN PLANTS

The biological activities of chrysoeriol in plants has not been elucidated, however, there is some work that indicates this flavone might serve as an insecticidal agent as detailed below. It is also likely that chrysoeriol serves antioxidant, anti-inflammatory, and UVB radiation protection functions, however, more research is needed to confirm these potential biological activities.

Insecticidal Agent: As isolated from the leaves and twigs of Melientha suavis, chrysoeriol was shown to be highly toxic against Spodoptera litura (the tobacco cutworm, a common agricultural pest), significantly decreasing the activities of both detoxification-related enzymes, glutathione S-transferase, and neurological enzymes[906].

USES IN INDUSTRY

Commercial preparations of chrysoeriol typically appear as pale yellow to yellow crystalline powders primarily used for research purposes. There is some interest in producing this flavone via means other than plant extraction, including synthetic production in engineered Nicotiana benthamiana leaves[907], and via in vivo biosynthesis in engineered Escherichia coli bacteria[908].

POTENTIAL USES IN MEDICINE

Chrysoeriol has the potential to be used in a variety of medica and therapeutic settings, including in the treatment of lung cancer and melanoma, as well as many different anti-inflammatory applications as indicated below:

Treatment for Brain Cancer: In a rat model of brain cancer, treatment with chrysoeriol significantly decreased cell viability and induced apoptosis by increasing the levels of Bax/Bcl-2 ratio and cleaved caspase-3/caspase-3 ratio, and by reducing the phosphorylation of PI3K, Akt, and mTOR expression[909].

Treatment for Lung Cancer: In A549 lung cancer cells, treatment with chrysoeriol induced autophagy by significantly upregulating the expression of LC3II and beclin-1, inducing sub-G1/G0 cell cycle arrest, and inhibiting cell migration and invasion by inhibiting the MAPK/ERK signaling pathway[910].

---

[906] Ruttanaphan, T., Thitathan, W., Piyasaengthong, N. et al. Chrysoeriol isolated from Melientha suavis Pierre with activity against the agricultural pest Spodoptera litura. Chem. Biol. Technol. Agric. 9, 21 (2022).

[907] Lee SB, Lee SE, Lee H, Kim JS, Choi H, Lee S, Kim BG. Engineering Nicotiana benthamiana for chrysoeriol production using synthetic biology approaches. Front Plant Sci. 2024 Dec 17;15:1458916.

[908] Wu, X., Yuwen, M., Pu, Z. et al. Engineering of flavonoid 3′-O-methyltransferase for improved biosynthesis of chrysoeriol in Escherichia coli. Appl Microbiol Biotechnol 107, 1663–1672 (2023).

[909] Wongkularb S, Limboonreung T, Tuchinda P, Chongthammakun S. Suppression of PI3K/Akt/mTOR pathway in chrysoeriol-induced apoptosis of rat C6 glioma cells. In Vitro Cell Dev Biol Anim. 2022 Jan;58(1):29-36.

[910] Wei W, He J, Ruan H, Wang Y. In vitro and in vivo cytotoxic effects of chrysoeriol in human lung carcinoma are facilitated through activation of autophagy, sub-G1/G0 cell cycle arrest, cell migration and invasion inhibition and modulation of MAPK/ERK signalling pathway. J BUON. 2019 May-Jun;24(3):936-942.

Treatment for Melanoma: In A375 and B16F10 melanoma cells, treatment with chrysoeriol inhibited the phosphorylation of STAT3 and downregulated the expression of STAT3-target genes involved in melanoma growth and metastasis, while in a murine model chrysoeriol restrained melanoma growth and tumor-related angiogenesis, and altered compositions of immune cells in the melanoma microenvironment[911].

Treatment for Rheumatoid Arthritis: Chrysoeriol has been shown to suppress hyperproliferation and evoke apoptosis in IL-6/sIL-6R-stimulated rheumatoid arthritis fibroblast-like synoviocytes, likely via the suppression of JAK2/STAT3 signaling[912].

Anti-inflammatory Agent: In 12-O-tetradecanoylphorbol-13-acetate-induced ear edema in mice, treatment with chrysoeriol ameliorated acute skin inflammation by reducing ear thickness, ear weight, and number of inflammatory cells in inflamed ear tissues[913]. As isolated from the leaves of Digitalis purpurea (foxglove), treatment with chrysoeriol potently inhibited the release of NO in Raw264.7 macrophages treated with lipopolysaccharide (LPS) and inhibited the LPS-induced inductions of iNOS gene[914].

Treatment for Microplastic Contamination/Injury: In polyethylene microplastics-induced testicular damage in rats, dietary supplementation with chrysoeriol significantly increased the activities of anti-oxidant enzymes, reduced levels of ROS and malondialdehyde, reduced the levels of inflammatory markers, dead sperm number, and sperm abnormality, increased the expressions of steroidogenic enzymes, and increased the levels of plasma testosterone, luteinizing and follicle stimulating hormone, and Bax and Caspase-3 expressions[915].

Treatment for Parkinsons Disease: In SH-SY5Y neuroblastoma cells, treatment with chrysoeriol significantly increased the viability of cells and decreased apoptosis by downregulating apoptotic cells, caspase-3 activity and antiapoptotic ratio via the activation of the PI3K/Akt pathway[916].

Bronchodilatory Agent: As the primary active constituent of an extract made from Rooibos tea, chrysoeriol exhibited significant selective bronchodilator effects, predominantly through KATP channel activation[917].

IMPLICATIONS FOR HUMAN HEALTH AND NUTRITION

Dietary acquisition of chrysoeriol is likely the only way to consume this flavone. When preparing, cooking, or serving food, consider working with herbs or spices including parsley, cardamom, sage, fennel, tarragon, thyme, and turmeric. You may also consider eating fruits like black huckleberry, loquat, pineapple, strawberry, or vegetables such as alfalfa, celery, soybean, corn, or Swiss chard, which are all known to contain chrysoeriol. Finally, snacking on nuts like hazelnuts or peanuts, or drinking teas made from lemongrass, chamomile, or rooibos may also allow dietary consumption of chrysoeriol.

[911] Liu YX, Chen YJ, Xu BW, Fu XQ, Ding WJ, Li SA, Wang XQ, Wu JY, Wu Y, Dou X, Liu B, Yu ZL. Inhibition of STAT3 signaling contributes to the anti-melanoma effects of chrysoeriol. Phytomedicine. 2023 Jan;109:154572.

[912] Wu JY, Chen YJ, Fu XQ, Li JK, Chou JY, Yin CL, Bai JX, Wu Y, Wang XQ, Li AS, Wong LY, Yu ZL. Chrysoeriol suppresses hyperproliferation of rheumatoid arthritis fibroblast-like synoviocytes and inhibits JAK2/STAT3 signaling. BMC Complement Med Ther. 2022 Mar 16;22(1):73.

[913] Wu JY, Chen YJ, Bai L, Liu YX, Fu XQ, Zhu PL, Li JK, Chou JY, Yin CL, Wang YP, Bai JX, Wu Y, Wu ZZ, Yu ZL. Chrysoeriol ameliorates TPA-induced acute skin inflammation in mice and inhibits NF-κB and STAT3 pathways. Phytomedicine. 2020 Mar;68:153173.

[914] Choi, DY., Lee, J.Y., Kim, MR. et al. Chrysoeriol potently inhibits the induction of nitric oxide synthase by blocking AP-1 activation. J Biomed Sci 12, 949–959 (2005).

[915] Ijaz MU, Saher F, Aslam N, Hamza A, Anwar H, Alkahtani S, Khan HA, Riaz MN. Evaluation of possible attenuative role of chrysoeriol against polyethylene microplastics instigated testicular damage: A biochemical, spermatogenic and histological study. Food Chem Toxicol. 2023 Oct;180:114043.

[916] Limboonreung T, Tuchinda P, Chongthammakun S. Chrysoeriol mediates mitochondrial protection via PI3K/Akt pathway in MPP+ treated SH-SY5Y cells. Neurosci Lett. 2020 Jan 1;714:134545.

[917] Khan, Au., Gilani, A.H. Selective bronchodilatory effect of Rooibos tea (Aspalathus linearis) and its flavonoid, chrysoeriol. Eur J Nutr 45, 463–469 (2006).

As of early 2025, cannabis and hemp plants are not known to contain chrysoeriol, except where the compound is a precursor in the biosynthesis of other molecules, notably cannflavin A and cannflavin B.

As always, exercise extreme caution when working with purified forms of chrysoeriol, as these preparations can be dangerous in some situations.

Chrysoeriol Review

Answer the following questions to test your knowledge of this flavonoid:

Question #1: What type of flavonoid is chrysoeriol?

    a. Flavone
    b. Flavorall
    c. Chalcone
    d. Flavonene

Question #2: The base flavonoid in the biosynthesis of chrysoeriol is:

    a. Naringeniis
    b. Apigenin
    c. Quercetin
    d. Luteolin

Question #3: What is the primary difference between chrysoeriol and luteolin?

    a. Position 3' and 5 Hydroxylation
    b. Two double bonded oxygen atoms
    c. Glycosylation
    d. B ring methylation

Question #4: What is the chemical formula for chrysoeriol?

    a. $C_{16}H_{12}O_6$
    b. $C_{15}H_{24}O_6$
    c. $C_{26}H_{28}O_6$
    d. $C_{21}H_{20}O_{10}$

Question #5: Name two types of variations of chrysoeriol:

1 _____ 2 _____

Question #6: Name two potential medical uses of chrysoeriol:

1 _____ 2 _____

For the answer key to Chrysoeriol, please visit www.cannabischemistry.org

KERACYANIN

Type: Anthocyanin
Chemical Formula: $C_{27}H_{31}ClO_{15}$
Molecular Weight: 630.98 g/mol
Flash Point: 32.00 °F (TGSC)[918]
Solubility: Soluble in water, methanol, ethanol, acetone
Oral LD50: 1,900 mg/kg (rat)
Biological Role: Unknown
Therapeutic Role: Anticancer agent
Commercial Use: Research chemical
Occurrence in Cannabis: Unknown, but possibly a major constituent

Occurs in Cannabis Strains: Keracyanin has been found in the leaves and flowers of two commercial hemp varieties, Fibrante and S1750, with researchers noting that the concentrations of this anthocyanin were higher than in other known natural sources, suggesting that cannabis biomass may be an excellent source of this compound[919].

INTRODUCTION

Also known as antirrhinin, cyaninoside, or sambucin, keracyanin is technically referred to as cyanidin 3-rutinoside, a name that indicates that a sugar group is attached to the third carbon of ring C. This anthocyanin has no known organoleptic properties, but it is likely responsible for the red and purple hues of cannabis varieties that contain keracyanin.

Interestingly, keracyanin is the only compound discussed in this book that does not contain only hydrogen, carbon, and oxygen atoms. Instead, this anthocyanin also features a chlorine atom, which changes the compound's features such as solubility, stability, biological function, human metabolism, and other attributes.

The scientific literature regarding keracyanin is light, with few studies available to document the compound's potential medical or therapeutic uses, and completely lacking in information about its biological activities in plants. Future editions of this text will be updated with emerging information on keracyanin, particularly now that it has been officially quantified in cannabis as a potentially major compound.

Keracyanin has been shown to be much more abundant in obviously pigmented cannabis varieties, and, interestingly, was the primary anthocyanin constituent in both hemp and drug types of cannabis, along with peonidin 3-O-rutinoside[920].

[918] The Good Scents Company, Data Sheet for Keracyanin. From: https://www.thegoodscentscompany.com/data/rw1699981.html. Accessed January 23, 2025.

[919] Bassolino L, Fulvio F, Pastore C, Pasini F, Gallina Toschi T, Filippetti I, Paris R. When Cannabis sativa L. Turns Purple: Biosynthesis and Accumulation of Anthocyanins. Antioxidants. 2023; 12(7):1393.

[920] Gagalova KK, Yan Y, Wang S, Matzat T, Castellarin SD, Birol I, Edwards D, Schuetz M. Leaf pigmentation in Cannabis sativa: Characterization of anthocyanin biosynthesis in colorful Cannabis varieties. Plant Direct. 2024 Nov 25;8(11):e70016.

CHEMICAL STRUCTURE

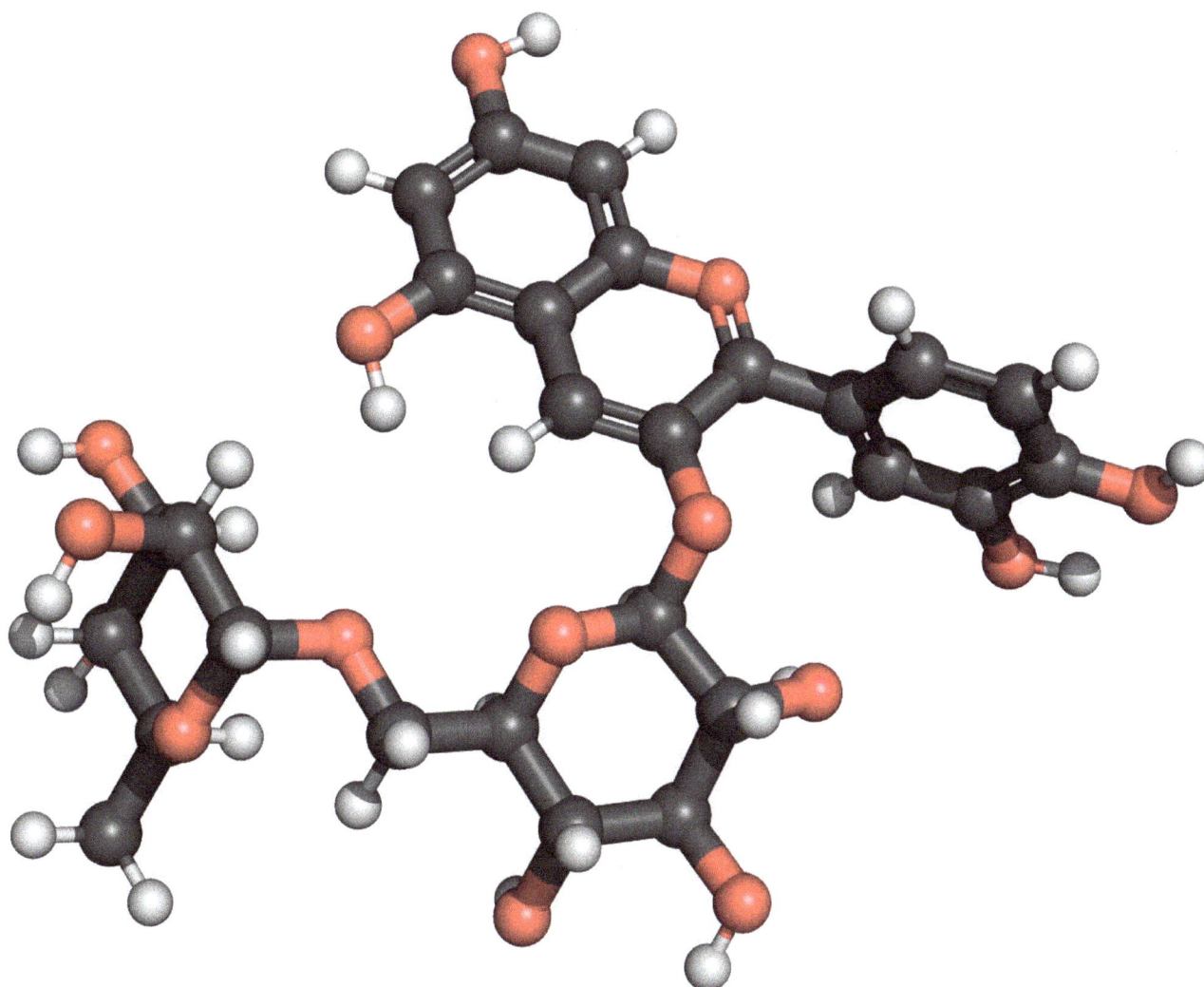

Keracyanin is biosynthesized in the phenylpropanoid pathway from phenylalanine, which is then converted to a flavanone via the chalcone synthase. The molecular structure of this compound features five rings. Rings A and B are benzene rings from the cyanidin core, ring C is a pyrilium ring (oxygen-containing heterocycle) in the cyanidin core containing the chlorine group, ring D is a pyranose ring with a glucose sugar group, and finally, ring E is a pyranose ring with a rhamnose sugar group.

The keracyanin molecule also contains seven endocyclic double bonds, one endocyclic double bonded oxygen atom, fifteen oxygen atoms total, a chlorine group, twenty-seven carbon atoms, and thirty-one hydrogen atoms, notated as $C_{27}H_{31}ClO_{15}$.

Variations of keracyanin include glycoside variations such as cyanidin 3-glucoside, cyanidin 3-galactoside, and cyanidin 3-sambubioside, substitution variations such as cyanidin 3,5-diglucoside and acylated cyanidin 3-rutinoside, methylation variations such as peonidin 3-rutinoside, as well as halogenated variations, and variants arising through polymerization.

207

OCCURRENCE IN PLANTS

Keracyanin is produced by a modest number of plants, making it reasonable to obtain this anthocyanin as part of a balanced diet that is rich in fruits and vegetables.

Asparagus	Cinnamon	Raspberry
Banana	Fig	Red currant
Black Currant	Gooseberry	Rhubarb
Black elder	Lychee	Rice
Cannabis	Mulberry	Rose
Cassava	Olive	Rubus spp.
Cherry	Pomegranate	Triticum spp.

BIOLOGICAL ACTIVITY IN PLANTS

At the time of publication of this text, there were no known publications or other literature that examined or discussed the biological activities of keracyanin in plants. However, based on the activities of other flavonoids, we can assume that this compound might serve antioxidant, UV protection, or antimicrobial roles, and/or it may act as an attractant to pollinators.

USES IN INDUSTRY

Keracyanin is primarily used for research and experimental purposes. There are no known commercial or household products that are manufactured using this anthocyanin.

POTENTIAL USES IN MEDICINE

Limited research has been conducted to study the medical and therapeutic potential of keracyanin. However, it seems likely that this anthocyanin could exhibit anti-inflammatory and antioxidant properties, which could have wide medical implications.

Treatment for Lung Cancer: In highly metastatic A549 human lung carcinoma cells, treatment with cyanidin 3-rutinoside and cyanidin 3-glucoside (as extracted from Morus alba L.) exerted a dose-dependent inhibitory effect on cell migration and invasion, without cytotoxicity to surrounding cells[921].

IMPLICATIONS FOR HUMAN HEALTH AND NUTRITION

Keracyanin and its degradation products have been shown to be biologically available in human test subjects within two hours of ingestion[922], making it relatively easy to obtain potentially therapeutic levels of this compound via a carefully chosen diet. Consider eating fruits that are known to contain keracyanin such as bananas, black currants, cherries, figs, gooseberries, mulberries, pomegranate, raspberries, or red currants. Vegetables that may contain this compound include asparagus, cassava, olives, rhubarb, or rice, or you can try eating foods that contain cinnamon or rose.

---

[921]Chen PN, Chu SC, Chiou HL, Kuo WH, Chiang CL, Hsieh YS. Mulberry anthocyanins, cyanidin 3-rutinoside and cyanidin 3-glucoside, exhibited an inhibitory effect on the migration and invasion of a human lung cancer cell line. Cancer Lett. 2006 Apr 28;235(2):248-59.

[922] Röhrig T, Kirsch V, Schipp D, Galan J, Richling E. Absorption of Anthocyanin Rutinosides after Consumption of a Blackcurrant (Ribes nigrum L.) Extract. J Agric Food Chem. 2019 Jun 19;67(24):6792-6797.

Interestingly, placing fruits like peaches in bags could significantly reduce levels of keracyanin[923]. This is an important consideration if you are eating fruits specifically to obtain dietary keracyanin.

Some cannabis varieties may contain keracyanin, but it is unlikely that laboratories are testing for this anthocyanin.

As always, use extreme caution when working with extracts or purified keracyanin, as these compounds can be dangerous in some situations.

---

[923] Wang Y, Yang C, Liu C, Xu M, Li S, Yang L, Wang Y. Effects of bagging on volatiles and polyphenols in "Wanmi" peaches during endocarp hardening and final fruit rapid growth stages. J Food Sci. 2010 Nov-Dec;75(9):S455-60.

Keracyanin Review

Answer the following questions to test your knowledge of this flavonoid:

Question #1: What type of flavonoid is keracyanin?

       a.   P-courmaryl
       b.   Anthocyanin
       c.   Chalcone
       d.   3-cyanidin

Question #2: Keracyanin is biosynthesized in what pathway?

       a.   Phelamine pathway
       b.   Isoprene pathway
       c.   Polyquercetin pathway
       d.   Phenylpropanoid pathway

Question #3: What is different about keracyanin than any other compound in this book?

       a.   Position 3' and 5 Hydroxylation
       b.   Chlorine atom
       c.   Glycosylation
       d.   B ring methylation

Question #4: What is the chemical formula for keracyanin?

       a. $C_{27}H_{31}ClO_{15}$
       b. $C_{15}H_{24}C1O_6$
       c. $C_{26}H_{28}O_6Cl_1$
       d. $C_{21}H_{20}O_{10}$

Question #5: Name two types of variations/modifications of keracyanin:

1 _____ 2 _____

Question #6: What properties does keracyanin likely to contribute to cannabis:

_____

For the answer key to Keracyanin, please visit www.cannabischemistry.org

BETA-SITOSTEROL

Type: Phytosterol [NOTE: *Not a flavonoid*]
Chemical Formula: $C_{29}H_{50}O$
Molecular Weight: 414.718 g/mol
Boiling Point: 498 °C (928.4 °F)
Flash Point: 220.40 °C (429 °F)
Melting Point: 140 °C (284 °F)
Solubility: Insoluble in water, soluble in acetone, ethanol, chloroform
Biological Role: Allelopathic agent, growth regulator
Therapeutic Role: Anticancer agent, antibacterial, anti-inflammatory
Commercial Use: Stabilizing agent, masking agent
Occurrence in Cannabis: Extremely common, likely in every chemovar

Occurs in Cannabis Strains: Cheese, Critical Dream, Fruit Punch, Huckleberry, Northern Lights, Purple Moose, Purple Punch, Skywalker, Space Candy, Strawberry Cough, Strawberry Banana, Super Lemon Haze, White Cookies

PRE-INTRODUCTION

During research for this book, the author found more than thirty cannabis websites and laboratories that listed beta-sitosterol as a flavonoid, including ACS Laboratory, Modern Canna Laboratory, WeedMaps, Alchimia, Merry Jane, and Green Relief Consulting. After a review of fifty-two published studies that focused on this compound, we confirmed that beta-sitosterol is true to its name; the compound is a plant sterol, not a flavonoid.

This situation is most likely caused by one of the most persistent and prevalent problems in the cannabis industry: the lack of scientific research, citations, and references in published materials, and the strong tendency for one cannabis website to copy the content and data of another.

The author sent fifteen unique organic cannabis strains for flavonoid testing to a laboratory within the US. The results showed that B-sitosterol was the top compound in all but one of the strains, often by double or more the listed ppm for the next flavonoid. This showed that, although B-sitosterol is not a flavonoid, it is clearly a prevalent and important compound in cannabis, thus its inclusion here. Interestingly, after removing B-sitosterol from our test results, it became clear that cannflavin A is the correct first flavonoid by concentration in cannabis.

INTRODUCTION

Beta-sitosterol (B-sitosterol) is the most common sterol in plants, and likely the top sterol by concentration in cannabis. This isoprene-based compound is closely related to cholesterol and is abundant in the human diet. B-sitosterol occurs naturally in dozens of plants and foods, exhibits significant anti-cancer properties, can be used to reduce cholesterol, and serves roles as an allelopathic agent in plants.

As mentioned previously, we tested fifteen strains for flavonoids and found that B-sitosterol occurred as the first compound by concentration in fourteen of them, while coming in third for the remaining tested strain.

Research carried out in 1974 found B-sitosterol as a primary compound in an Indian variety of cannabis sativa L.[924], while a 2018 study showed that B-sitosterol was a primary constituent in the hexane extract made from the roots of a high-CBD cannabis variety grown by the University of Mississippi[925].

---

[924] Mole, M. Leonard, and Carlton E. Turner. Phytochemical screening of cannabis sativa L. I: Constituents of an indian variant. Journal of Pharmaceutical Science 63.1 (1974): 154-156.

[925] Elhendawy M, A, Wanas A, S, Radwan M, M, Azzaz N, A, Toson E, S, ElSohly M, A: Chemical and Biological Studies of Cannabis sativa Roots. Med

CHEMICAL STRUCTURE

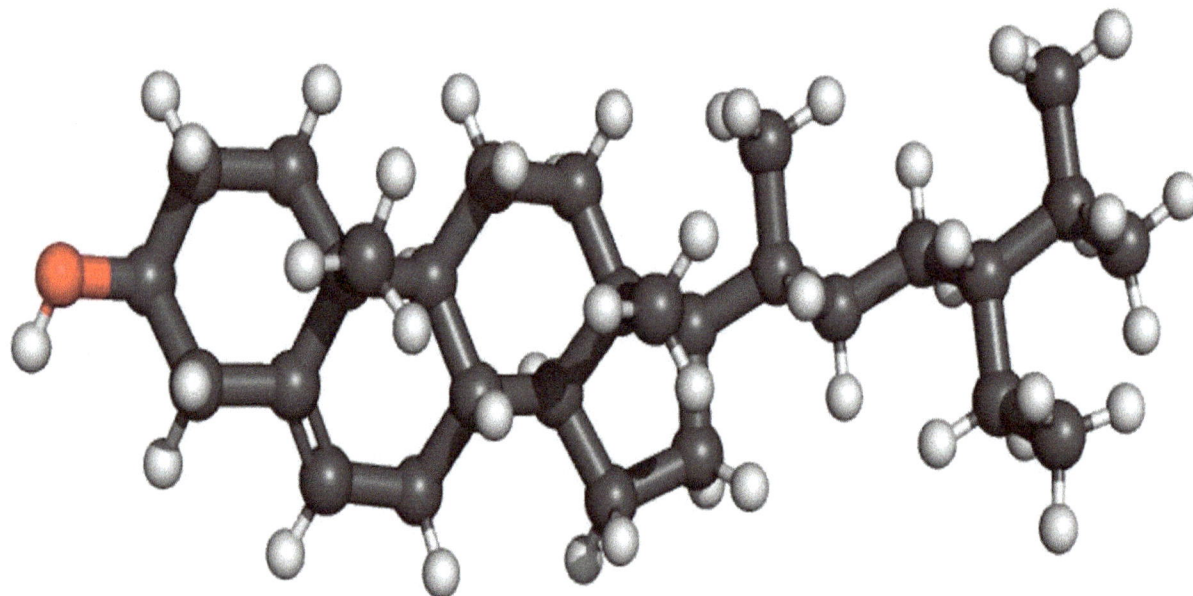

Like the terpenes, B-sitosterol is formed initially from isoprene, then cycloarteol[926], via the mevalonate pathway. This sterol consists of twenty-nine carbon atoms, fifty hydrogen atoms, and one oxygen atom notated as $C_{29}H_{50}O$. The molecular skeleton of this compound is comprised of four rings; two rings of benzene and one of pyrane, while the fourth ring is likely comprised of diphenylpropane. This latter ring is fused to ring A, with ring C being oxygenated and ring B containing an endocyclic double- bond.

OCCURRENCE IN PLANTS

B-sitosterol is found in many herbs, spices, seeds, roots and tubers, vegetables, and fruits; we have created a small non-exhaustive list.

Allspice	Almond	Anise
Apple	Apricot	Avocado
Baboon	Bamboo	Banana
Barley	Basil	Basswood
Black bean	Black elder	Black pepper
Bell pepper	Beet	Bilberry
Broccoli	Brussels sprouts	Buckthorn
Buckwheat	Burdock	Cabbage
Caper	Cannabis	Caraway
Cardamom	Carrot	Cashew
Cayenne	Celery	Chamomile
Cherry	Chicory	Cinnamon
Clove	Coconut	Coffee
Coriander	Corn	Cotton
Cress	Cucumber	Cumin
Dandelion	Date palm	Dill

Cannabis Cannabinoids 2018;1:104-111.

[926] Soodabeh Saeidnia, Azadeh Manayi, Ahmad R. Gohari, and Mohammad Abdollahi. The Story of Beta sitosterol. A Review European Journal of Medicinal Plants 4(5): 590-609, 2014.

Fennel	Ginger	Ginkgo biloba
Goosefoot	Grape	Guava
Hazelnut	Horsemint	Horseradish
Hyssop	Kiwi	Leek
Lemon	Lemongrass	Lettuce
Loquat	Lovage	Mandarin
Marango	Marjoram	Melon
Mugwort	Mustard	Nutmeg
Oat	Oregano	Papaya
Parsley	Pea	Peanut
Persimmon	Pineapple	Pistachio
Plum	Pomegranate	Poppy
Potato	Purslane	Rapini
Raspberry	Rice	Roselle
Rosemary	Sage	Salvia
Sesame	Shitake	Soybean
Spearmint	Spinach	Sunflower
Tamarind	Tarragon	Thyme
Tomato	Turmeric	Walnut
Watermelon	Wheat germ	

BIOLOGICAL ACTIVITY IN PLANTS

Few studies have been conducted on the biological roles of B-sitosterol in plants, but some research shows that the compound exhibits allelopathic properties and plant growth regulator activity:

Allelopathic Agent: As a primary constituent in a methanolic extract of euphorbia heterophylla (Japanese poinsettia), B-sitosterol contributed to potent allelopathic activity, inhibiting the germination, root, and shoot growth of sorghum and lettuce by 100%[927]. As the primary active constituent in a methanolic extracts of hibiscus sabdariffa (roselle), B-sitosterol inhibited the growth of cress, lettuce, alfalfa, timothy, ryegrass, crabgrass, buckwheat, Chinese sprangletop, jungle rice, barnyard grass, and sand fescue[928].

Plant Growth Regulator: B-sitosterol has been shown to be a plant growth regulator in white clover exposed to harsh conditions. This compound was found to increase OH (hydroxyl) scavenging in white clover by more than seven times when exposed to water stress, and increased antioxidant capacity by more than 190%, while also increasing total amino and organic acids by more than 40%[929].

USES IN INDUSTRY

B-sitosterol is commonly derived from African star grass or pine species, and appears as a white, waxy powder or crystalline solid in commercial samples, which are typically mixed with other phytosterols such as stigmasterol and campesterol. B-sitosterol is widely distributed in the plant kingdom and naturally occurs in many foods, but it is

---

[927] da Silva UP, Furlani GM, Demuner AJ, da Silva OLM, Varejao EVV. Allelopathic activity and chemical constituents of extracts from roots of Euphorbia heterophylla L. Nat Prod Res. 2019 Sep;33(18):2681-2684.

[928] Piyatida Pukclai, Masashi Sato, F. Kimura, H. Kato-Noguchi Tsolation of B-sitosterol from Hibiscus sabdariffa L Allelopathy Journal 32(2):289-300. October 2013.

[929] Li, Zhou, Bizhen Cheng, Bin Yong, Ting Liu, Yan Peng, Xinquan Zhang, Xiao Ma, Linkai Huang, Wei Liu, and Gang Nie. Metabolomics and physiological analyses reveal B-sitosterol as an important plant growth regulator inducing tolerance to water stress in white clover. Planta 250.6 (2019): 2033-2046.

also added to prepared and packaged foods, especially sauces, salad dressings, and oils. In Spain, B-sitosterol is the primary ingredient in an immunomodulator feed supplement for pigs[930].

Industrial uses of this compound include use as emulsion stabilizers, masking agents, skin conditioners, and stabilizing agents in other formulations. The Good Scents Company notes that this compound is "not for fragrance use, not for flavor use."

POTENTIAL USES IN MEDICINE

B-sitosterol has the potential to be used in the treatment and prevention of serious diseases like breast, colon, renal, and other cancers, Alzheimer's disease, pneumonia, and diabetes, as well as use in reducing cholesterol or blood lipid levels.

Treatment & Prevention of Colon & Colorectal Cancer: Antioxidant and other effects of B-sitosterol were shown to help lower liver lipid peroxides and protect against depletion of antioxidants in rats with DMH-induced colon cancer[931]. As isolated from the leaves of asclepias curassavica, B-sitosterol was shown to inhibit the growth of and induce apoptosis in human colon cancer cells, while simultaneously exhibiting chemopreventive effects attributed to its radical quenching ability in vitro, with minimal toxicity to normal cells[932]. As extracted from sweet potato, B-sitosterol reversed a tumor-inducing critical decrease in the diversity of gut bacteria in mice with induced colorectal cancer[933].

Treatment for Breast Cancer: B-sitosterol has been shown to ameliorate oxidative stress, increase saturated and unsaturated fatty acids, and downregulate NF-kB (a protein complex with numerous roles) expression in rats with induced mammary gland carcinoma[934]. B-sitosterol also induces G1 arrest and causes depolarization of mitochondrial membrane potential in human breast adenocarcinoma cells[935].

Treatment for Renal Cancer: In a rat model of experimental renal cancer, B-sitosterol was shown to both induce apoptosis and inhibit proliferation[936] of malignant cells.

Treatment for Diabetic Neuropathy: As the primary active constituent in the biologically guided fractionation of moringa oleifera seeds, B-sitosterol significantly attenuated the hyperalgesia and tactile allodynia associated with painful diabetic neuropathy in alloxan-induced diabetic mice[937].

[930] Fraile L, Crisci E, Cordoba L, Navarro MA, Osada J, Montoya M. Immunomodulatory properties of beta- sitosterol in pig immune responses. Int Immunopharmacol. 2012 Jul;13(3):316-21.

[931] Baskar AA, Al Numair KS, Gabriel Paulraj M, Alsaif MA, Muamar MA, Ignacimuthu S. B-sitosterol prevents lipid peroxidation and improves antioxidant status and histoarchitecture in rats with 1,2-dimethylhydrazine-induced colon cancer. J Med Food. 2012 Apr;15(4):335-43.

[932] Baskar, A.A., Ignacimuthu, S., Paulraj, G.M. et al. Chemopreventive potential of B-Sitosterol in experimental colon cancer model - an In vitro and In vivo study. BMC Complement Altern Med 10, 24 (2010).

[933] Ma, Hang, Yang Yu, Meimei Wang, Zhaoxing Li, Heshan Xu, Cheng Tian, Jian Zhang, Xiaoli Ye, and Xuegang Li. Correlation between microbes and colorectal cancer: tumor apoptosis is induced by sitosterols through promoting gut microbiota to produce short-chain fatty acids. Apoptosis 24.2 (2018): 168-183.

[934] Manral, C., Roy, S., Singh, M. et al. Effect of B-sitosterol against methyl nitrosourea-induced mammary gland carcinoma in albino rats. BMC Complement Altern Med 16, 260 (2016).

[935] Vundru, S.S., Kale, R.K. & Singh, R.P. B-sitosterol induces G1 arrest and causes depolarization of mitochondrial membrane potential in breast carcinoma MDA-MB-231 cells. BMC Complement Altern Med 13, 280 (2013).

[936] Sharmila, Ramalingam, and Ganapathy Sindhu. Modulation of Angiogenesis, Proliferative Response and Apoptosis by B-Sitosterol in Rat Model of Renal Carcinogenesis. Indian journal of clinical biochemistry: IUCB 32.2 (2018): 142-152.

[937] Raafat K, Hdaib F. Neuroprotective effects of Moringa oleifera: Bio-guided GC-MS identification of active compounds in diabetic neuropathic pain model. Chin U Integr Med. 2017 Dec 12.

Reduce Cholesterol and Blood Lipid Levels: B-sitosterol has been found to significantly lower LDL cholesterol levels[938], and, in a murine model of induced hyperlipidemia, exhibited blood-lipid-lowering effects, as well as the potential to reduce the weight of liver fat, perirenal fat, and epididymal fat[939]. However, some researchers have noted that more work on this compound is needed, as it is poorly absorbed.

Prevention of Alzheimer's Disease: Because high cholesterol may be linked to Alzheimer's disease, B-sitosterol's blood lipid and cholesterol lowering effects have been studied in relation to this condition. It was found that B-sitosterol effectively inhibits high cholesterol-driven platelet AB release and prevents high cholesterol-induced increase of activities of B- and y-secretase[940], both of which are essential in the pathogenesis of Alzheimer's.

Anti-inflammatory Agent: As a primary constituent in various extracts made from the leaves of ficus radicans, B-sitosterol contributed to the significant inhibition of inflammation in induced ear oedema in mice, and induced paw oedema in rats[941].

Treatment for Colitis: B-sitosterol has been shown to inhibit colon shortening, lower myeloperoxidase activity, and inhibit the expression of proinflammatory cytokines in induced colitis in mice. Researchers reasoned that B-sitosterol may ameliorate colitis by inhibiting the NF-kB pathway[942].

Antileishmanial Agent: As a primary constituent in an ethanolic extract made from the dried and powdered bark of cornus florida, B-sitosterol exhibited promising in vitro antileishmanial activity[943].

Antibacterial Agent: As a major compound in an ethanolic extract made from lyophilized aloe vera gel, B-sitosterol contributed to the significant growth inhibition of nine common bacteria, including S. typhi, E. coli, A. salmonicida, streptococcus spp., and S. griseus[944]. As a primary constituent in a methanolic extract made from the leaves of elaeophorbia drupiferan (used in West African religious ceremonies), B- sitosterol exhibited antimicrobial activities against twenty-eight of thirty-three gram-negative and gram-positive bacteria (84.8%), including multidrug resistant phenotypes[945].

Treatment for Streptococcus Pneumoniae Infections: B-sitosterol has been shown to prevent S. pneumoniae infection by protecting against cell lysis caused by pneumolysin, inducing pneumolysin oligomerization, and protecting cells from damage by other cholesterol-dependent toxins[946].

[938] Fernandez ML, Vega-Lopez S. Efficacy and safety of sitosterol in the management of blood cholesterol levels. Cardiovasc Drug Rev. 2005 Spring;23(1):57-70.

[939] Yuan, C., Zhang, X., Long, X. et al. Effect of B-sitosterol self-microemulsion and B-sitosterol ester with linoleic acid on lipid-lowering in hyperlipidemic mice. Lipids Health Dis 18, 157 (2019).

[940] Chun Shi, Jun Liu, Fengming Wu, Xiao Ming Zhu, David T. Yew, Jie Xu. B-sitosterol inhibits high cholesterol- induced platelet B-amyloid release Journal of Bioenergetics and Biomembranes December 2011, Volume 43, Issue 6, p 691-697.

[941] Naressi, Maria Augusta, Marcos Alessandro dos Santos Ribeiro, Ciomar Aparecida Bersani-Amado, Maria Lucilia M Zamuner, Willian Ferreira da Costa, Clara M Abe Tanaka, and Maria Helena Sarragiotto. Chemical composition, anti-inflammatory, molluscicidal and free-radical scavenging activities of the leaves of Ficus radicans 'Variegata' (Moraceae). Natural product research 26.4 (2012): 323-330.

[942] Lee, In-Ah, Eun-Jin Kim, and Dong-Hyun Kim. Inhibitory effect of B-sitosterol on TNBS-induced colitis in mice. Planta medica 78.9 (2012): 896-8.

[943] Graziose, Rocky, Patricio Rojas-Silva, Thirumurugan Rathinasabapathy, Carmen Dekock, Mary H Grace, Alexander Poulev, Mary Ann Lila, Peter Smith, and Ilya Raskin. Antiparasitic compounds from Cornus florida L. with activities against Plasmodium falciparum and Leishmania tarentolae. Journal of ethnopharmacology 142.2 (2012): 456-461.

[944] Bawankar, Raksha, V. C. Deepti, Pooja Singh, Rathinasamy Subashkumar, Govindasamy Vivekanandhan, and Subramanian Babu. Evaluation of Bioactive Potential of an Aloe vera Sterol Extract. Phytotherapy Research 27.6 (2013): 864-868.

[945] Igor K. Voukeng, Blaise K. Nganou, Louis P. Sandjo, Ilhami Celik, Veronique P. Beng, Pierre Tane & Victor Kuete. Antibacterial activities of the methanol extract, fractions and compounds from Elaeophorbia drupifera (Thonn.) Stapf. (Euphorbiaceae) BMC Complementary and Alternative Medicine volume 17, Article number: 28 (2017).

[946] Li, H., Zhao, X., Wang, J. et al. B-sitosterol interacts with pneumolysin to prevent Streptococcus pneumoniae infection. Sci Rep 5, 17668 (2015).

Analgesic Agent: As the primary active component in petroleum ether extracts made from the leaves of nyctanthes arbortristis, B-sitosterol exhibited significant analgesic effects in acetic acid induced and carrageenan induced writhing tests in mice, with measured analgesic effects comparable to that of ibuprofen and Pentazocine[947].

IMPLICATIONS FOR HUMAN HEALTH & NUTRITION

Increasing dietary consumption of B-sitosterol is possible by consuming foods that are high in this phytosterol. For example, when preparing, cooking, or serving food, consider working with herbs and spices like allspice, anise, barley, basil, black pepper, caraway, cardamom, coriander, cumin, dill, fennel, garlic, ginger, marjoram, nutmeg, onion, oregano, parsley, rosemary, sage, spearmint, tarragon, or thyme.

Fruits that contain this compound include apples, apricots, bananas, cantaloupe, cherries, figs, grapes, guava, kiwi, lemons, mandarins, melons, oranges, papaya, pineapples, plums, pomegranates, raspberries, and watermelon. Vegetables to consider include avocados, black beans, bell peppers, beets, broccoli, Brussels sprouts, cabbage, capers, carrots, celery, corn, cucumbers, horseradish, lettuce, peas, arugula, soybean, spinach, and tomatoes. Starches that contain B-sitosterol include rice, potatoes, and shitake mushrooms, while several nuts also feature this sterol: almonds, hazelnuts, peanuts, pistachios, and walnuts.

While it is likely that most or all cannabis chemovars contain B-sitosterol, we have confirmed this compound in the following strains: Cheese, Critical Dream, Fruit Punch, Huckleberry, Northern Lights, Purple Moose, Purple Punch, Skywalker, Space Candy, Strawberry Cough, Strawberry Banana, Super Lemon Haze, and White Cookies

---

[947] Nirmal, Sunil, Subodh Pal, Subhash Mandal, and Anuja Patil. Analgesic and anti-inflammatory activity of B- sitosterol isolated from Nyctanthes arbortristis leaves. Tnflammopharmacology 20.4 (2011): 219-224.

FINAL EXAM

Congratulations on completing The Formidable Book of Flavonoids! Now, test the knowledge that you have gained and retained by taking this thirty-one question final exam.

Question #1: What is the general biosynthesis pathway for flavonoids?

        a.   Mevalonate pathway
        b.   Phenylbarbitol pathway
        c.   Phenylpropanoid pathway
        d.   O-methylpropane pathway

Question #2: What are the six major subclasses of flavonoids?

_____, _____, _____, _____,

_____, _____

Question #3: Name two primary differences between flavonoids and terpenes:

_____, _____

Question #4: Name three biological activities of flavonoids in plants:

_____, _____, _____

Question #5: Name six flavonoids that are common to cannabis:

_____, _____, _____, _____,

_____, _____

Question #6: According to the author, what is the best way to obtain flavonoids from cannabis?

_____

Question #7: Who coined the term "Vitamin P" for flavonoids?

_____

Question #8: Name three other Friends of Flavonoids (other than the person above):

_____, _____, _____

Question #9: Common pollinators such as bees, flies, wasps, butterflies, moths, and beetles do not show a preference for the color:

    a. Red
    b. Purple
    c. Green
    d. Yellow
    e. Blue

Question #10: What is the chemical notation for the base flavonoid compound?

_____

Question #11: Describe glycosylation:

_____

Question #12: Can a single flavonoid exhibit opposing behavior, such as pro-oxidant AND anti-oxidant activity?

    a. Yes
    b. No

Question #13: What type of molecule is beta-sitosterol?

    a. Terpenoid
    b. Flavonoid
    c. Flavan-3-ol
    d. Plant sterol

Question #14: Flavonoid content in cannabis is higher in:

    a. Flowers
    b. Seeds
    c. Stems
    d. Bracts
    e. Leaves

Question #15: Flavonoids undergo extensive metabolic transformations in the human body, which influences:

    a. Bioavailability and biological effects
    b. Coloration and genome
    c. C-ring structural attachment
    d. Size and optical rotation

Question #16: Excessive consumption of flavonoid-rich supplements may increase the risk of:

    a. Cardiac arrest
    b. Liver toxicity
    c. Blood infection
    d. Neurological disorder

Question #17: Who coined the term "the entourage effect?"

    a. Li Tian
    b. Ethan Russo
    c. Shimon Ben-Shabat
    d. Mark J. Miller

Question #18: Explain how Synergy is different than the Entourage Effect:

_____

Question #19: What do the health implications of synergy depend on?

    a. Desired outcome
    b. Total flavonoid intake
    c. Monotherapy efficacy
    d. Treatment modality

Question #20: Name three types of synergy:

_____ , _____ , _____

Question #21: Accumulation of cannflavin A in cannabis is determined primarily by genetics, and accumulation can be increased by:

    a. High humidity
    b. High air temperature
    c. Low humidity
    d. Low air temperature

Question #22: Quercetin is closely related to which other flavonoid:

    a. Rutin
    b. Kaempferol
    c. Orientin
    d. Cannflavin A

Question #23: Cannflavin C was discovered long after Cannflavins A and B. In what year was Cannflavin C discovered?

    a. 2005
    b. 1981
    c. 2006
    d. 2008

Question #24: Name two flavonoids that are widely used in traditional Chinese medicine:

_____ , _____

Question #25: Fisetin is a well-known:

      a.   Anti-diarrheal agent
      b.   Anti-aging agent
      c.   Anti-malarial agent
      d.   Anti-antibody agent

Question #26: How many rings is the baicalin molecule comprised of?

      a.   Two
      b.   Three
      c.   Four
      d.   Five
      e.   Six

Question #27: This flavonoid is often the precursor chemical to many other flavonoids:

      a.   Chrysin
      b.   Naringenin
      c.   Hesperiden
      d.   Orientin

Question #28: Apigetrin is also known as:

      a.   Cosmosioside
      b.   Cosmosiin
      c.   Cosmetin
      d.   All of the above

Question #29: Researchers have been able to produce many flavonoids in engineered:

      a.   E. Coli
      b.   Thale cress
      c.   Amanita muscaria
      d.   Dihydrogen monoxide

Question #30: Name the only major anthocyanin to be found in cannabis as of early 2025:

      a.   Isovitexin
      b.   Keracyanin
      c.   Bergamotene
      d.   O-periden

Question #31: Which is the only flavonoid described in this book to feature a chlorine atom?

      a.   Cannflavin B
      b.   Rutin
      c.   Silymarin
      d.   Keracyanin
      e.   Wogonin

Acknowledgements

PubChem

One of the best resources for information about flavonoids is PubChem, which has a massive database of information on these and other molecules, including structural and other molecular/atomic data, biological activity, toxicity, patents, and references to thousands of studies carried out with each compound. PubChem is an open chemistry database at the National Institutes of Health. The author thanks PubChem for providing such a valuable resource to the public, and refers readers to contribute to and utilize the data found there at: https://pubchem.ncbi.nlm.nih.gov/.

The Good Scents Company

For more than four decades, The Good Scents Company has been the leading source of information about volatile molecules for use in the foods, flavors, and fragrances industries. The website consists of a huge database describing the appearance, smell and taste profile of various small molecules, their plant sources, and commercial usage data. The Good Scents Company often features valuable information that is not available on PubChem or other scientific sources. The author expresses significant appreciation for the work carried out by TGSC and refers readers to the site for supplemental or other interesting data not reported here. http://www.thegoodscentscompany.com/index.htm

About the Author

Russ Hudson is an international cannabis consultant, researcher, and author of several books about cannabis. Russ has dedicated more than 30 years to cannabis, working with the private social clubs of Spain, the coffeeshops and suppliers of the Netherlands, and cannabis, hemp, and CBD producers in Germany, Switzerland, Portugal, Belgium, and other EU countries, while also specializing as a cannabis licensing, compliance, and regulatory affairs expert in more than a dozen US states. Russ has been arrested and jailed for cultivation and trafficking, he has been the subject of a Vice documentary, he has written a children's book about cannabis, and he currently researches and collaborates with the world's leading cannabinoid, terpene, flavonoid, and other scientists. The author of this textbook and The Big Book of Terps, Russ intends to publish the next textbook in the Cannabis Chemistry Collection soon, this time likely about alkaloids, or lipids. If you have a preference or another suggestion with data to support it, please contact the author: info@cannabischemistry.com

www.ingramcontent.com/pod-product-compliance
Lightning Source LLC
Chambersburg PA
CBHW041603260326
41914CB00011B/1371